SOFT MATTER PHYSICS

Soft Matter Physics

MASAO DOI

Emeritus Professor, University of Tokyo
Director of the Center of Soft Matter Physics and its Applications, Beihang University

OXFORD
UNIVERSITY PRESS

Great Clarendon Street, Oxford, OX2 6DP,
United Kingdom

Oxford University Press is a department of the University of Oxford.
It furthers the University's objective of excellence in research, scholarship,
and education by publishing worldwide. Oxford is a registered trade mark of
Oxford University Press in the UK and in certain other countries

First Edition published in 2013

Published in the United States of America by Oxford University Press
198 Madison Avenue, New York, NY 10016, United States of America

British Library Cataloguing in Publication Data

Data available

Library of Congress Control Number: 2013938495

ISBN 978–0–19–965295–2

Printed and bound by
CPI Group (UK) Ltd, Croydon, CR0 4YY

Dedication

To my father and mother

Preface

Soft matter is a class of materials which include polymers, colloids, surfactants, and liquid crystals. They are indispensable in our everyday life and in modern technology. For example, gels and creams are soft matter we use in everyday life. Plastics, textiles, and rubbers are soft matter used in our houses, cars, and many other industrial products. Liquid crystals are used in display devices, and IC chips are made with the use of photo-reactive polymers. Furthermore, our body is made of soft matter and a variety of soft matter is used in medical treatments.

Soft matter does not belong to the category of conventional fluids or solids in a simple sense. For example, bubble gum, a soft matter made of polymers, can be stretched indefinitely like a fluid, but it behaves like a rubber when it is stretched and released quickly. Gel is a soft solid which can contain very large amounts of fluids. A liquid crystal is a fluid which has optical anisotropy like a crystal. These properties are important in applications of soft matter. Plastic bottles are made by blowing a liquid of polymers (like a bubble gum), gels are used in many industrial processes, and liquid crystals are used in optical devices.

This book is an introduction to this important class of materials from the viewpoint of physics. The properties of soft matter are complex, but they can be understood in terms of physics. The aim of this book is to provide a base for such physical thinking for soft matter.

Soft matter has the common characteristic that it consists of large elements such as macromolecules, colloidal particles, molecular assemblies, or ordered molecules. The size we need to discuss is between 0.01 μm and 100 μm. This is much larger than the scales of atoms and electrons, but still smaller than the scales of macroscopic mechanics. Physics on that scale is now attracting considerable interest in other areas of science and engineering. I believe soft matter is a good starting point in learning the physics of topics such as correlation, phase transitions, fluctuations, non-equilibrium dynamics, etc., since soft matter is the material that we see and play with in everyday life. I have tried to connect the topics discussed in this book with the phenomena we see or have seen as much as possible.

This book was first published in Japanese by Iwanami, but considerable rewriting has been done both for the overall structure and for individual sections. I thank Iwanami for kindly accepting this project.

Acknowledgements

Many people helped me in writing this book. I have had continual encouragement from Prof. Koji Okano (Tokyo) and Prof. Kunihiro Osaki (Kyoto). Professor Hiroshi Maeda (Kyushu) gave me useful comments for the revision of Chapter 10.

This book emerged from lectures I gave at Tokyo University. I have benefited from students who read the early version of the manuscript, made comments, and corrected errors.

I have also benefited from long-term stays at the Institute of Academia Sinica (Taiwan), the Kavli Institute of Theoretical Physics (Beijing), and the Kavli Institute of Theoretical Physics (Santa Barbara). I thank Prof. Chin-Kun Hu (Taipei), Prof. Wenbing Hu (Nanjin), and Prof. Kurt Kremer (Mainz) who gave me the opportunities of staying there and allowing me to give lectures. The experience helped me enormously in restructuring and revising the presentation.

Contents

What is soft matter?

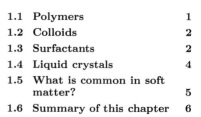

Soft matter includes a large variety of materials, typically composed of polymers, colloids, liquid crystals, surfactants, and other mesoscopic constituents. In this chapter, we shall discuss (i) what are these materials, (ii) what are the common characteristics of these materials, and (iii) what is the physics that gives such characteristics to soft matter.

1.1 Polymers

Polymers are long string-like molecules made of certain chemical units called monomers. The monomers are connected sequentially as shown in Fig. 1.1. The number of monomers in a polymer is typically several thousand, and can be as large as tens of millions. Polymers are indispensable materials in modern technology; they are used as plastics, rubbers, films, and textiles.

Polymers are also the basic molecules of life. The machinery of life is realized by proteins which are natural polymers made of amino acids. The genetic information of life is inscribed in another important class of bio-polymer, DNA.

In polymers, a variety of materials can be produced simply by changing the types of monomers and the way they are connected. Polymers are usually string-like as shown in Fig. 1.2(a), but branches can be introduced as in Fig. 1.2(b) or the strings can be cross-linked as in Fig. 1.2(c). Plastics are usually made of string-like polymers (Fig. 1.2(a)). They can be brought into a liquid state which can be easily molded into any desired shapes. On the other hand, rubbers are made of cross-linked polymers (Fig. 1.2(c)). Since the molecules are all connected, rubbers cannot flow. They change their shape under external forces, but recover their original shape when the forces are removed. The branched polymers (Fig. 1.2(b)) are intermediate, and behave like a rubbery material which can flow. The adhesives used in adhesive tapes are made of branched polymers which are intermediate between the liquid state and rubbery state.

Depending on the chemical structure of monomers, polymeric materials can be very soft, like rubbers in rubber bands, or can be very hard, like plastics in car bumpers. Materials which are hard for small strain, but become plastic for large strain have been made by mixing these monomers. New types of polymers, electro-conductive polymers, or photo-reactive polymers have been developed and are used extensively in modern technologies.

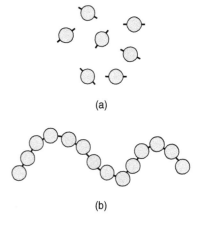

Fig. 1.1 (a) Monomer; (b) polymer.

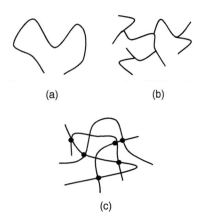

Fig. 1.2 Various polymers made of the same monomer, but by different monomer connections. (a) Linear polymer; (b) branched polymer; (c) network polymer.

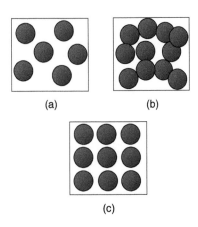

Fig. 1.3 Various phases of spherical colloids. (a) Sol; (b) gel; (c) crystal.

1.2   Colloids

Colloids are small solid particles or fluid droplets dispersed in other fluids. The diameter of colloidal particles is typically between 1 [nm] and 1 [µm] and is much larger than atomic size.

Colloids are seen in everyday life. Milk is a colloidal solution with nutritious particles (made of fat and proteins) of about 0.1 µm dispersed in water. Water colour is a colloidal solution of coloured solid particles (called pigments) dispersed in water.

Colloidal suspensions made of solid particles are fluids and can flow if the particle concentration is low, but become solid-like and cease to flow when the particle concentration becomes large. For example, water colour is a fluid when dissolved in water, but becomes solid as it dries up. The fluid state is called a sol, and the solid state is called a gel (see Fig. 1.3).

The transition from sol to gel can be induced without changing the concentration of particles. By the addition of certain chemicals to a colloidal solution, the interaction between colloidal particles can be changed, leading to a transformation from sol to gel. For example, when vinegar is added to milk, the milk becomes quite viscous and loses fluidity. This is caused by the aggregation of milk particles whose interaction becomes attractive when vinegar is added.

The solidification of colloids usually occurs by random aggregation of particles as shown in Fig. 1.3(b). If the particles are carefully made to have the same size and shape, they can form a regular structure, very much like the usual atomic crystal as shown in Fig. 1.3(c). Such colloidal crystals show strong diffraction of light in the same way as X-ray diffraction in atomic crystals. Since the lattice constants of colloidal crystals are of the order of 0.5 [µm], the diffraction is seen as an iridescent colour scattered by these crystals.

There are a variety of colloidal particles, of various shapes (spherical, rod-like, or plate-like particles), and of various surface properties which are used in daily life as well as in industrial applications.

1.3   Surfactants

Oil and water are a typical example of antagonistic pairs. Usual materials are classified into either hydrophilic (water-loving) or oleophilic (oil-loving). Materials which dissolve in oil usually do not dissolve in water, i.e., oleophilic materials are usually hydrophobic (water hating), and vice versa. Surfactants are a special class of so-called amphiphilic materials which can dissolve in both water and oil. Surfactant molecules are made of two parts, one part is hydrophilic, and the other part is hydrophobic (i.e., oleophilic) (see Fig. 1.4(a)). Surfactant molecules can dissolve in water by forming the structure shown in Fig. 1.4(b), where the hydrophilic part is exposed to water encapsulating the hydrophobic part. Such a molecular assembly is called a micelle. On the other hand, surfactant

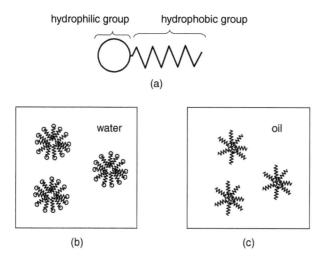

hydrophilic group   hydrophobic group

(a)

water   oil

(b)   (c)

Fig. 1.4 (a) Surfactant molecule; (b) micelles formed in water, (c) inverse micelles formed in oil.

molecules can dissolve in oil by exposing the hydrophobic group to oil concealing the hydrophilic group in an inverted micelle (Fig. 1.4(c)).

Depending on the surfactants, micelles can take various shapes including spheres, cylinders, or lamellae (see Fig. 1.5). Spherical micelles typically consist of several tens of molecules, but cylindrical and lamellar micelles can be much larger than this, as large as millions or billions of molecules. Although the size and shape of the micelles are primarily determined by the structure of the surfactant molecules, they can vary depending on the environment, solvent type, temperature, and surfactant concentration, etc.

Surfactants are used in soaps and detergents in our daily life. Due to their amphiphilic molecular nature, surfactant molecules prefer to sit at the interface between oil and water. Therefore, if a surfactant is added to the mixture of oil and water, surfactant molecules tend to assemble at the oil–water interface. To accommodate these amphiphilic molecules, the interfacial area must increase. Accordingly, the oil becomes encapsulated in small droplets coated by surfactant, and is dispersed in water. This is the basic mechanism for the function of detergents (see Fig. 1.6). The term surfactant comes from this function; it is an abbreviation of 'surface activating agents'.

Surfactants are important in dispersion technology. This is the technology of using surfactants to make seemingly homogeneous materials from components that do not form stable mixtures in usual conditions, for

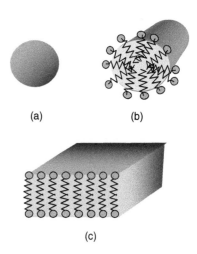

(a)   (b)

(c)

Fig. 1.5 Various structures of micelles formed by surfactants. (a) A spherical micelle; (b) a cylindrical micelle; (c) a lamellar bilayer.

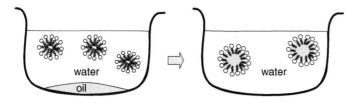

water   water
oil

Fig. 1.6 Dispersion of oil droplets in water by the action of surfactants.

example oil and water, or inorganic particulates and water. The stand-ard way is to break the material into fine grains (particles or droplets) in other materials with the aid of surfactants. The surfactants coat the surface of the dispersed materials and prevent their aggregation. This technology is important in many applications (for example, cosmetics and food) as well as in creating new functional materials.

1.4   Liquid crystals

Condensed states of matter are usually classified into two states: the crystalline state where the molecules are ordered, and the liquid state where the molecules are disordered. For certain materials, molecules form a semi-ordered state between crystal and liquid. Such materials are called liquid crystals.

Consider, for example, a material made of rod-like molecules. In the crystalline state, the molecules are regularly placed on lattice sites and are completely aligned as in Fig. 1.7(a). In this state, there is complete order in position and orientation. In the liquid state, the position and orientation of molecules are both random (Fig. 1.7(d)). In the liquid crystalline state, molecules are intermediately ordered between these two states.

A class of liquid crystals, nematics, is illustrated in Fig. 1.7(b). Here the molecules retain the orientational order, while there is no positional order. The nematic liquid crystal is a fluid which has anisotropy: it flows when the container is tilted, but it is anisotropic and shows birefringence. Another class of liquid crystals, smectics, is illustrated in Fig. 1.7(c). Here, in addition to the orientational order, there is a partial positional order: the molecules are regularly placed along a certain direction (the z-axis in Fig. 1.7(c)), but their positions are random in the x–y plane.

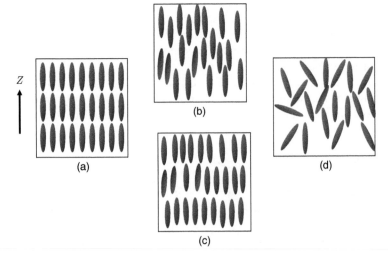

Fig. 1.7 Various phases of materials made of rod-like molecules. (a) Crystal; (b) nematic liquid crystal; (c) smectic liquid crystal; (d) isotropic liquid.

Nematic liquid crystals are extensively used in display devices since their optical properties are easily controlled by an electric field. The changes in the optical properties of liquid crystals are also used as sensors. Strong fibres (used in bullet-proof jackets) are made from liquid crystalline polymers since polymer chains are strongly oriented in liquid crystals.

1.5   What is common in soft matter?

As we have seen, soft matter includes a large class of materials.[1] A natural question is then what is common in these materials, and why are they discussed in the framework of 'soft matter'.

What is common in the above materials is that they all consist of structural units that are much larger than atoms. Polymer molecules typically consist of millions of atoms. Colloidal particles of diameter 0.1 [μm] involve billions of atoms. Molecules constituting surfactants and liquid crystals are not very large (including only tens or hundreds of atoms), but they form ordered structures and move together in a large unit: surfactant molecules form micelles which move as a collective entity, and liquid crystalline molecules move together when they rotate.

The fact that soft matter consists of large molecules or assemblies of molecules which move collectively gives two characteristics to soft matter:

(a) Large and nonlinear response. Soft matter shows a large response to weak forces. Polymer molecules consisting of tens of thousands of atoms are easily deformed, and this gives the softness of rubbers and gels. Colloidal particles form a very soft solid, which is used in cosmetics and paints. The optical properties of liquid crystals are easily changed by an electric field as we have seen. Such large responses cannot be described by linear relations between the force and the response. For example, rubbers can be elongated by several hundred percent of their initial length and their mechanical responses cannot be described by a linear relation between the stress and strain. The nonlinear response is quite important in soft matter.

(b) Slow and non-equilibrium response. The collectivity of soft matter slows down their dynamics. The response time of simple liquids is of the order of 10^{-9} [s], while it can be billions or more times longer (1 [s] to 10^4 [s]) in solutions of polymers and colloids. Consequently, the properties of the non-equilibrium state, or the dynamics in the non-equilibrium state, are quite important in soft matter.

These characteristics are the results of the fact that the fundamental structural units of soft matter are very large. The reason can be understood by the following examples.

Consider a tube containing a solution in which the density of solute is slightly larger than that of solvent. If the tube is rotated as shown in

[1] The materials discussed here do not cover the whole class of soft matter. Other materials, such as granular materials and glassy materials, are often included in soft matter, but will not be discussed in this book.

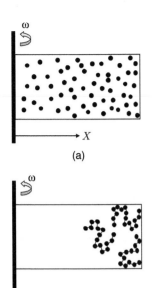

(a)

(b)

Fig. 1.8 When a tube containing a solution is rotated, the heavier component (assumed to be the solute) is pushed outward by the centrifugal force, and the concentration becomes non-uniform. In a solution of monomers, the concentration changes little as in (a). On the other hand, if these monomers are connected to form large polymers, a large concentration gradient is created as in (b).

<hr/>

[2] m can be expressed as $m = \Delta \rho v$, where $\Delta \rho$ is the difference in the density of solute and solvent, and v is the volume of the solute molecule. Equation (1.1) is derived from the fact that the centrifugal force acting on the particle at x is $m\omega^2 x$.

Fig. 1.8, the solute will be pushed outward by the centrifugal force. This effect is expressed by the potential energy for the solute molecule

$$U(x) = -\frac{1}{2}m\omega^2 x^2 \tag{1.1}$$

where m is the effective mass of the solute molecule,[2] ω is the angular velocity, and x is the distance from the rotation centre. At equilibrium, the probability of finding the particle at x is given by the Boltzmann distribution

$$P(x) \propto \exp\left(-\frac{U(x)}{k_B T}\right) \propto \exp\left(\frac{m\omega^2 x^2}{2k_B T}\right) \tag{1.2}$$

For small solute molecules, the factor $m\omega^2 x^2/k_B T$ is small, and hence the solution remains essentially homogeneous in the container. Now suppose that the monomers are transformed to polymers consisting of N monomers. Since the mass of the polymer is Nm, the probability changes as

$$P'(x) \propto \exp\left(\frac{Nm\omega^2 x^2}{2k_B T}\right) \tag{1.3}$$

For large N, the factor $Nm\omega^2 x^2/k_B T$ becomes non-negligible, and the polymer moves to the outer end of the tube. Indeed, polymer molecules are often separated from polymer solutions by centrifugal forces. This method is not practical for monomer solutions since the angular velocity needed to separate monomers is very large.

The same considerations apply in controlling the molecular orientation in liquid crystals. The orientation of the molecules constituting liquid crystals can be controlled by an electric or magnetic field. In the isotropic phase, the field needed to change the molecular orientation is very large, but in the liquid crystalline phase, the required field becomes small since the molecules in the liquid crystalline phase rotate collectively.

In the following chapters, we shall first discuss the equilibrium properties of several typical soft matter systems, polymers, colloids, surfactants, and liquid crystals. Next, we shall discuss non-equilibrium phenomena such as relaxation, diffusion, permeation, flow, and deformation. The last chapter is devoted to ionic soft matter. Through these discussions, we would like to demonstrate that the principles of physics are working in these seemingly complex systems.

1.6   Summary of this chapter

Soft matter includes a large class of materials (polymers, colloids, surfactants, and liquid crystals, etc.). A common feature of these materials is they all consist of large structural units. This gives two characteristics to soft matter:

(1) large and nonlinear responses;

(2) slow and non-equilibrium responses.

Further reading

(1) *Fragile Objects: Soft Matter, Hard Science and the Thrill of Discovery*, Pierre Gilles de Gennes, Jacques Badoz, translated by Axel Reisinger, Copernicus (1996).

(2) *Introduction to Soft Matter: Synthetic and Biological Self-Assembling Materials*, Ian W. Hamley, Wiley (2007).

(3) *Soft Condensed Matter*, Richard A. L. Jones, Oxford University Press (2002).

2 Soft matter solutions

Many soft matter systems exist as solutions. Solutions are made by dissolving a material in a liquid. The dissolved material is called the solute and the liquid is called the solvent. The main player in solutions is the solute, but the solvent also plays an important role. The effective interaction between solute molecules can be controlled by the solvent: solute molecules attract or repel each other depending on the solvent. By properly changing the solvent conditions (e.g., by changing the temperature or composition of the solvent), one can induce various orderings (crystallization, phase separation) of solute molecules.

In this chapter, we shall first discuss the thermodynamics of solutions, summarizing basic concepts of solutions such as mixing free energy, osmotic pressure, chemical potential, etc., and discussing how these quantities are related to the miscibility of solute and solvent. We then discuss two soft matter solutions: polymer solutions and colloidal solutions. The essential characteristic of these solutions is that the size of solute molecule is much larger than that of the solvent molecule. The main theme here is how the size affects the solution properties.

2.1 Thermodynamics of solutions

2.1.1 Free energy of solutions

Consider a homogeneous solution made of two components, solute and solvent. The thermodynamic state of a two-component solution can be specified by four parameters: temperature T, pressure P, and two other parameters which specify the amount of solute and solvent in the solution. A natural choice for these parameters is N_p and N_s, the number of solute and solvent molecules, respectively.[1] Then the Gibbs free energy of the solution is written as

$$G = G(N_p, N_s, T, P) \tag{2.1}$$

The solution state can be specified by other parameters, such as the volume V of the solution, and the solute concentration.

There are many ways to represent the solute concentration. A commonly used quantity is weight concentration, the weight of solute molecules per unit volume (typically in units of kg/m^3)

$$c = \frac{m_p N_p}{V} \tag{2.2}$$

[1] In this book, we shall use the symbol p to denote solute (meaning *polymer* or *particle*) and the symbol s to denote solvent.

where m_p is the mass of a solute molecule. The solute concentration can be represented by the molar fraction x_m, or the mass fraction ϕ_m of solute, each being defined by

$$x_m = \frac{N_p}{N_p + N_s} \tag{2.3}$$

$$\phi_m = \frac{m_p N_p}{m_p N_p + m_s N_s} \tag{2.4}$$

In the literature of soft matter, the solute concentration is often represented by the volume fraction ϕ defined by

$$\phi = \frac{v_p N_p}{v_p N_p + v_s N_s} \tag{2.5}$$

where v_p and v_s are specific volumes defined by

$$v_p = \left(\frac{\partial V}{\partial N_p}\right)_{T,P,N_s}, \qquad v_s = \left(\frac{\partial V}{\partial N_s}\right)_{T,P,N_p} \tag{2.6}$$

The specific volume satisfies the following equation.[2]

$$V = v_p N_p + v_s N_s \tag{2.8}$$

Equation (2.8) indicates that v_p and v_s correspond to the volume of solute molecule and the volume of solvent molecule, respectively.

In general, v_p and v_s are functions of T, P, and solute concentration. However, in many soft matter solutions (especially in organic solvents), v_p and v_s change little with these parameters. Therefore, in this book, we shall assume that v_p and v_s are constants.[3] Such a solution is called incompressible. In an incompressible solution, c and ϕ are related by

$$c = \frac{m_p}{v_p}\phi = \rho_p \phi \tag{2.9}$$

where $\rho_p = m_p/v_p$ is the density of pure solute.

Given the Gibbs free energy $G(N_p, N_s, T, P)$, the volume of the solution is given by

$$V = \frac{\partial G}{\partial P} \tag{2.10}$$

Since V is constant, $G(N_p, N_s, T, P)$ can be written as

$$G(N_p, N_s, T, P) = PV + F(N_p, N_s, T) \tag{2.11}$$

The function $F(N_p, N_s, T)$ represents the Helmholtz free energy of the solution. Usually the Helmholtz free energy is written as a function of N_p, N_s, T, and V. However in an incompressible solution, V is given by eq. (2.8), and cannot be an independent variable.

[2] Equation (2.8) is derived as follows. If N_p and N_s are increased by a factor α under constant T and P, the volume of the system also increases by the factor α. Therefore

$$V(\alpha N_p, \alpha N_s, T, P) = \alpha V(N_p, N_s, T, P) \tag{2.7}$$

Differentiating both sides of eq. (2.7) with respect to α, and using eq. (2.6), we have eq. (2.8).

[3] If we use the mass fraction ϕ_m instead of the volume fraction ϕ, we do not need the assumption of constant v_p and v_s since m_p and m_s are constant. Here we use the volume fraction ϕ to keep the notational consistency of the book.

The Helmholtz free energy $F(N_p, N_s, T)$ is an extensive quantity and therefore must satisfy the scaling relation $F(\alpha N_p, \alpha N_s, T) = \alpha F(N_p, N_s, T)$ for any number α. Hence $F(N_p, N_s, T)$ is written as[4]

$$F(N_p, N_s, T) = V f(\phi, T) \qquad (2.14)$$

where $f(\phi, T)$ represents the Helmholtz free energy per unit volume of the solution, and is called the Helmholtz free energy density. By eq. (2.11), the Gibbs free energy is written as

$$G(N_p, N_s, T, P) = V[P + f(\phi, T)] \qquad (2.15)$$

In eqs. (2.14) and (2.15), V and ϕ are expressed in terms of N_p and N_s using eqs. (2.8) and (2.5).

[4] Setting $\alpha = v_p/V$ in the equation $F(\alpha N_p, \alpha N_s, T) = \alpha F(N_p, N_s, T)$, we have

$$F\left(\frac{v_p N_p}{V}, \frac{v_p N_s}{V}, T\right) = \frac{v_p}{V} F(N_p, N_s, T) \qquad (2.12)$$

Hence

$$\begin{aligned} F(N_p, N_s, T) &= \frac{V}{v_p} F\left(\frac{v_p N_p}{V}, \frac{v_p N_s}{V}, T\right) \\ &= \frac{V}{v_p} F\left(\phi, \frac{v_p}{v_s}(1-\phi), T\right) \end{aligned} \qquad (2.13)$$

The right-hand side can be written in the form of eq. (2.14)

2.1.2   Mixing criterion

If a drop of benzene is mixed with a large amount of water, it dissolves in the water completely. If the amount of water is not sufficient, the mixing is not complete, and undissolved droplets of benzene remain in the solution. Whether two liquids 1 and 2 mix completely (i.e., to form a homogeneous solution) or not depends on the liquids, their volume and temperature. The mixing behaviour of liquids can be discussed if the free energy density $f(\phi, T)$ is known.

Suppose that a solution having volume and concentration (V_1, ϕ_1) is mixed with another solution of (V_2, ϕ_2) and a homogeneous solution is made (see Fig. 2.1(a)).

If the two solutions mix completely, the final solution has volume $V_1 + V_2$ and volume fraction

$$\phi = \frac{\phi_1 V_1 + \phi_2 V_2}{V_1 + V_2} = x\phi_1 + (1-x)\phi_2 \qquad (2.16)$$

where x is defined by

$$x = \frac{V_1}{V_1 + V_2} \qquad (2.17)$$

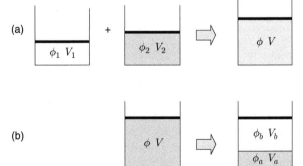

Fig. 2.1 (a) The situation of mixing: two solutions having concentration ϕ_1 and ϕ_2 are mixed and produce a homogeneous solution of concentration ϕ. (b) The situation of demixing: a homogeneous solution of concentration ϕ separates into two solutions of concentration ϕ_a and ϕ_b.

Before mixing, the free energy of the system is equal to $V_1 f(\phi_1) + V_2 f(\phi_2)$. (Here the argument T is omitted for simplicity.) On the other hand, the free energy of the final solution is given by $(V_1 + V_2) f(\phi)$. In order to have a homogeneous solution, the free energy after mixing must be less than that before mixing, i.e.,

$$(V_1 + V_2) f(\phi) < V_1 f(\phi_1) + V_2 f(\phi_2) \qquad (2.18)$$

or

$$f(x\phi_1 + (1 - x)\phi_2) < x f(\phi_1) + (1 - x) f(\phi_2) \qquad (2.19)$$

If the two solutions mix homogeneously at any volume ratio V_1/V_2, eq. (2.19) has to be satisfied for any x in the region of $0 \leq x \leq 1$. This condition is equivalent to the condition that $f(\phi)$ is upper concave in the region $\phi_1 < \phi < \phi_2$ (see Fig. 2.2(a)), i.e.,

$$\frac{\partial^2 f}{\partial \phi^2} > 0, \qquad \text{for} \quad \phi_1 < \phi < \phi_2 \qquad (2.20)$$

On the other hand, if $f(\phi)$ has an upper convex part in some region between ϕ_1 and ϕ_2 as in Fig. 2.2(b), the solution ceases to stay homogeneous. As explained in the caption of Fig. 2.2, the system can lower its free energy by separating into two solutions with concentration ϕ_a and ϕ_b. Such phenomena are called demixing or phase separation, and will be discussed further in later sections.

2.1.3 Osmotic pressure

The thermodynamic force which tends to mix solute and solvent can be measured by osmotic pressure. Osmotic pressure is the force which appears when a solution is brought into contact with pure solvent across a semi-permeable membrane (see Fig. 2.3). A semi-permeable membrane allows solvent to pass through, but prohibits solute to pass through. In the case that the free energy of the system is lowered if solvent and solute mix, solvent molecules tend to flow into the solution crossing the semi-permeable membrane. If the semi-permeable membrane could move freely, all solvent molecules would move into the solution region by pushing the semi-permeable membrane to the right, eventually making a homogeneous solution. In order to prevent this from happening, a force has to be applied to the semi-permeable membrane as shown in Fig. 2.3. Osmotic pressure Π is the force, per unit area of the semi-permeable membrane, needed to keep the solution volume constant.

Let $F_{tot}(V)$ be the free energy of the whole system consisting of solution (of volume V) and pure solvent (of volume $V_{tot} - V$). If we move the semi-permeable membrane and change the solution volume by dV, we do work $-\Pi dV$ to the system. This work is equal to the change of the free energy $dF_{tot}(V)$ of the whole system. Therefore Π is expressed as

$$\Pi = -\frac{\partial F_{tot}(V)}{\partial V} \qquad (2.21)$$

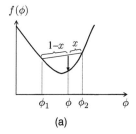

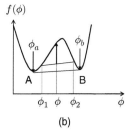

Fig. 2.2 Free energy of a homogeneous solution is plotted as a function of solute concentration. (a) The case that the solute and solvent can mix at any composition. (b) The case that phase separation takes place. If a solution of concentration ϕ_1 is mixed with the other solution of concentration ϕ_2 with the ratio $x : (1 - x)$ and forms a homogeneous solution, the free energy of the system changes from $x f(\phi_1) + (1-x) f(\phi_2)$ to $f(x\phi_1 + (1-x)\phi_2)$. The change of the free energy is indicated by the arrows in the figure. The free energy decreases in the case of (a), but increases in the case of (b).

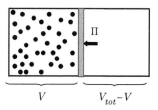

Fig. 2.3 Definition of osmotic pressure. When a solution is brought into contact with pure solvent across a semi-permeable membrane, the solvent tends to move into the solution and to increase the volume of the solution. In order to keep the volume of the solution at a fixed value V, a force has to be applied to the membrane. The force per unit area of the membrane is the osmotic pressure $\Pi(V)$ of the solution.

On the other hand, $F_{tot}(V)$ can be expressed by the free energy density $f(\phi)$ as

$$F_{tot} = Vf(\phi) + (V_{tot} - V)f(0) \tag{2.22}$$

From eqs. (2.21) and (2.22), it follows that[5]

$$\Pi(\phi) = -f(\phi) + \phi f'(\phi) + f(0) \tag{2.23}$$

Here $f'(\phi)$ stands for $\partial f / \partial \phi$.

The osmotic pressure represents the strength of the tendency that solvent and solute mix together. If the osmotic pressure is large, solvent tends to permeate strongly into the region of solution, or, equivalently, solute tends to diffuse strongly into the region of solvent. As will be shown in Chapters 7 and 8, osmotic pressure is the driving force for the mixing phenomena of solute and solvent such as diffusion and permeation.

2.1.4   Chemical potential

The thermodynamic force for mixing can also be expressed in terms of the chemical potential. The chemical potential of solute μ_p and that of solvent μ_s are defined by

$$\mu_p = \left(\frac{\partial G}{\partial N_p} \right)_{N_s, T, P}, \qquad \mu_s = \left(\frac{\partial G}{\partial N_s} \right)_{N_p, T, P} \tag{2.24}$$

Using $G = (N_p v_p + N_s v_s)[P + f(\phi, T)]$ and $\phi = N_p v_p / (N_p v_p + N_s v_s)$, μ_p is written (after some calculation) as

$$\mu_p(\phi, T, P) = v_p \left[P + f(\phi, T) + (1 - \phi) f'(\phi, T) \right] \tag{2.25}$$

Similarly

$$\mu_s(\phi, T, P) = v_s \left[P + f(\phi, T) - \phi f'(\phi, T) \right] \tag{2.26}$$

Using eq. (2.23), eq. (2.26) can be written as

$$\mu_s(\phi, T, P) = v_s[P - \Pi(\phi, T) + f(0, T)] = v_s[P - \Pi(\phi, T)] + \mu_s^{(0)}(T) \tag{2.27}$$

where $\mu_s^{(0)}(T)$ is a function of temperature only.

If one differentiates both sides of (2.25) and (2.26) with respect to ϕ, one gets

$$\frac{\partial \mu_p}{\partial \phi} = v_p (1 - \phi) \frac{\partial^2 f}{\partial \phi^2} \qquad \frac{\partial \mu_s}{\partial \phi} = -v_s \phi \frac{\partial^2 f}{\partial \phi^2} \tag{2.28}$$

Therefore, in a homogeneous solution where eq. (2.20) is satisfied, $\partial \mu_p / \partial \phi$ is always positive, and the chemical potential of solute increases with the increase of solute concentration. On the other hand, the chemical potential of solvent decreases with the increase of solute concentration. In general, molecules migrate from a high chemical potential region to a low chemical potential region. Therefore if there is a gradient of solute concentration, solute molecules tend to migrate from high concentration regions to low concentration regions. On the other hand, solvent molecules move from low solute concentration regions to high concentration regions. Both migrations of solute and solvent reduce the concentration gradient and make the solution more uniform. This is the thermodynamic origin of the diffusion phenomena taking place in solution.

2.1.5 Dilute solution

When the solute concentration is sufficiently low, the effect of interaction between solute molecules can be ignored, and the osmotic pressure is given by van't Hoff's law: the osmotic pressure is proportional to the number density $n = N_p/V = \phi/v_p$ of solute molecules:

$$\Pi = \frac{N_p k_B T}{V} = \frac{\phi k_B T}{v_p} \tag{2.29}$$

The interaction between solute molecules gives correction terms to van't Hoff's law. At low concentration, the correction terms are written as a power series in ϕ:

$$\Pi = \frac{\phi k_B T}{v_p} + A_2 \phi^2 + A_3 \phi^3 + \cdots \tag{2.30}$$

The coefficients A_2 and A_3 are called the second and third virial coefficients, respectively. The virial coefficients are expressed in terms of the effective interaction potential between solute molecules. If the interaction is repulsive, A_2 is positive, while if the interaction is attractive, A_2 can be negative.

If the osmotic pressure $\Pi(\phi)$ is known as a function of ϕ, the free energy density $f(\phi)$ can be calculated. Equation (2.23) is written as

$$\Pi = \phi^2 \frac{\partial}{\partial \phi} \left(\frac{f}{\phi} \right) + f(0) \tag{2.31}$$

$f(\phi)$ can be obtained by integrating (2.31). If one uses eq. (2.30) for $\Pi(\phi)$, one finally gets

$$f(\phi) = f_0 + k_0 \phi + \frac{k_B T}{v_p} \phi \ln \phi + A_2 \phi^2 + \frac{1}{2} A_3 \phi^3 + \cdots \tag{2.32}$$

where f_0 represents $f(0)$ and k_0 is a constant independent of ϕ. Notice that the coefficients f_0, k_0, A_2, and A_3 are functions of temperature T. Using eqs. (2.26) and (2.27), the chemical potentials of solvent and solute are written as

$$\mu_s(\phi) = \mu_s^0 + P v_s - \frac{v_s}{v_p} k_B T \phi - v_s (A_2 \phi^2 + A_3 \phi^3 + \cdots) \tag{2.33}$$

$$\mu_p(\phi) = \mu_p^0 + Pv_p + k_BT \ln \phi$$
$$+ v_p \left[\left(2A_2 - \frac{k_BT}{v_p} \right) \phi + \left(\frac{3}{2} A_3 - A_2 \right) \phi^2 + \cdots \right] \quad (2.34)$$

2.2 Phase separation

2.2.1 Coexistence of two phases

As discussed in Section 2.1.2, if the free energy density $f(\phi)$ has an upper convex part where $\partial^2 f/\partial \phi^2 < 0$, a homogeneous solution cannot be stable. Suppose that by a change of external parameters (such as temperature and pressure), a homogeneous solution is brought into such a state. Then the solution starts to separate into two regions, a high concentration region and a low concentration region. The high concentration region is called the concentrated phase, and the low concentration region is called the dilute phase, and the phenomenon is called phase separation.

The phenomenon of phase separation in solution is analogous to the phenomenon of the gas–liquid transition in a one-component system. For example, if a vapour of water is cooled, a certain part of the water condenses and forms a liquid phase (high-density phase), while the rest remains as a gas phase (low-density phase). Likewise, the phase separation of a solution results in the coexistence of a concentrated phase and a dilute phase: the density and the pressure in the gas–liquid transition correspond to the concentration and the osmotic pressure, respectively, in the phase separation of solutions.

Thermodynamic criteria for the phase separation can be obtained in the same way as the criterion of mixing. Suppose that a homogeneous solution of volume V and concentration ϕ separates into two solutions, one having volume V_1 and concentration ϕ_1 and the other having volume V_2 and concentration ϕ_2. The volumes V_1 and V_2 are determined by the conservation of the solution volume, $V = V_1 + V_2$, and the conservation of the solute volume, $\phi V = \phi_1 V_1 + \phi_2 V_2$. This gives

$$V_1 = \frac{\phi_2 - \phi}{\phi_2 - \phi_1} V, \qquad V_2 = \frac{\phi - \phi_1}{\phi_2 - \phi_1} V \quad (2.35)$$

Therefore, the free energy of the phase-separated state is given by

$$F = V_1 f(\phi_1) + V_2 f(\phi_2) = V \left[\frac{\phi_2 - \phi}{\phi_2 - \phi_1} f(\phi_1) + \frac{\phi - \phi_1}{\phi_2 - \phi_1} f(\phi_2) \right] \quad (2.36)$$

As shown in Fig. 2.2(a), the expression in the brackets [ ] corresponds to the line connecting the two points $P_1(\phi_1, f(\phi_1))$ and $P_2(\phi_2, f(\phi_2))$ in the free energy curve. Therefore to minimize (2.36), one needs to seek the two points P_1 and P_2 on the curve $f(\phi)$ so that the height of the line $P_1 P_2$ at ϕ is minimized. Such a line is given by the common tangent for the curve $f(\phi)$ (see Fig. 2.4(a)). Let ϕ_a and ϕ_b be the concentrations at the tangent points. A homogeneous solution which has concentration

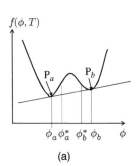

(a)

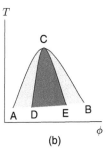

(b)

Fig. 2.4 Free energy and phase diagram of a solution. (a) ϕ_a and ϕ_b are the concentrations at which the line connecting the points P_a and P_b on the curve is tangent to the free energy curve $f(\phi)$. ϕ_a^* and ϕ_b^* are the concentrations at which $\partial^2 f(\phi)/\partial \phi^2$ becomes equal to zero. A solution of concentrations ϕ ($\phi_a < \phi < \phi_b$) can minimize its free energy by phase separating into two solutions of concentrations ϕ_a and ϕ_b. The solution is unstable if $\phi_a^* < \phi < \phi_b^*$, and is locally stable if $\phi_a < \phi < \phi_a^*$ or $\phi_b^* < \phi < \phi_b$. (b) Phase diagram obtained by plotting $\phi_a, \phi_b, \phi_a^*, \phi_b^*$ as a function of temperature. Solutions in the dark grey region are unstable and solutions in the light grey region are metastable.

ϕ in the region $\phi_a < \phi < \phi_b$ becomes most stable (i.e., its free energy becomes a minimum) if it separates into two solutions of concentrations ϕ_a and ϕ_b.

The condition that the line $P_a P_b$ is the common tangent for the curve $f(\phi)$ can be written as

$$f'(\phi_a) = f'(\phi_b), \qquad f(\phi_a) - f'(\phi_a)\phi_a = f(\phi_b) - f'(\phi_b)\phi_b \qquad (2.37)$$

It is easy to confirm that the condition (2.37) is equivalent to the condition that the chemical potential of the concentrated phase and the dilute phase are both equal to each other for solute and solvent ($\mu_p(\phi_a) = \mu_p(\phi_b)$ and $\mu_s(\phi_a) = \mu_s(\phi_b)$). Under this condition, the osmotic pressures of both phases are also equal to each other ($\Pi(\phi_a) = \Pi(\phi_b)$).

2.2.2   Phase diagram

The concentration region $\phi_a < \phi < \phi_b$ can be further divided into two regions (see Fig. 2.4). If $\partial^2 f/\partial\phi^2 < 0$, the solution is unstable. The graphical construction explained in Fig. 2.2 indicates that any small deviation from the homogeneous state lowers the free energy. Therefore the system is unstable. On the other hand, if $\partial^2 f/\partial\phi^2 > 0$, any small deviation from the homogeneous state increases the free energy. In other words, the system can remain homogeneous as long as the deviation from the original state is small. Such a state is called locally stable, or metastable. Let ϕ_a^*, ϕ_b^* be the concentrations at which $\partial^2 f/\partial\phi^2$ is equal to zero. They correspond to the points of zero curvature in the graph of $f(\phi)$. In the region of $\phi_a^* < \phi < \phi_b^*$, the solution is unstable, and in the regions of $\phi_a < \phi < \phi_a^*$ and $\phi_b^* < \phi < \phi_b$, the solution is metastable.

Since $\phi_a, \phi_b, \phi_a^*, \phi_b^*$ are functions of temperature, they can be drawn in the $\phi - T$ plane. An example is shown in Fig. 2.4(b). The curves AC and BC denote the lines $\phi_a(T)$ and $\phi_b(T)$, and the curves DC and EC denote the lines $\phi_a^*(T)$ and $\phi_b^*(T)$. The curve connecting A, C, and B is called the coexistence curve or the binodal line. The solution in the state below this curve can phase separate into two phases having concentrations $\phi_a(T)$ and $\phi_b(T)$. On the other hand, the curve connecting D, C, and E stands for the boundary between the metastable state and unstable state, and is called the stability boundary or the spinodal line.

The top of the spinodal line is called the critical point. Since the two concentrations $\phi_a^*(T)$ and $\phi_b^*(T)$ which satisfy $\partial^2 f/\partial\phi^2 = 0$ merge at the critical point, $\partial^3 f/\partial\phi^3$ must be equal to zero at the critical point. Therefore the critical point C is determined by the following two equations

$$\frac{\partial^2 f}{\partial\phi^2} = 0, \qquad \frac{\partial^3 f}{\partial\phi^3} = 0 \qquad \text{at the critical point} \qquad (2.38)$$

The coexistence curve and the spinodal curve merge (having the same tangent) at the critical point.

2.3   Lattice model

2.3.1   Molecules on a lattice

The free energy curve $f(\phi)$ can be obtained experimentally (e.g., by using eq. (2.31)). Alternatively, $f(\phi)$ can be calculated theoretically if the Hamiltonian of the system is known. In this section we shall show such a calculation using a simple model called the lattice model.

The lattice model is shown in Fig. 2.5(a). Here the microscopic configuration of molecules in the solution is represented by an arrangement of black and white circles placed on a lattice: the black circle represents a solute molecule and the white circle represents a solvent molecule. For simplicity, we assume that the solute molecule and the solvent molecule have the same volume v_c, and that each cell in a lattice is occupied by either a solute molecule or a solvent molecule. The solution volume V and the solute volume fraction ϕ are given by

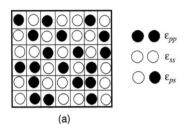

$$V = v_c N_{tot}, \qquad \phi = \frac{N_p}{N_{tot}} \tag{2.39}$$

where $N_{tot} = N_p + N_s$ is the total number of cells.

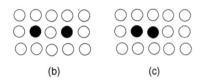

To take into account the interaction between the molecules, we assume that there is an interaction energy between the neighbouring molecules. Let ϵ_{pp}, ϵ_{ss}, and ϵ_{ps} be the interaction energy acting between the solute–solute, solvent–solvent, and solute–solvent pairs located at neighbouring sites in the lattice model. Since neighbouring molecules usually attract each other by van der Waals forces, ϵ_{pp} ϵ_{ss}, and ϵ_{ps} are usually negative. The total energy of the system is given by the sum of all such energies. The energy E_i for a certain configuration i is calculated by

$$E_i = \epsilon_{pp} N_i^{(pp)} + \epsilon_{ss} N_i^{(ss)} + \epsilon_{ps} N_i^{(ps)} \tag{2.40}$$

where $N_i^{(pp)}$ stands for the number of neighbouring solute–solute pairs in the configuration i, and $N_i^{(ss)}$ and $N_i^{(ps)}$ are defined in the same way.

The partition function of this system is calculated by

$$Z = \sum_i \exp(-E_i/k_B T) \tag{2.41}$$

where the summation is taken for all possible configurations of the molecules. The free energy density $f(\phi, T)$ is then obtained by

$$f(\phi, T) = -\frac{k_B T}{V} \ln Z \tag{2.42}$$

Fig. 2.5 Lattice model for symmetric solutions. Solute and solvent molecules are represented by black and white circles, respectively.

2.3.2   Effective interaction between solute molecules

Rigorous calculation of the summation in eq. (2.41) is difficult. As an approximation, we replace E_i in the summation by the mean value $\overline{E}$. This is called the mean field approximation. In this approximation, all terms in the summation are constant independent of i, and therefore the summation gives

$$Z \simeq W \exp(-\overline{E}/k_B T) \qquad (2.43)$$

where W is the number of terms in the summation. W is equal to the number of ways of placing N_p molecules on $N_{tot} = N_p + N_s$ cells, and is given by

$$W = \frac{(N_p + N_s)!}{N_p! N_s!} \qquad (2.44)$$

The mean energy $\overline{E}$ can be estimated as follows. Each cell in the lattice has z neighbouring cells (z is called the coordination number). Among the z cells, $z\phi$ cells are, on average, occupied by solute molecules, and the remaining $z(1-\phi)$ cells are occupied by solvent molecules. Therefore the average number of neighbouring solute pairs is estimated by

$$\overline{N_{pp}} = (1/2)z\phi N_p = z N_{tot}\phi^2/2 \qquad (2.45)$$

Similarly, the average number of N_{ss} and N_{ps} are given by

$$\overline{N_{ss}} = z N_{tot}(1 - \phi)^2/2 \qquad (2.46)$$

$$\overline{N_{ps}} = z N_{tot}\phi(1 - \phi) \qquad (2.47)$$

Consequently, the average energy $\overline{E}$ is

$$\begin{aligned}
\overline{E} &= \epsilon_{pp}\overline{N_{pp}} + \epsilon_{ps}\overline{N_{ps}} + \epsilon_{ss}\overline{N_{ss}} \\
&= \frac{1}{2}N_{tot}z\left[\epsilon_{pp}\phi^2 + 2\epsilon_{ps}\phi(1-\phi) + \epsilon_{ss}(1-\phi)^2\right] \\
&= \frac{1}{2}N_{tot}z\Delta\epsilon\phi^2 + C_0 + C_1\phi \qquad (2.48)
\end{aligned}$$

where

$$\Delta\epsilon = \epsilon_{pp} + \epsilon_{ss} - 2\epsilon_{ps} \qquad (2.49)$$

and C_0 and C_1 are constants independent of ϕ.

The energy $\Delta\epsilon$ represents the effective interaction between the solute molecules in the solution. Notice that this energy depends on the inter-action energy between all pairs. The reason is explained in Fig. 2.5(b) and (c). Consider a pair of solute molecules placed apart from each other in the sea of solvent molecules (Fig. 2.5(b)). When the pair is brought into contact with each other as in Fig. 2.5(c), two pairs of solute–solvent molecules disappear, and two new pairs (a solute–solute pair and a solvent–solvent pair) appear. The energy change associated with this recombination is $\Delta\epsilon$.

The above argument indicates that whether solute molecules like each other or not in solutions is not determined by ϵ_{pp} alone. It depends on the interaction energy of other pairs. For example, the pair of solute molecules which attract each other in vacuum (i.e., ϵ_{pp} is negative) may repel each other in solutions if the attractive interaction between solute and solvent molecules is stronger than the attractive interaction between solute molecules. The effective interaction between solute molecules is determined by $\Delta\epsilon$. If $\Delta\epsilon$ is positive ($\Delta\epsilon > 0$), solute molecules tend to stay away, and the solution is homogeneous. On the other hand, if $\Delta\epsilon$

is negative ($\Delta\epsilon < 0$), solute molecules attract each other, and if their attraction is strong enough, phase separation takes place.

The above argument can be stated in a different way. The energy $\Delta\epsilon$ can be regarded as representing the affinity between solute and solvent. If $\Delta\epsilon$ is positive, solute and solvent attract each other, and therefore they tend to mix. On the other hand, if $\Delta\epsilon$ is negative, solute and solvent repel each other, and if $\Delta\epsilon$ is negative and large, phase separation takes place.

2.3.3   Free energy

Using eqs. (2.43)–(2.48), the free energy of the solution is obtained as

$$F = -k_B T \ln Z = -k_B T \ln W + \overline{E} \tag{2.50}$$

The first term on the right hand side represents the entropic contribution. The mixing entropy is given by

$$S_{mix} = k_B \ln W = k_B[\ln(N_p + N_s)! - \ln N_p! - \ln N_s!] \tag{2.51}$$

By using Stirling's formula ($\ln N! = N \ln N - N$), this is written as

$$S_{mix} = k_B[(N_p + N_s)\ln(N_p + N_s) - N_p \ln N_p - N_s \ln N_s]$$
$$= k_B\left[-N_p \ln\left(\frac{N_p}{N_p + N_s}\right) - N_s \ln\left(\frac{N_s}{N_p + N_s}\right)\right]$$
$$= k_B N_{tot}[-\phi \ln\phi - (1 - \phi)\ln(1 - \phi)] \tag{2.52}$$

Hence the free energy of the solution is given by.

$$F = N_{tot}\left\{k_B T\left[\phi \ln\phi + (1 - \phi)\ln(1 - \phi)\right] + \frac{z}{2}\Delta\epsilon\phi^2\right\} \tag{2.53}$$

Here we have dropped the terms which are represented as a linear function of ϕ (such as $(C_0 + C_1\phi$ and $PV)$) since they do not affect whether the solution remain homogeneous or phase separated.[6] Therefore the free energy density $f(\phi) = F/V$ is given by

$$f(\phi) = \frac{1}{v_c}\left\{k_B T\left[\phi \ln\phi + (1 - \phi)\ln(1 - \phi)\right] + \frac{1}{2}z\Delta\epsilon\phi^2\right\} \tag{2.54}$$

This is often rewritten in the following form

$$f(\phi) = \frac{k_B T}{v_c}\left[\phi \ln\phi + (1 - \phi)\ln(1 - \phi) + \chi\phi(1 - \phi)\right] \tag{2.55}$$

where χ is defined by

$$\chi = -\frac{z}{2k_B T}\Delta\epsilon \tag{2.56}$$

In going from eq. (2.54) to eq. (2.55), a linear term in ϕ has been added. This is for the convenience of subsequent analysis. The function $f(\phi)$ defined by eq. (2.55) has a mirror symmetry with respect to the axis of $\phi = 1/2$.

[6] Whether the solution remains homogeneous or not is determined by the difference in the free energy on the left- and right-hand sides of eq. (2.19), i.e., $\Delta f = f(x\phi_1 + (1 - x)\phi_2) - xf(\phi_1) - (1 - x)f(\phi_2)$. The linear term in $f(\phi)$ does not affect the value of Δf.

The osmotic pressure of the solution is calculated by eq. (2.23)

$$\Pi = -f(\phi) + \phi f'(\phi) + f(0) = \frac{k_B T}{v_c}\left[-\ln(1-\phi) - \chi\phi^2 \right] \quad (2.57)$$

In dilute solutions ($\phi \ll 1$), this can be written as

$$\Pi = \frac{k_B T}{v_c}\left[\phi + \left(\frac{1}{2} - \chi\right)\phi^2 \right] \quad (2.58)$$

The first term corresponds to van't Hoff's law. The second term corresponds to the second virial coefficient A_2 in eq. (2.30).

The second virial coefficient A_2 represents the effective interaction between solute molecules. If A_2 is positive, the net interaction is repulsive, and if A_2 is negative, the net interaction is attractive. Notice that in the lattice model, there are two interactions considered. One is the excluded volume interaction which arises from the constraint that solute molecules cannot occupy the same lattice sites. This gives a repulsive contribution and is represented by the positive constants $1/2$ in eq. (2.58). The other is the energetic interaction acting between the neighbouring molecules. This can be attractive or repulsive depending on the sign of $\Delta\epsilon$ or χ. It is attractive when $\delta\epsilon < 0$ or $\chi > 0$. The second virial coefficients A_2 represents whether solute molecules repel or attract each other as a sum of these two effects.

2.3.4 Phase separation

Figure 2.6(a) shows the schematic shape of the free energy function $f(\phi)$ for various values of χ. $f(\phi)$ has a mirror symmetry with respect to the line $\phi = 1/2$. If χ is less than a certain critical value χ_c, $f(\phi)$ has only one minimum at $\phi = 1/2$, while if χ is larger than χ_c, $f(\phi)$ has two local minima. The value of χ_c is determined by the condition that the curvature at $\phi = 1/2$ changes sign; i.e., $\partial^2 f/\partial\phi^2 = 0$ at $\phi = 1/2$. This condition gives the critical point

$$\chi_c = 2, \qquad \phi_c = \frac{1}{2} \quad (2.59)$$

The spinodal line is determined by $\partial^2 f/\partial\phi^2 = 0$, and is given by

$$\chi = \frac{1}{2}\frac{1}{\phi(1-\phi)} \quad (2.60)$$

The coexistence curve is obtained by the common tangent construction discussed in Section 2.3.4. Since the free energy $f(\phi)$ curve has a mirror symmetry with respect to the line at $\phi = 1/2$, the common tangent line is given by the line connecting the two local minima of $f(\phi)$. Thus the concentrations $\phi_a(T)$ and $\phi_b(T)$ in the coexistence region are given by the two solutions of the equation $\partial f/\partial\phi = 0$. By eq. (2.55), this is given by

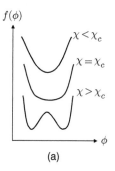

(a)

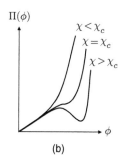

(b)

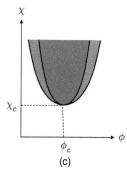

(c)

Fig. 2.6 (a) The Free energy $f(\phi)$ for a symmetric solution is plotted against solute volume fraction ϕ for various values of χ. (b) Osmotic pressure is plotted against ϕ. (c) Phase diagram of the solution. The dark shaded region is the unstable region, and the light shaded region is the metastable region.

$$\chi = \frac{1}{1 - 2\phi} \ln\left(\frac{1-\phi}{\phi}\right) \tag{2.61}$$

Figure 2.6(b) shows the osmotic pressure as a function of ϕ. For $\chi > \chi_c$, the curve has a part where $\partial\Pi/\partial\phi < 0$. This corresponds to the unstable region. Fig. 2.6(c) is a schematic phase diagram of the symmetric solution.

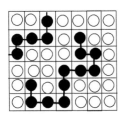

2.4   Polymer solutions

2.4.1   Lattice model of polymer solutions

In the previous section, solute molecules and solvent molecules are assumed to have the same size. In solutions of soft matter, e.g., polymer solutions and colloidal solutions, solute molecules (or particles) are much larger than solvent molecules. Let us now consider how this asymmetry in size affects the solution properties.

The size effect of polymer solutions can be discussed by the lattice model shown in Fig. 2.7, where a polymer is represented by N segments (represented by the black circles) connected by bonds. The segment corresponds to a monomer before the polymerization reaction. Here, for simplicity, it is assumed that the segment and solvent molecules have the same size. For such a model, the free energy density is now given by

Fig. 2.7 Lattice model for polymer solutions. A polymer is represented by N dark circles connected in series. The building block of the polymer (i.e., the dark circle) is called the segment.

$$f(\phi) = \frac{k_B T}{v_c}\left[\frac{1}{N}\phi\ln\phi + (1-\phi)\ln(1-\phi) + \chi\phi(1-\phi)\right] \tag{2.62}$$

The difference between eq. (2.62) and eq. (2.55) is the factor $1/N$ in front of $\phi\ln\phi$. This factor comes from the fact that the mixing entropy per segment of a polymer molecule is now given by $\phi\ln\phi/N$ since N segments are connected and cannot be placed independently.

For the free energy (2.62), the osmotic pressure is given by

$$\Pi = \frac{k_B T}{v_c}\left[\frac{\phi}{N} - \ln(1-\phi) - \phi - \chi\phi^2\right] \tag{2.63}$$

For $\phi \ll 1$, this is written as

$$\Pi = \frac{k_B T}{v_c}\left[\frac{\phi}{N} + \left(\frac{1}{2} - \chi\right)\phi^2\right] \tag{2.64}$$

Equation (2.64) gives van't Hoff's law $\Pi = k_B T\phi/(Nv_c)$ in the limit of $\phi \to 0$. However, for this behaviour to be observed, ϕ has to be extremely small. Due to the factor $1/N$, the first term is usually negligibly small compared with the second term. Therefore the osmotic pressure of a polymer solution is usually written as

$$\Pi = A_2\phi^2 \tag{2.65}$$

Figure 2.8(a) shows the schematic curve of the osmotic pressure as a function of ϕ.

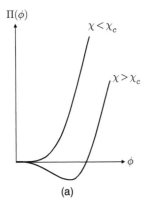

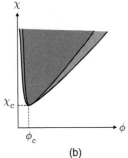

Fig. 2.8 (a) Osmotic pressure of polymer solutions is plotted against ϕ. (b) Phase diagram of polymer solutions. The dark shaded region is the unstable region, and the light shaded region is the metastable region.

The polymer size effect also appears in the phase diagram, in the form of an asymmetric shape. The spinodal line is calculated by $\partial^2 f / \partial \phi^2 = 0$, and is given by

$$\chi = \frac{1}{2} \left[\frac{1}{1 - \phi} + \frac{1}{N\phi} \right] \qquad (2.66)$$

This is illustrated in Fig. 2.8(b). The bottom of the spinodal line gives the critical point (ϕ_c, χ_c), which is now given by

$$\phi_c = \frac{1}{1 + \sqrt{N}}, \qquad \chi_c = \frac{1}{2} \left(1 + \frac{1}{\sqrt{N}} \right)^2 \qquad (2.67)$$

For large N, χ_c is equal to $1/2$. This is the value of χ at which the second virial coefficient A_2 changes from positive to negative.

In the case of symmetric solutions, phase separation does not take place at the value of χ where the second virial coefficient A_2 becomes zero. This is because there is a gain of entropy by mixing. When solvent and solute mix with each other, the entropy of the system increases, and there is a gain of free energy even if the effective interaction A_2 is zero. In the case of polymer solutions, the effect of entropy is small, and therefore phase separation takes place as soon as the second virial coefficient becomes negative.

2.4.2   Correlation effect in polymer solutions

The theory presented above ignores an important effect in polymer solutions, the correlation effect in the segment density. In the above theory, segment density is assumed to be uniform in the system. (This assumption has been used, for example, in evaluating the mean energy $\overline{E}$.) In reality, this assumption is not correct.

Figure 2.9 shows a schematic picture of polymer solutions. In a very dilute solution (Fig. 2.9(a)), polymer chains are well separated from each other: each chain occupies a spherical region of radius R_g. In such situations, the density of polymer segments is not homogeneous. The density is high (of the order of N/R_g^3) inside the region of the polymer coil, but is zero outside the region. Such a dilute solution is realized when the weight concentration c satisfies[7]

$$\frac{4\pi}{3} R_g^3 \frac{c}{m_p} < 1 \qquad (2.68)$$

The polymer coils start to overlap each other when the left-hand side of this equation exceeds one. The concentration

$$c^* = \frac{m_p}{\frac{4\pi}{3} R_g^3} \qquad (2.69)$$

is called the overlap concentration.

Below the overlap concentration c^*, or near c^*, the distribution of segments is strongly correlated: the segment density around a given segment is higher than the average value. It has been shown that if the

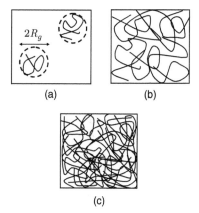

Fig. 2.9 Polymer solutions at various concentrations. (a) In a very dilute solution, polymers are separated from each other and there is a strong correlation in the segment density. (b) As concentration increases, polymers starts to overlap, but the correlation effect is still there. (c) At high concentration, the correlation effect becomes weak, and the segment density is nearly homogeneous.

[7] Note that c/m_p represents the number density of the polymer coil.

polymer is very large (i.e., if $N \gg 1$) this correlation effect is important, and gives a fundamental change in the theoretical results. For example, in a dilute solution with $\chi < \chi_c$, the osmotic pressure is not proportional to ϕ^2 as in eq. (2.65), but to a higher power ϕ^α, with exponent α between 2.2 and 2.3.[8] The correlation effect becomes less important as the polymer concentration increases, and above a certain concentration (say $\phi > 0.25$), the polymer solution can be described by the mean field theory discussed in Section 2.4.1.

[8] See the book by de Gennes, *Scaling Concepts in Polymer Physics*, listed in Further reading.

2.4.3   Polymer blends

The size effect is more dramatic in the case that both solute and solvent are polymers. Consider a mixture of two kinds of polymers A and B. Let N_A and N_B be the number of segments in the respective polymers. The free energy density for such a mixture can be written as a function of the volume fraction of A segments ϕ

$$f(\phi) = \frac{k_B T}{v_c} \left[\frac{\phi}{N_A} \ln \phi + \frac{1 - \phi}{N_B} \ln(1 - \phi) + \chi \phi(1 - \phi) \right] \qquad (2.70)$$

The spinodal line for such a mixture is now given by

$$\chi = \frac{1}{2} \left(\frac{1}{\phi N_A} + \frac{1}{(1 - \phi)N_B} \right) \qquad (2.71)$$

For large polymers of $N_A \gg 1$ and $N_B \gg 1$, the right-hand side of eq. (2.71) is very small. In this case eq. (2.71) is equivalent to the condition that $\chi < 0$, or $\Delta\epsilon > 0$. In other words, for two polymers A and B to mix together, segments A and B have to attract each other. This condition is satisfied only for very special pairs of polymers. Usually different polymers do not mix because the entropy gain by mixing is very small in polymers.

2.5   Colloidal solutions

2.5.1   Interaction potential between colloidal particles

The size effect is also important in colloidal solutions. The entropic effect of mixing is of the order of $k_B T$ per particle and is independent of the size of the particle, while the interaction energy between particles increases with the size of the particle. Therefore whether the particles are dispersed homogeneously or not is primarily determined by the interaction energy between the colloidal particles.

Consider two colloidal particles placed in a sea of solvent. The effective interaction energy $U(r)$ between the particles is defined as the work needed to bring the particles from infinity to the configuration in which the centre-to-centre distance is r. Alternatively, $U(r)$ can be defined as the difference in the free energy between the two states; one is the state where the distance is r, and the other is the state where the distance is infinity.

The energy $U(r)$ represents the effective interaction between solute with the effect of solvent included. Such an energy can be defined for any objects (molecules, molecular assemblies, particles) in solution, and is called the potential of mean force. The effective interaction energy $\Delta\epsilon$ in Section 2.3 corresponds to $U(r)$ for neighbouring solute molecules.

Theoretically, if $U(r)$ is given, colloidal solutions can be discussed in the same theoretical framework as that for solutions of small molecules. However, this approach is not usually taken in colloid science. This is because there are important differences in the interaction potential $U(r)$ for colloidal particles and for small molecules. The differences are of a quantitative nature, but they need to be noted.

(a) Interaction range: In solutions of small molecules, the range of interaction is comparable with the size of the molecules, while in colloidal solutions, the interaction range (typically of the order of nm) is much smaller than the size of the particles (typically of the order of 0.1 µm). Since the interaction potential changes significantly in a length-scale much smaller than the particle radius R, the interaction potential U in colloidal particles is usually expressed as a function of the gap distance $h = r - 2R$, not r itself (see Fig. 2.10).

(b) Magnitude of interaction: In solutions of small molecules, the magnitude of the interaction energy is less than $k_B T$, while in colloidal solutions, the interaction energy is usually much larger (typically tens of times larger) than $k_B T$ since it involves a large number of atoms.

An example of the colloidal interaction potential is shown in Fig. 2.10(b), where the energy is now shown as a function of the gap distance $h = r - 2R$. The potential $U(h)$ has a strong attractive part for very small h due to the van der Waals attraction. The depth of the attractive part is usually much larger than $k_B T$. Therefore if two particles are brought into contact, they are unable to separate. If the interaction potential has only an attractive part, particles in solution will aggregate quickly. For the solution to be stable, the interaction potential requires a repulsive part as shown in the figure.

Various methods have been developed to modify the interaction potential between particles in solutions in order to stabilize colloidal dispersions. In fact, the form of the interaction potential is a central issue in colloid science. This problem will be discussed in more detail in Chapter 4.

If the system is at equilibrium, it is possible to discuss the stability of a colloidal solution using the same theoretical framework as that of phase separation. Again such an approach is not usually taken in colloid science. The reason is that the colloidal solution is quite often not at true equilibrium.[9] Consider a colloidal solution in which the particles are interacting via the potential denoted as stable in Fig. 2.10. The

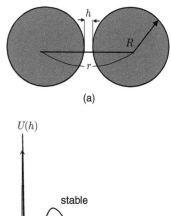

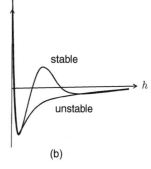

Fig. 2.10 (a) Interaction between two colloidal particles. (b) An example of the effective interaction potential between colloidal particles. Here 'unstable' indicates the potential for the unstable dispersion in which particles aggregate quickly, while 'stable' indicates the potential for the stable dispersions in which aggregation takes place so slowly that the dispersion *looks* stable.

[9] The equilibrium state of colloidal dispersions can be realized if the interaction potential has no attractive part. In fact, colloidal dispersions have been used as a model of a system of hard spheres or hard rods.

second virial coefficient for particles interacting by such a potential is negative due to the deep potential valley at the closest distance. Therefore if we apply the thermodynamic criteria for the stability of colloidal dispersions, we find that most colloidal dispersions are not stable.

For practical purposes, whether the colloidal solution is thermodynamically stable (i.e., absolutely stable) or not does not matter so much. As long as a colloidal dispersion remains homogeneous before it is used up, the colloidal dispersion can be regarded as stable. Therefore the stability of colloidal dispersions is usually discussed in terms of the aggregation speed, or the interaction potential. This is an example of the cultural gap between the theory of solutions and the theory of colloidal dispersions.

Further discussion on the stability of colloidal dispersions is given in Chapter 4.

2.6   Multi-component solutions

So far, we have been considering two-component systems, namely solutions made of one solute in a single solvent. However, solutions in soft matter usually have many components. For example, in simple polymer solutions, say a solution of polystyrene in benzene, the polystyrene molecules do not all have the same molecular weight, and therefore they need to be treated as different components. Also, mixed solvents, for instance mixture of water and acetone, are often used. Moreover, in electrolyte solutions, there are always many ionic components involved.

Quite often, these multi-component aspects of solutions are ignored. For instance, solutions of polystyrene in a solvent are usually regarded as two-component solutions made of monodisperse polystyrene and solvent, and homogeneously mixed binary solvents are regarded as a single-component solvent. Though such simplification is useful and necessary to advance our thinking, it is important to remember that it is only an approximation to reality.

Thermodynamic equations for many-component solutions can be derived in the same way as for two-component solutions (see problems (2.4) and (2.7)). Here we mention a different aspect that is specific to soft matter.

In general, if a mixed solvent is used to dissolve large solute (polymers or colloids), their composition will vary in space. For example, in dilute polymer solutions, the solvent composition inside the polymer coil region will generally be different from that of the outer region as certain components prefer to stay in the polymer coil region while other components prefer to stay away from the polymer. This effect can be seen on a macroscopic scale. If a network of polymer (i.e., polymer gel) is placed in a mixed solvent, the gel absorbs solvent for preference, and the solvent composition inside the gel is different from that outside. Further discussion of this aspect is given in Chapter 10.

2.7 Summary of this chapter

Solute and solvent mix together to form a homogeneous solution if the free energy of the mixed state is lower than the free energy of the unmixed state. The mixing behaviour is entirely characterized by the free energy density $f(\phi, T)$ expressed as a function of solute concentration (or the solute volume fraction ϕ). $f(\phi, T)$ can be obtained from the osmotic pressure $\Pi(\phi, T)$, or the chemical potentials $\mu_p(\phi, T)$ or $\mu_s(\phi, T)$.

The thermodynamic force for mixing has two origins. One is entropic, which always favours mixing. The other is energetic, which usually disfavours mixing.

In solutions of small molecules, solute and solvent do mix even if they do not like each other due to the entropic effect. On the other hand, in typical soft matter solutions (polymer solutions and colloidal dispersions), the solute is much larger than the solvent. Therefore, the entropic effect is negligible, and the energetic effect dominates. Accordingly, polymers and colloids are much more difficult to dissolve in solvent.

Further reading

(1) *Introduction to Polymer Physics*, Masao Doi, Oxford University Press (1996).

(2) *Scaling Concepts in Polymer Physics*, Pierre-Gilles Gennes, Cornell University Press (1979).

(3) *Polymer Physics*, Michael Rubinstein and Ralph H. Colby, Oxford University Press (2003).

Exercises

(2.1) 10 g of sugar is dissolved in 200 ml of water. Assuming that the sugar consists of molecules of molecular weight 500 g/mol, answer the following questions.

   (a) Calculate the weight concentration c [g/cm^3], weight fraction ϕ_m, and molar fraction x_m of sugar in the solution.

   (b) Estimate the osmotic pressure Π of the solution.

(2.2) Polystyrene of molecular weight 3×10^7 g/mol has a radius of gyration R_g of about 100 nm in benzene. Answer the following questions.

   (a) Estimate the overlap concentration at which the polymer molecules start to overlap each other.

   (b) Estimate the osmotic pressure at the overlap concentration.

(2.3) A solution is held in a cylinder which is sealed by two pistons, and divided into two chambers by a semi-permeable membrane (see Fig. 2.11) Initially, the concentrations of the two chambers are the same at number density n_0, and the chambers have the same volume hA (h and A are the height and the cross-section of the chamber). A weight W is placed on top of the cylinder, causing the piston to move down. Assuming that the solution is ideal, answer the following questions.

   (a) Calculate the displacement x of the piston at equilibrium ignoring the density of the solution.

   (b) Calculate the displacement x of the piston taking into account of the density of the solution ρ.

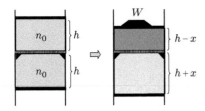

Fig. 2.11

(2.4) Consider an $(n + 1)$-component solution. Let N_i be the number of molecules of component i $(i = 0, 1, 2, \ldots, n)$ in the solution, and v_i be the specific volume of each component. The total volume of the solution is given by

$$V = \sum_{i=0}^{n} v_i N_i \qquad (2.72)$$

Assuming that v_i are all constant, answer the following questions.

(a) Show that the Gibbs free energy of the solution is given by

$$\begin{aligned} & G(N_0, \ldots, N_n, T, P) \\ & = PV + V f(\phi_1, \phi_2, \ldots, \phi_n, T) \end{aligned} \qquad (2.73)$$

where $\phi_i = v_i N_i / V$ is the volume fraction of the component i.

(b) Show that the osmotic pressure Π across a semi-permeable membrane which allows only the component 0 to pass through is given by

$$\Pi = -f + \sum_{i=1}^{n} \phi_i \frac{\partial f}{\partial \phi_i} + f(0) \qquad (2.74)$$

Also show that the chemical potential of the component i is given by

$$\begin{aligned} \mu_i &= v_i \left[P + f + \sum_{j=1}^{n} (\delta_{ij} - \phi_j) \frac{\partial f}{\partial \phi_j} \right] \\ &= v_i \left[\frac{\partial f}{\partial \phi_i} + P - \Pi + f(0) \right], \\ & \qquad\qquad \text{for } i = 1, \ldots n. \quad (2.75) \end{aligned}$$

$$\mu_0 = v_0 [P - \Pi + f(0)] \qquad (2.76)$$

(c) Prove that the chemical potentials μ_i $(i = 0, 1, \ldots, n)$ satisfy the following identity

$$\sum_{i=0}^{n} \frac{\phi_i}{v_i} \mu_i = f + P \qquad (2.77)$$

(d) Consider the situation that the volume fraction ϕ_i $(i = 1, 2, \ldots, n)$ is much smaller than ϕ_0. In this situation, the osmotic pressure is written as

$$\Pi = \sum_{i=1}^{n} \frac{\phi_i}{v_i} k_B T + \sum_{i=1}^{n} \sum_{j=1}^{n} A_{ij} \phi_i \phi_j \qquad (2.78)$$

Show that the chemical potentials of the minor components $(i = 1, 2, \ldots, n)$ are given by

$$\begin{aligned} \mu_i &= \mu_i^0(T) + P v_i + k_B T \ln \phi_i \\ &\quad + \sum_{j=1}^{n} \left(2 v_i A_{ij} - \frac{v_i}{v_j} k_B T \right) \phi_j \qquad (2.79) \end{aligned}$$

and that the chemical potential of the major component is

$$\mu_0 = \mu_0^0(T) + v_0 [P - \Pi] \qquad (2.80)$$

(2.5) For the osmotic pressure given by the lattice model (eq. (2.57)), answer the following questions.

(a) Plot $\Pi(\phi)$ as a function of ϕ for $\chi = 0, 1, 2, 4$.

(b) Draw the spinodal line in the ϕ–χ plane, and indicate the critical point.

(c) Draw the binodal line in the ϕ–χ plane.

(2.6) For the free energy expression for a polymer solution (eq. (2.62)), answer the following questions.

(a) Plot $f(\phi)$ as a function of ϕ for $\chi = 0.4, 0.5, 0.6$ and $N = 100$.

(b) Draw the spinodal line in the ϕ–χ plane, and indicate the critical point.

(c) Show that for large positive χ, the binodal curve is approximately given by $\phi_a = e^{-N(\chi-1)}$, and $\phi_b = 1 - e^{-(\chi+1)}$.

(2.7) *(a) Consider the $(n + 1)$-component solution whose free energy density is given by $f(\phi_1, \phi_2, \ldots, \phi_n)$, where ϕ_i denote the volume fraction of component i. Show that the spinodal line, i.e., the boundary between the stable region and the unstable region, is given by

$$\det \left(\frac{\partial^2 f}{\partial \phi_i \partial \phi_j} \right) = 0 \qquad (2.81)$$

(b) Consider a polymer solution which contains polymers consisting of the same types of segments, but having different numbers of segments, N_1 and N_2. The free energy density of the system is written as

$$f(\phi) = \frac{k_B T}{v_c}\left[(1-\phi)\ln(1-\phi) + \sum_{i=1,2}\frac{1}{N_i}\phi_i\ln\phi_i\right.$$

$$\left. - \chi\left(\sum_{i=1,2}\phi_i\right)^2\right] \tag{2.82}$$

Show that the spinodal line of such a solution is given by

$$\chi = \frac{1}{2}\left[\frac{1}{1-\phi} + \frac{1}{N_w\phi}\right] \tag{2.83}$$

where

$$\phi = \sum_{i=1,2}\phi_i \tag{2.84}$$

is the volume fraction of polymer, and N_w is defined by

$$N_w = \frac{1}{\phi}\sum_{i=1,2}\phi_i N_i \tag{2.85}$$

3

Elastic soft matter

[1] In this book, we call a network of polymer containing a solvent a gel, and that not containing solvent a rubber. There are rubbers which contain solvent (or additives of small molecules). The distinction between rubber and gel is not clear, and it seems that the distinction is not in the structure of the material. Rather, it is in the usage of the material. If the mechanical property of the material is utilized, it is called a rubber, and if the ability to keep fluids in an elastic state is utilized, it is called a gel. Different definitions are used for gels in other materials. For example, in colloid science, gel is defined as a disordered state of colloidal dispersions which do not have fluidity (see Section 1.2).

Elastic soft matter is familiar to us in daily life as rubbers and gels. Rubbers and gels are composed of network polymers, as shown in Fig. 1.2(c). Rubbers usually do not contain solvent, while gels contain solvent.[1] The essential characteristic of these materials is that they have a well-defined, conserved shape: they deform when external forces are applied, but recover the original shape when the forces are removed. This is distinct from the liquid state: liquid does not have its own shape and can be deformed without any work.

Materials which have their own shapes are called elastic materials. Metals and glasses are typical hard elastic materials, while rubbers and gels are soft elastic materials. Though classified in elastic materials, rubbers and gels have unique properties quite different from metals and glasses. They can be deformed by up to tens or hundreds of percent without breaking. This difference has a molecular origin.

The elasticity of metals and glasses comes from the fact that the atoms in the material are trapped in the state of local energy minimum, and resist any change from the state. Rubbers and gels are distinct from these materials in the sense that atoms are moving vigorously, very much like in the fluid state (polymer melts and polymer solutions). Rubbers and gels are made by introducing cross-links in the fluid state of polymers. The fraction of cross-links is usually very small (less than a few percent), and most of the material remains in the same state as that of fluids. Nevertheless, a small number of cross-links is often sufficient to change the fluid into an elastic material. In this chapter, we shall discuss the equilibrium properties of this unique class of materials.

3.1 Elastic soft matter

3.1.1 Polymer solutions and polymer gels

Figure 3.1 shows a schematic comparison between the fluid state (solution) and the elastic state (gel) of polymers. In polymer solutions, long-chain-like molecules are entangled with each other. Therefore, a polymer solution is a viscous material. It is a material that resists flow, but it can flow. Indeed, a polymer solution in a container can be poured on a table if sufficient time is given. On the other hand, in polymer gels, the polymers are chemically cross-linked and form a large molecular network. Such a material cannot flow, and cannot be poured: if the container is placed upside down, the material may fall out, but will not spread on the table.

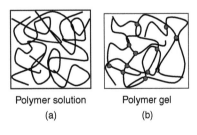

Polymer solution   Polymer gel
(a)   (b)

Fig. 3.1 Structural difference of polymeric material in a fluid state (a) and in an elastic state (b).

A fluid does not have a well-defined shape, and can be deformed in-definitely. On the other hand, an elastic material has its own shape and can be reversibly deformed to a finite extent. Though this distinction may seem very clear, it is not so in many soft matter systems.

Consider for example the making of gelatin jelly. To make a gelatin jelly, one first dissolves gelatin (a natural polymer) in hot water, and cools it in a refrigerator. In the refrigerator, the polymer forms a network and solidifies. If the cooling time in the refrigerator is not sufficient, one gets a viscous fluid. As time goes on, the viscosity increases, and finally one gets a gel, but it is not obvious to tell precisely when the polymer solution has changed from fluid to elastic.

The mechanical properties of soft matter are generally quite complex, and are the subject of rheology, the science of flow and deformation of matter. This subject will be discussed in Chapter 9. Here we give a brief introduction to certain concepts needed to discuss the equilibrium properties of elastic soft matter.

3.1.2   Viscosity and elasticity

Whether a material is a fluid or an elastic material can be studied using the experimental setup shown in Fig. 3.2(a). Here the sample is set between two horizontal plates. The bottom plate is fixed, while the top plate moves horizontally pulled by a string. Suppose that a weight is hung on the string at time $t = 0$, and the string is cut at a later time t_0. This operation gives a constant force F applied on the top plate for the time between 0 and t_0. The mechanical properties of the material can be characterized by the behaviour of the displacement of the top plate, $d(t)$. If the material is a fluid, the top plate keeps moving as long as the force is applied, and remains displaced from the original position when the force is set to zero. On the other hand, if the material is elastic, the

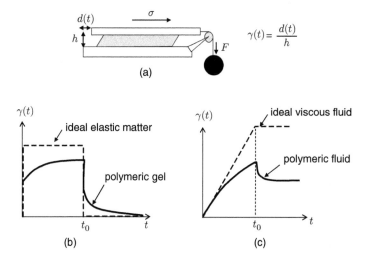

Fig. 3.2 (a) An experimental setup to test the mechanical properties of a material. The sample is set between two parallel plates, and the horizontal displacement of the top plate is measured when a constant force F is applied on the top plate at time $t = 0$ and is removed at time t_0. (b) Response of solid-like materials. (c) Response of fluid-like materials.

top plate moves to the position of mechanical equilibrium, and returns to the original position when the force is set to zero.

To analyse the result, we define two quantities: the shear stress σ and the shear strain γ. The shear stress σ is the shearing force acting per unit area of the sample surface. For the geometry shown in Fig. 3.2(a), σ is given by

$$\sigma = \frac{F}{S} \qquad (3.1)$$

where S is the area of the top surface of the material. The shear strain γ is given by

$$\gamma = \frac{d}{h} \qquad (3.2)$$

where h is the distance between the plates. If the material is uniform, the relation between σ and γ is independent of the sample size.[2]

If the material is an ideal elastic material (called Hookean elastic), the shear stress is proportional to the shear strain, and is written as

$$\sigma = G\gamma \qquad (3.3)$$

where G is a material constant called the shear modulus. For such an ideal material, the shear strain $\gamma(t)$ behaves as shown by the dashed curve in Fig. 3.2(b), i.e., $\gamma(t)$ is equal to σ/G for $t < t_0$ and is zero for $t > t_0$.

On the other hand, if the material is an ideal viscous fluid (often called a Newtonian fluid), the shear stress is proportional to the time derivative of the shear strain:

$$\sigma = \eta\dot{\gamma} \qquad (3.4)$$

where $\dot{\gamma}$ represents the time derivative of $\gamma(t)$, and η is another material constant called the viscosity (or shear viscosity). For such a material, the response of the shear strain is shown by the dashed line in Fig. 3.2(c), i.e., $\gamma(t)$ increases with time as $(\sigma/\eta)t$ for $t < t_0$, and stays at a constant value for $t > t_0$.

For polymeric materials, the time dependence of $\gamma(t)$ is more complex as indicated in Fig. 3.2(b) and (c). Polymeric materials usually have long relaxation times, and exhibit complex transient behaviour before reaching steady states. The complex time dependence of materials is known as viscoelasticity. This subject will be discussed in Chapter 9. In this chapter, we shall restrict the discussion to the equilibrium state of elastic materials. In this case, we can discuss the mechanical properties of a material quite generally in terms of the free energy.

3.1.3   Elastic constants

If the deformation of an elastic material is small, the relation between stress and strain is linear, and the mechanical property of the material is completely characterized by the elastic constants. For isotropic

[2] The shear stress is a particular component of a tensorial quantity called the stress tensor. Likewise, the shear strain is a particular component of the strain tensor. Complete characterization of the mechanical properties of a material generally requires a relation between such tensorial quantities. A detailed discussion is given in Appendix A. Here we continue the discussion restricting ourselves to shear deformation.

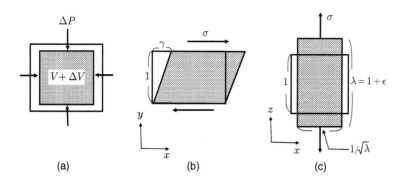

Fig. 3.3 Elastic constants obtained by various deformations. (a) Bulk modulus, (b) shear modulus, (c) elongational modulus (Young's modulus).

materials, there are two elastic constants; one is the shear modulus G defined above, and the other is the bulk modulus K which is defined below.

Suppose that a material is compressed isotropically by an extra pressure ΔP, and, in response, the volume of the material changes from V to $V + \Delta V$. (Notice that $\Delta V < 0$ if $\Delta P > 0$, see Fig. 3.3(a).) For small volume change, ΔP is proportional to the ratio $\Delta V/V$, which is called the volume strain, and is written as

$$\Delta P = -K \frac{\Delta V}{V} \tag{3.5}$$

The constant K is called the bulk modulus. In the definition of eq. (3.5), a minus sign is included to make K positive.

The elastic constant of the material for other types of deformation can be expressed in terms of K and G. For example, consider that a rectangular material is stretched uniformly along the z-direction as shown in Fig. 3.3(c). Let λ be the elongational ratio, i.e., $\lambda = L'_z/L_z$, where L_z and L'_z are the length of the material in the z-direction before and after the deformation. The elongational strain ϵ is defined by

$$\epsilon = \lambda - 1 \tag{3.6}$$

The force acting per unit area on the top surface is called the elongational stress, and the ratio between the elongational stress σ and the elongational strain ϵ is called Young's modulus E:

$$\sigma = E\epsilon \tag{3.7}$$

As shown in Appendix A, Young's modulus is expressed in terms of K and G as

$$E = \frac{9KG}{3K + G} \tag{3.8}$$

If the material is elongated in the z-direction, it usually shrinks in the x- and y-directions. The strain in the x-direction is defined by $\epsilon_x = L'_x/L_x - 1$, and the ratio ϵ_x/ϵ defines Poisson's ratio ν:

$$\nu = -\frac{\epsilon_x}{\epsilon} \tag{3.9}$$

ν is expressed in terms of K and G as

$$\nu = \frac{3K - 2G}{2(3K + G)} \tag{3.10}$$

A derivation of these equations is given in Appendix A (see also problem (3.2)).

The bulk modulus represents the restoring force of an isotropically compressed material, and it can be defined both for fluids and elastic materials. The magnitude of the bulk modulus is essentially determined by the density of the material, and its value does not change significantly between fluids and elastic materials. For example, the bulk modulus of rubber and polymer melts is of the order of $1\,\mathrm{GPa}$, which is about the same as the bulk modulus of metals.

On the other hand, the shear modulus represents the restoring force of a deformed material which tends to recover its own shape. Such a force does not exist in fluids, but does exist in elastic materials. When cross-links are introduced to polymer fluids, the shear moduli start to deviate from zero, and increase as the cross-linking reaction proceeds.

It must be noted that the softness of soft matter is due to the smallness of the shear modulus: the bulk modulus of soft matter is of the same order as that of metals, and is much larger than the shear modulus. Therefore, in discussing the deformation of soft matter, we may assume that the bulk modulus K is infinitely large, i.e., the volume change does not take place when the material is deformed by external forces. This assumption, called the incompressible assumption, is valid for typical stress levels (less than $10\,\mathrm{MPa}$) in both the fluid and the elastic states of soft matter. Throughout this book, we shall make this assumption. For an incompressible material, eq. (3.8) becomes

$$E = 3G \tag{3.11}$$

3.1.4 Continuum mechanics for elastic materials

Free energy density of deformation

Let us now consider how an elastic material deforms under external forces. As long as the system is at equilibrium, the mechanical behaviour of a material is completely characterized by a function called the deformation free energy. To define this, we choose an equilibrium state of the material with all external forces removed, and define this as the reference state. We then define the deformation free energy as the difference in the free energy between the deformed state and the reference state. The deformation free energy per unit volume in the reference state is called the deformation free energy density.

Before discussing the explicit form of the deformation free energy, let us first consider how to specify the deformation of a material.

Orthogonal deformation

A simple example of deformation is the orthogonal deformation shown in Fig. 3.4(a), where a cubic material is stretched by factors λ_1, λ_2, and

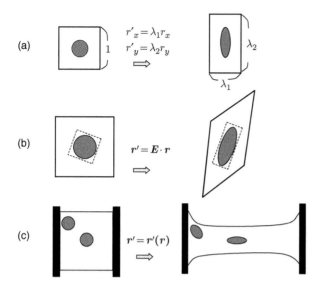

(a)

$r'_x = \lambda_1 r_x$
$r'_y = \lambda_2 r_y$

λ_2

λ_1

(b)

$r' = E \cdot r$

(c)

$r' = r'(r)$

Fig. 3.4 Deformation of an elastic material. (a) Orthogonal deformation: the material is stretched (or compressed) along three orthogonal directions. (b) Uniform deformation: the mapping from r to r' is linear. This deformation is expressed as a combination of rotation and orthogonal deformation. (c) General deformation: the mapping from r to r' is nonlinear. On a local scale, however, the mapping $r \to r'(r)$ can be regarded as the same as the linear mapping $r \to r' = E \cdot r$.

λ_3 along the three orthogonal directions.[3] The deformation displaces the point which was located at (r_x, r_y, r_z) before the deformation to the position

$$r'_x = \lambda_1 r_x, \qquad r'_y = \lambda_2 r_y, \qquad r'_z = \lambda_3 r_z \qquad (3.12)$$

Let $f(\lambda_1, \lambda_2, \lambda_3)$ be the deformation free energy density for this deformation. For simplicity, we shall denote $f(\lambda_1, \lambda_2, \lambda_3)$ by $f(\lambda_i)$. As we show in the following, in isotropic materials, $f(\lambda_i)$ is the only quantity we need to know about the material to calculate the deformation of the material under external forces.

Uniform deformation

The deformation of an elastic material can be described by a mapping $r \to r'(r)$, where r and r' are the positions of a material point (the point moving with the material) before and after the deformation. If r' is a linear transformation of r, i.e., if r' is written as

$$r' = E \cdot r \qquad (3.13)$$

with a constant tensor E, the deformation is called uniform since all volume elements in the material are deformed in the same way.

In this book, we shall denote the x, y, z components of vectors and tensors by Greek suffices $\alpha, \beta, \ldots$, and use the Einstein notation that summation is taken for the indices that appear twice in a term. In this notation, eq. (3.13) is written as

$$r'_\alpha = E_{\alpha\beta} r_\beta \qquad (3.14)$$

The right-hand side of this equation is identical to $\sum_{\beta=x,y,z} E_{\alpha\beta} r_\beta$. The tensor E in eq. (3.14) is called the deformation gradient tensor since $E_{\alpha\beta}$ is written as

$$E_{\alpha\beta} = \frac{\partial r'_\alpha}{\partial r_\beta} \tag{3.15}$$

The free energy $f(\boldsymbol{E})$ for the deformation $\boldsymbol{E}$ can be expressed in terms of $f(\lambda_i)$. The reason is as follows. If the material is deformed uniformly, a sphere is mapped to an ellipsoid as shown in Fig. 3.4(b). Therefore the sphere is elongated along three orthogonal directions. Mathematically, it can be proven that any tensor $\boldsymbol{E}$ can be written as $\boldsymbol{E} = \boldsymbol{Q}{\cdot}\boldsymbol{L}$, where $\boldsymbol{L}$ is a diagonal tensor, and $\boldsymbol{Q}$ is an orthogonal tensor (the tensor which satisfies $Q_{\alpha\gamma}Q_{\beta\gamma} = \delta_{\alpha\beta}$).[4] Physically this means that any uniform deformation is equivalent to a combination of orthogonal stretching (represented by $\boldsymbol{L}$) and rotation (represented by $\boldsymbol{Q}$, see Fig. 3.4(b)). Let λ_i $(i = 1, 2, 3)$ be the values of the diagonal elements of the tensor $\boldsymbol{L}$. It can be shown that λ_i^2 are the eigenvalues of the symmetric tensor

$$B_{\alpha\beta} = E_{\alpha\mu}E_{\beta\mu} \tag{3.18}$$

and satisfy the following identities (see problem (3.3)).

$$\lambda_1^2 + \lambda_2^2 + \lambda_3^2 = E_{\alpha\beta}^2 = B_{\alpha\alpha} \tag{3.19}$$

$$\lambda_1\lambda_2\lambda_3 = \det(\boldsymbol{E}) \tag{3.20}$$

Therefore $f(\boldsymbol{E})$ is obtained from $f(\lambda_i)$.

[4] Since the tensor $\boldsymbol{B} = \boldsymbol{E} \cdot \boldsymbol{E}^t$ is a symmetric tensor, it can be diagonalized with a proper orthogonal tensor $\boldsymbol{Q}$:

$$\boldsymbol{L}^2 = \boldsymbol{Q}^t \cdot \boldsymbol{B} \cdot \boldsymbol{Q} \tag{3.16}$$

Hence

$$\boldsymbol{Q} \cdot \boldsymbol{L}^2 \cdot \boldsymbol{Q}^t = \boldsymbol{B} = \boldsymbol{E} \cdot \boldsymbol{E}^t \tag{3.17}$$

Therefore $\boldsymbol{E}$ is written as $\boldsymbol{Q} \cdot \boldsymbol{L}$.

General deformation

The deformation free energy for a general deformation can be calculated from $f(\boldsymbol{E})$. As shown in Fig. 3.4(c), in any deformation of $\boldsymbol{r} \to \boldsymbol{r'} = \boldsymbol{r'}(\boldsymbol{r})$, the local deformation is always uniform since a small vector dr_α is mapped into $dr'_\alpha = (\partial r'_\alpha/\partial r_\beta)dr_\beta$ by the deformation. The deformation gradient tensor at point $\boldsymbol{r}$ is given by $E_{\alpha\beta}(\boldsymbol{r}) = \partial r'_\alpha/\partial r_\beta$. Therefore the total free energy of deformation is given by

$$F_{tot} = \int d\boldsymbol{r} f(\boldsymbol{E}(\boldsymbol{r})) \tag{3.21}$$

The equilibrium state of the material is given by minimizing F_{tot} under proper boundary conditions. Examples will be given later.

The equilibrium state can also be obtained by considering the force balance in the material. This approach is more general since it can account for the viscous stress which cannot be derived from the free energy. This subject will be discussed in Chapter 9 and also in Appendix A.

3.2   Elasticity of a polymer chain

3.2.1   Freely jointed chain model

Having discussed the elasticity of materials from a macroscopic viewpoint, we now discuss the elasticity from a molecular viewpoint.

The elasticity of rubbers and gels comes from the elasticity of polymer molecules.

Polymers are flexible molecules, and can be easily stretched by weak forces. This can be demonstrated by the simple polymer model shown in Fig. 3.5(a), where a polymer molecule is represented by N segments of constant length b. The segments are connected by flexible joints and can point in any direction independently of other segments. Such a model is called a freely jointed chain.

To discuss the elasticity of a freely jointed chain, we shall consider the relation between the vector $\boldsymbol{r}$ joining the two ends of the chain, and the force $\boldsymbol{f}$ acting at the chain ends. Before studying this problem, we first study the case of $\boldsymbol{f} = 0$, when there is no external force acting on the chain ends.

If there is no external forces acting on the chain ends, the distribution of the vector $\boldsymbol{r}$ is isotropic. Therefore the average $\langle \boldsymbol{r} \rangle$ is equal to zero. Let us therefore consider the average $\langle \boldsymbol{r}^2 \rangle$.

Let $\boldsymbol{b}_n$ $(n = 1, 2, \ldots, N)$ be the end-to-end vector of the n-th segment. Then $\boldsymbol{r}$ is written as

$$\boldsymbol{r} = \sum_{n=1}^{N} \boldsymbol{b}_n \tag{3.22}$$

Therefore the mean square of $\boldsymbol{r}$ is calculated as

$$\langle \boldsymbol{r}^2 \rangle = \sum_{n=1}^{N} \sum_{m=1}^{N} \langle \boldsymbol{b}_n \cdot \boldsymbol{b}_m \rangle \tag{3.23}$$

For a random distribution of segments, the $\boldsymbol{b}_n$ are independent of each other. Thus

$$\langle \boldsymbol{b}_n \cdot \boldsymbol{b}_m \rangle = \delta_{nm} b^2 \tag{3.24}$$

Therefore

$$\langle \boldsymbol{r}^2 \rangle = \sum_{n=1}^{N} \langle \boldsymbol{b}_n^2 \rangle = Nb^2 \tag{3.25}$$

The average size of a polymer in a force free state can be estimated by $\bar{r} = \sqrt{\langle \boldsymbol{r}^2 \rangle}$, which is equal to $\sqrt{N}b$. On the other hand, the maximum value of the end-to-end distance r of the freely jointed chain is $r_{max} = Nb$. Hence polymer chains can be extended, on average, by a factor $r_{max}/\bar{r} = \sqrt{N}$. Since N is large ($N = 10^2 \simeq 10^6$), polymer molecules are very stretchable.

3.2.2 Equilibrium distribution of the end-to-end vector

Let us consider the statistical distribution of $\boldsymbol{r}$ in the force free state. Since $\boldsymbol{r}$ is expressed as a sum of independent variables, its distribution is given by a Gaussian distribution. This is a result of the central limit theorem in statistics. According to the central limit theorem, if x_n $(n = 1, 2, \ldots, N)$ are statistical variables independent of each other and

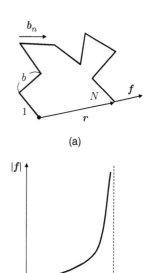

Fig. 3.5 (a) Freely jointed chain model for a polymer. (b) The relation between the end-to-end distance $|\boldsymbol{r}|$ of a polymer chain and the force $|\boldsymbol{f}|$ needed to extend the chain.

obeying the same distribution, the distribution of the sum $X = \sum_{n=1}^{N} x_n$ approaches to the following Gaussian distribution for $N \gg 1$:

$$P(X) = \frac{1}{(2\pi N\sigma^2)^{1/2}} \exp\left[-\frac{(X - N\bar{x})^2}{2N\sigma^2}\right] \tag{3.26}$$

where $\bar{x}$ and σ are the mean and the variance of x

$$\bar{x} = \langle x \rangle, \qquad \sigma^2 = \langle (x - \bar{x})^2 \rangle \tag{3.27}$$

In the present problem, x_n corresponds to the x, y, z components of the vector $\boldsymbol{b}_n$. The average of $b_{n\alpha}$ is equal to zero, and the variance is given by

$$\langle b_{nx}^2 \rangle = \langle b_{ny}^2 \rangle = \langle b_{nz}^2 \rangle = \frac{b^2}{3} \tag{3.28}$$

Hence, the distribution of $\boldsymbol{r}$ is given by

$$P(\boldsymbol{r}) = \left(\frac{3}{2\pi N b^2}\right)^{3/2} \exp\left(-\frac{3r^2}{2Nb^2}\right) \tag{3.29}$$

3.2.3   Elasticity of a polymer chain

Let us now study the elasticity of a freely jointed chain. For this purpose, we suppose that the end-to-end vector of the chain is fixed at $\boldsymbol{r}$. Let $U(\boldsymbol{r})$ be the free energy of such a chain.[5] Given $U(\boldsymbol{r})$, the average force $\boldsymbol{f}$ that we need to apply at the chain end to fix the position is calculated as follows. If we change the end-to-end vector from $\boldsymbol{r}$ to $\boldsymbol{r} + d\boldsymbol{r}$, we do work $\boldsymbol{f} \cdot d\boldsymbol{r}$ to the chain. This work is equal to the change of $U(\boldsymbol{r})$. Hence $dU = \boldsymbol{f} \cdot d\boldsymbol{r}$, or

$$\boldsymbol{f} = \frac{\partial U(\boldsymbol{r})}{\partial \boldsymbol{r}} \tag{3.30}$$

Now if the chain end is not fixed, the end-to-end vector $\boldsymbol{r}$ can take various values. The distribution of $\boldsymbol{r}$ is related to $U(\boldsymbol{r})$ by the Boltzmann distribution

$$P(\boldsymbol{r}) \propto \exp\left(-\frac{U(\boldsymbol{r})}{k_B T}\right) \tag{3.31}$$

On the other hand, the distribution of $\boldsymbol{r}$ of a free chain is given by eq. (3.29). Comparing eq. (3.29) with eq. (3.31), we have

$$U(\boldsymbol{r}) = \frac{3k_B T}{2Nb^2} r^2 \tag{3.32}$$

where terms which do not depend on $\boldsymbol{r}$ have been dropped.

From eqs. (3.30) and (3.32), the force acting on the chain end is given by

$$\boldsymbol{f} = \frac{3k_B T}{Nb^2} \boldsymbol{r} \tag{3.33}$$

Equation (3.33) indicates that as far as the relation between the end-to-end vector $\boldsymbol{r}$ and the force $\boldsymbol{f}$ is concerned, the polymer molecule can be

[5] $U(\boldsymbol{r})$ is the free energy of a chain which is subject to the constraint that the end-to-end vector is fixed at $\boldsymbol{r}$. Such free energy is generally called restricted free energy. A general discussion of the restricted free energy is given in Appendix B.

regarded as a molecular spring having spring constant $k = 3k_BT/Nb^2$. With the increase of N, the spring constant decreases, i.e., the polymer chain becomes softer.

It is important to note that the origin of the spring force is the thermal motion of the chain. At temperature T, the polymer chain is moving randomly. The force $\boldsymbol{f}$ represents the force needed to fix the end-to-end vector at $\boldsymbol{r}$. This force has the same origin as the pressure of a gas. The pressure of a gas is the force needed to keep the gas molecules within a fixed volume. More precisely, both the force $\boldsymbol{f}$ in a polymer chain and the pressure in a gas have their origin in the entropy change. When a chain is stretched, the entropy of the chain is reduced. Likewise, when a gas is compressed, the entropy of the gas is reduced. Both forces arise from the change of the entropy. The characteristic of such entropy-originating forces (or entropic force) is that they are proportional to the absolute temperature T. This is indeed seen in the force $\boldsymbol{f}$ of a polymer chain, and in the pressure of a gas.

According to eq. (3.33), the end-to-end vector $\boldsymbol{r}$ can increase indefinitely with the increase of $\boldsymbol{f}$. This is not true for the freely jointed chain: the maximum value of $|\boldsymbol{r}|$ of the freely jointed chain is Nb. In fact, eq. (3.33) is only valid for the situation of $|\boldsymbol{r}|/Nb \ll 1$, or $fb/k_BT \ll 1$. The exact relation between $\boldsymbol{f}$ and $\boldsymbol{r}$ can be calculated as follows.

Suppose that an external force $\boldsymbol{f}$ is applied at the chain end. Then the average of the end-to-end vector $\boldsymbol{r}$ is not zero. We shall calculate the average of $\boldsymbol{r}$ in such a situation, and obtain the relation between $\boldsymbol{f}$ and $\boldsymbol{r}$.

Under the external force $\boldsymbol{f}$, the chain has the potential energy

$$U_f(\{\boldsymbol{b}_n\}) = -\boldsymbol{f} \cdot \boldsymbol{r} = -\boldsymbol{f} \cdot \sum_n \boldsymbol{b}_n \qquad (3.34)$$

The probability of finding the chain in the configuration $\{\boldsymbol{b}_n\} = (\boldsymbol{b}_1, \boldsymbol{b}_2, \ldots, \boldsymbol{b}_N)$ is given by

$$P(\{\boldsymbol{b}_n\}) \propto e^{-\beta U_f} = \Pi_{n=1}^{N} \, e^{\beta \boldsymbol{f} \cdot \boldsymbol{b}_n} \qquad (3.35)$$

where $\beta = 1/k_BT$. Therefore the distributions of $\boldsymbol{b}_n$ are independent of each other, and the mean of $\boldsymbol{b}_n$ is calculated as

$$\langle \boldsymbol{b}_n \rangle = \frac{\int_{|b|=b} d\boldsymbol{b} \; \boldsymbol{b} e^{\beta \boldsymbol{f} \cdot \boldsymbol{b}}}{\int_{|b|=b} d\boldsymbol{b} \; e^{\beta \boldsymbol{f} \cdot \boldsymbol{b}}} \qquad (3.36)$$

The subscript $|\boldsymbol{b}| = b$ attached to the integral means that the integral is taken over the surface of the sphere of radius b. The integral can be performed analytically, and the result is[6]

$$\langle \boldsymbol{b}_n \rangle = b\frac{\boldsymbol{f}}{|\boldsymbol{f}|} \left[\coth(\xi) - \frac{1}{\xi} \right] \qquad (3.37)$$

where

$$\xi = \frac{|\boldsymbol{f}|b}{k_BT} \qquad (3.38)$$

[6] Let θ be the angle that the vector $\boldsymbol{b}$ makes against $\boldsymbol{f}$. The integral in the denominator is then written as

$$\int_{|b|=b} d\boldsymbol{b} \; e^{\beta \boldsymbol{f} \cdot \boldsymbol{b}} = b^2 \int_0^{\pi} d\theta 2\pi \sin\theta e^{\xi \cos\theta}$$
$$= 4\pi b^2 \sinh\xi$$

A similar calculation for the numerator gives eq. (3.37).

[7] The left-hand side of eq. (3.39) is actually the average $\langle r \rangle$, but the symbol $\langle \cdots \rangle$ is omitted here since under the external force f, the deviation of r from $\langle r \rangle$ is negligibly small for $N \gg 1$.

Hence the end-to-end vector is given by[7]

$$r = N\langle b_n \rangle = Nb\frac{f}{|f|}\left[\coth(\xi) - \frac{1}{\xi}\right] \qquad (3.39)$$

The relation between r and f is plotted in Fig. 3.5(b).

In the weak force limit ($\xi \ll 1$), eq. (3.39) reduces to

$$r = Nb\frac{f}{|f|}\frac{\xi}{3} = \frac{Nb^2}{3k_BT}f \qquad (3.40)$$

which agrees with eq. (3.33). On the other hand, in the strong force limit ($\xi \gg 1$), eq. (3.39) gives

$$r = Nb\frac{f}{|f|}\left(1 - \frac{1}{\xi}\right) \qquad (3.41)$$

or

$$|f| = \frac{k_BT}{b}\frac{1}{\left(1 - \frac{|r|}{Nb}\right)} \qquad (3.42)$$

Therefore the force diverges as $|r|$ approaches the maximum length Nb.

3.3 Kuhn's theory for rubber elasticity

3.3.1 Free energy of deformed rubber

Having seen the elasticity of a polymer chain, let us now consider the elasticity of a polymer network, i.e., a rubber. Consider the model shown in Fig. 3.6. The polymer network consists of cross-links (denoted by filled circles), and the polymer chains connecting them. We shall call the part of the polymer chain between neighbouring cross-links the subchain.

Now suppose that the rubber is in the reference state. As we have seen in Section 3.2.3, if the subchain consists of N segments and its end-to-end vector is r, its free energy contribution is given by $(3k_BT/2Nb^2)r^2$. If the interaction between the subchains is ignored, the free energy of the polymer network (per unit volume) can be written as an integral over the population of these subchains:

$$\tilde{f}_0 = n_c \int dr \int_0^\infty dN \ \Psi_0(r, N)\frac{3k_BT}{2Nb^2}r^2 \qquad (3.43)$$

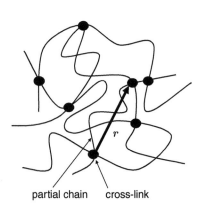

partial chain   cross-link

Fig. 3.6 Polymer network in a rubber.

where n_c is the number of subchains in unit volume and $\Psi_0(r, N)$ is the probability that a subchain consists of N segments and has the end-to-end vector r in the reference state.

The probability distribution function $\Psi_0(r, N)$ depends on how the polymers are cross-linked. Here we proceed by making a simple assumption. We assume that the distribution of the end-to-end vector r of a subchain in the reference state is the same as that of a free chain at equilibrium: if the chain consists of N segments, the distribution of r is given by eq. (3.29). Therefore

$$\Psi_0(r, N) = \left(\frac{3}{2\pi Nb^2}\right)^{3/2} \exp\left(-\frac{3r^2}{2Nb^2}\right)\Phi_0(N) \qquad (3.44)$$

where $\Phi_0(N)$ represents the distribution of N, and satisfies

$$\int_0^\infty dN \Phi_0(N) = 1 \qquad (3.45)$$

Now suppose that the polymer network is deformed uniformly, and each material point is displaced from r to $r' = E \cdot r$. We assume that when the network is deformed, the end-end-vector r of a subchain changes in the same way as the material point, i.e., the end-to-end vector changes from r to $E \cdot r$. Then the free energy of deformation can be written as

$$f(E) = \tilde{f}(E) - \tilde{f}_0 = n_c \int dr \int_0^\infty dN\ \Psi_0(r, N)\frac{3k_BT}{2Nb^2}\left[(E \cdot r)^2 - r^2\right]$$
$$(3.46)$$

The integral on the right-hand side can be calculated as

$$\int dr\Psi_0(r, N)(E \cdot r)^2 = E_{\alpha\beta}E_{\alpha\gamma}\int dr\ \Psi_0(r, N)r_\beta r_\gamma$$
$$= E_{\alpha\beta}E_{\alpha\gamma}\frac{Nb^2}{3}\delta_{\beta\gamma}\Phi_0(N) \qquad (3.47)$$

Using eqs. (3.47) and (3.45) in eq. (3.46), we finally get the following expression for the free energy of deformation

$$f(E) = \frac{1}{2}n_ck_BT\left[(E_{\alpha\beta})^2 - 3\right] \qquad (3.48)$$

This can be rewritten, by use of eq. (3.19), as

$$f(\lambda_i) = \frac{1}{2}n_ck_BT\left[\sum_i \lambda_i^2 - 3\right] \qquad (3.49)$$

Materials described by this model are called neo-Hookean.

3.3.2 Stress–strain relation for typical deformations

Shear deformation

Let us now study the mechanical response of a rubber for typical deformations.

First, let us consider the shear deformation shown in Fig. 3.3(b). In this deformation the material point located at (x, y, z) is displaced to

$$x' = x + \gamma y , \quad y' = y , \quad z' = z \qquad (3.50)$$

The deformation gradient tensor is now given by

$$(E_{\alpha\beta}) = \begin{pmatrix} 1 & \gamma & 0 \\ 0 & 1 & 0 \\ 0 & 0 & 1 \end{pmatrix} \tag{3.51}$$

Hence the elastic energy of deformation is calculated by eq. (3.48) as

$$f(\gamma) = \frac{1}{2} n_c k_B T \gamma^2 \tag{3.52}$$

To calculate the shear stress σ, we consider that the shear strain is increased from γ to $\gamma + d\gamma$. The work done to the material of unit volume is $\sigma d\gamma$. This work is equal to the change of the deformation free energy density df. Hence the shear stress is given by

$$\sigma = \frac{\partial f}{\partial \gamma} \tag{3.53}$$

For eq. (3.52), this gives

$$\sigma = n_c k_B T \gamma \tag{3.54}$$

Therefore the shear modulus G is given by

$$G = n_c k_B T \tag{3.55}$$

The number density of the subchain n_c is expressed by using the average molecular weight between the cross-links M_x as $n_c = \rho/(M_x/N_{Av})$, where ρ is the density of the rubber, and N_{Av} is the Avogadro number. Hence eq. (3.55) is rewritten as

$$G = \frac{\rho N_{Av} k_B T}{M_x} = \frac{\rho R_G T}{M_x} \tag{3.56}$$

where $R_G = N_{Av} k_B$ is the gas constant.

Equation (3.56) has the same form as the bulk modulus of an ideal gas:

$$K_{gas} = n k_B T = \frac{\rho R_G T}{M} \tag{3.57}$$

where n is the number density of molecules in the gas and M is the molecular weight of the molecule. The shear modulus of a rubber is the same as the bulk modulus of a gas which has the same number density of molecules as that of the subchain. This gives an idea of why rubber is soft. In the case of a gas, the bulk modulus is small since the density ρ is small. In the case of a rubber, the shear modulus is small since M_x is large. As long as the polymer maintains the network structure, M_x can be made very large by decreasing the cross-link density. For example, for $M_x = 10^4$, G is about $0.25\,\mathrm{MPa}$, which is about the same as the bulk modulus of air. In principle, we can make an elastic material which has nearly zero shear modulus.

Uniaxial elongation

Next, let us consider the elongational deformation shown in Fig. 3.3(c). If an incompressible sample is stretched in the z-direction by a factor λ, the sample will shrink in the x- and y-directions by a factor $1/\sqrt{\lambda}$ in order to keep the volume constant. Therefore the deformation gradient tensor is given by

$$(E_{\alpha\beta}) = \begin{pmatrix} 1/\sqrt{\lambda} & 0 & 0 \\ 0 & 1/\sqrt{\lambda} & 0 \\ 0 & 0 & \lambda \end{pmatrix} \tag{3.58}$$

The elastic energy of deformation becomes

$$f(\lambda) = \frac{1}{2}n_c k_B T \left(\lambda^2 + \frac{2}{\lambda} - 3 \right) = \frac{1}{2}G \left(\lambda^2 + \frac{2}{\lambda} - 3 \right) \tag{3.59}$$

Consider a cubic sample of rubber (of unit volume) stretched to λ in the z-direction. Let $\sigma(\lambda)$ be stress for this deformation (i.e., the force acting on the unit area of the sample surface normal to the z-axis). Since the surface area is now $1/\lambda$, the total force acting on the sample surface is $\sigma(\lambda)/\lambda$, and the work needed to stretch the sample further by $d\lambda$ is $(\sigma(\lambda)/\lambda)d\lambda$. Equating this with the free energy change df, we have[8]

$$\sigma = \lambda\frac{\partial f}{\partial \lambda} \tag{3.60}$$

Using eq. (3.59),

$$\sigma = G \left(\lambda^2 - \frac{1}{\lambda} \right) \tag{3.61}$$

For small deformation, λ can be written as $\lambda = 1 + \epsilon$ ($\epsilon \ll 1$). Then

$$\sigma = 3G\epsilon \tag{3.62}$$

Hence Young's modulus E is given by $3G$, which is a general result for incompressible materials (eq. (3.11)).

According to eq. (3.60), λ increases indefinitely with the increase of σ. This is not true for real rubbers. Real rubbers cannot be extended beyond a certain maximum value λ_{max}. This is due to the finite extensibility of polymer molecules. According to the freely jointed chain model, the maximum value of the end-to-end distance of a subchain is Nb. Therefore, λ_{max} is estimated as $Nb/\sqrt{N}b = \sqrt{N}$.

We have calculated the stress for two deformations, shear and elongation. The stress tensor for a general deformation can be calculated from $f(\boldsymbol{E})$. This is discussed in Appendix A (see eq. (A.24)).

3.3.3   Balloon inflation

As an example of the nonlinear analysis of the deformation of rubbers, we consider the inflation of a rubber balloon (see Fig. 3.7(a)). We assume that the balloon is a thin membrane arranged as a spherical shell. Let h

[8] The stress $\sigma(\lambda)$ represents the force per unit area of the *current* state. On the other hand, the force per unit area of the *reference* state is given by $\sigma_E = \partial f/\partial\lambda$, which is often called the engineering stress.

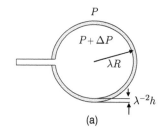

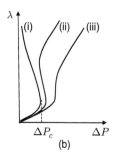

Fig. 3.7 (a) Inflating a rubber balloon. (b) The relation between the expansion of the balloon λ and the applied pressure ΔP.

be the thickness of the membrane and let R be the radius of the sphere in the force free state. Suppose that we inflate the balloon and make the pressure inside the balloon higher than that outside by ΔP. We will calculate the balloon radius R' as a function of ΔP.

If the radius is changed by a factor $\lambda = R'/R$, the rubber membrane is stretched by a factor λ in each orthogonal direction in the plane, and compressed by a factor λ^{-2} along the direction normal to the plane due to the incompressible condition. Therefore the free energy of deformation per unit volume is $(G/2)(2\lambda^2 + \lambda^{-4} - 3)$, and the total free energy of the system is given by

$$F_{tot} = 4\pi R^2 h \frac{1}{2} G \left(2\lambda^2 + \frac{1}{\lambda^4} - 3 \right) - \frac{4\pi}{3} R^3 \Delta P (\lambda^3 - 1) \qquad (3.63)$$

The second term represents the work needed to change the air volume in the balloon. The equilibrium value of λ is given by the condition $\partial F_{tot}/\partial \lambda = 0$. This gives

$$4\pi R^2 h \frac{1}{2} G \left(4\lambda - \frac{4}{\lambda^5} \right) - \frac{4\pi}{3} R^3 \Delta P 3\lambda^2 = 0 \qquad (3.64)$$

or

$$\frac{R\Delta P}{Gh} = 2 \left(\frac{1}{\lambda} - \frac{1}{\lambda^7} \right) \qquad (3.65)$$

Figure 3.7(b) shows the relation between λ and ΔP. The curve denoted by (i) in the figure indicates eq. (3.65). According to this equation, ΔP takes a maximum ΔP_c at $\lambda_c = 7^{1/6} \cong 1.38$. If ΔP is less than ΔP_c, λ increases with the increase of ΔP. However if ΔP exceeds ΔP_c, there is no solution. If a pressure higher than ΔP_c is applied, the balloon will expand indefinitely and will rupture. In real rubbers, this does not happen due to the finite extensibility of the chain. If the effect of finite extensibility is taken into account, the relation between λ and ΔP becomes as shown by curves (ii) or (iii) in Fig. 3.7(b). In the case of (ii), the balloon radius changes discontinuously at a certain pressure. In the case of (iii), there is no discontinuity, but the balloon radius changes drastically at a certain pressure. Such phenomena are indeed seen in real balloons.

3.4   Polymer gels

3.4.1   Deformation free energy

A polymer gel is a mixture of a polymer network and a solvent. Such a gel is elastic like a rubber, but unlike rubber, gel can change its volume by taking in (or expelling) solvent. For example, when a dried gel is placed in a solvent, the gel swells by absorbing the solvent from the surroundings. This phenomenon is called swelling. Conversely, if the gel is placed in air, the solvent evaporates, and the gel shrinks. The volume change of a gel can also be caused by external forces. If a weight is placed on top of a gel, solvent is squeezed out from the gel, and the gel shrinks.

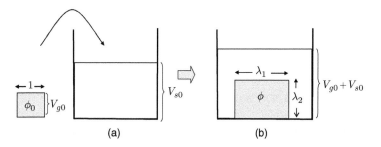

Fig. 3.8 Definition of the deformation free energy density for a gel. (a) The state before the deformation, i.e., the reference state. (b) The state after the deformation. The deformation free energy density is defined as the difference in the free energy between the two states divided by V_{g0}.

To discuss the equilibrium state, we define the deformation free energy $f_{gel}(\lambda_i)$ for a gel. This is defined as follows. We consider a gel which has a cubic shape of volume V_{g0} in the reference state (see Fig. 3.8(a)). Suppose that this gel is immersed in a pure solvent of volume V_{s0}, and is stretched along three orthogonal directions by factors λ_1, λ_2, and λ_3 (see Fig. 3.8(b)). The volume of the gel is now $\lambda_1\lambda_2\lambda_3 V_{g0}$, and the total volume of the system is $V_{g0} + V_{s0}$. The deformation free energy density $f_{gel}(\lambda_i)$ is defined as the difference in the free energy between the two states (b) and (a) divided by V_{g0}. As in the case of rubber, the equilibrium state of the gel under external forces can be calculated if $f_{gel}(\lambda_i)$ is known.

Let us consider the actual form of $f_{gel}(\lambda_i)$. The deformation free energy of a gel consists of two parts. One is the elastic energy $f_{ela}(\lambda_i)$, the energy needed to deform the polymer network, and the other is the mixing energy $f_{mix}(\lambda_i)$, the energy needed to mix the polymer and solvent:

$$f_{gel}(\lambda_i) = f_{ela}(\lambda_i) + f_{mix}(\lambda_i) \tag{3.66}$$

The elastic energy $f_{ela}(\lambda_i)$ has the same origin as that of rubber and can be written as

$$f_{ela}(\lambda_i) = \frac{G_0}{2}\left(\lambda_1^2 + \lambda_2^2 + \lambda_3^2 - 3\right) \tag{3.67}$$

where G_0 is the shear modulus of the gel in the *reference state*.

The mixing energy $f_{mix}(\lambda_i)$ has the same origin as that of polymer solutions discussed in Section 2.4, and may be written in the same form. If the volume fraction of polymer in the reference state is ϕ_0, the volume fraction in the current state is

$$\phi = \frac{\phi_0}{\lambda_1\lambda_2\lambda_3} \tag{3.68}$$

When the volume fraction changes from ϕ_0 to ϕ, a polymer solution of concentration ϕ_0 and volume V_{g0} is mixed with pure solvent of volume $(\lambda_1\lambda_2\lambda_3 - 1)V_{g0}$, and becomes a solution of concentration ϕ and volume $\lambda_1\lambda_2\lambda_3 V_{g0}$. The change in the free energy for this operation is given by

$$V_{g0}f_{mix}(\lambda_i) = \lambda_1\lambda_2\lambda_3 V_{g0}f_{sol}(\phi) - V_{g0}f_{sol}(\phi_0) - (\lambda_1\lambda_2\lambda_3 - 1)V_{g0}f_{sol}(0) \tag{3.69}$$

where $f_{sol}(\phi)$ is the free energy density of the polymer solution of concentration ϕ. Equation (3.69) is written, by use of eq. (3.68), as

$$f_{mix}(\lambda_i) = \frac{\phi_0}{\phi}[f_{sol}(\phi) - f_{sol}(0)] - [f_{sol}(\phi_0) - f_{sol}(0)] \qquad (3.70)$$

The last term does not depend on λ_i, and will be dropped in the subsequent calculation.

The free energy of the polymer solution $f_{sol}(\phi)$ is given by eq. (2.62). Since the number of polymer segments, N, in the polymer network can be regarded as infinite, $f_{sol}(\phi)$ can be written as

$$f_{sol}(\phi) = \frac{k_B T}{v_c}\left[(1-\phi)\ln(1-\phi) + \chi\phi(1-\phi)\right] \qquad (3.71)$$

Therefore, the free energy density of the gel is given by

$$f_{gel}(\lambda_i) = \frac{G_0}{2}(\lambda_1^2 + \lambda_2^2 + \lambda_3^2 - 3) + \frac{\phi_0}{\phi}f_{sol}(\phi) \qquad (3.72)$$

where we have used the fact that $f_{sol}(0) = 0$ for the expression (3.71).

We shall now study the equilibrium state of the gel using this expression.

3.4.2   Swelling equilibrium

Consider the equilibrium state of the gel shown in Fig. 3.8. If there is no force acting on the gel, the gel expands isotropically, so we may set $\lambda_1 = \lambda_2 = \lambda_3 = \lambda$. Now λ can be expressed by ϕ and ϕ_0 as $\lambda = (\phi_0/\phi)^{1/3}$. Therefore, eq. (3.72) gives

$$f_{gel}(\phi) = \frac{3G_0}{2}\left[\left(\frac{\phi_0}{\phi}\right)^{2/3} - 1\right] + \frac{\phi_0}{\phi}f_{sol}(\phi) \qquad (3.73)$$

The equilibrium state of the gel is obtained by minimizing eq. (3.73) with respect to ϕ. The condition $\partial f_{gel}/\partial \phi = 0$ gives

$$G_0\left(\frac{\phi}{\phi_0}\right)^{1/3} = \phi f'_{sol} - f_{sol} \qquad (3.74)$$

The right-hand side is equal to the osmotic pressure $\Pi_{sol}(\phi)$ of the polymer solution of concentration ϕ. Therefore eq. (3.74) can be rewritten as

$$\Pi_{sol}(\phi) = G_0\left(\frac{\phi}{\phi_0}\right)^{1/3} \qquad (3.75)$$

The left-hand side of the equation represents the force that drives polymers to expand and mix with the solvent, while the right-hand side represents the elastic restoring force of the polymer network which resists the expansion. The equilibrium volume of the gel is determined by the balance of these two forces.

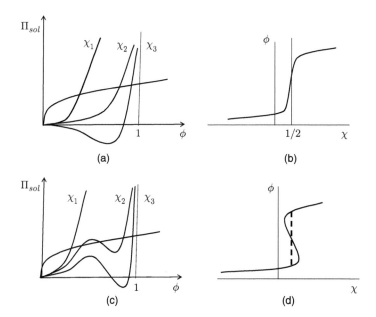

Fig. 3.9 (a) Graphical solution of eq. (3.75), where $\chi_1 < \chi_2 < \chi_3$. (b) The solution of eq. (3.75) for ϕ is plotted against χ. In this case, the equilibrium volume of the gel changes continuously as a function of χ. Similar graphs are shown in (c) and (d) for the situation that $\Pi_{sol}(\phi)$ has a local maximum and minimum. In this case, the equilibrium volume of the gel changes discontinuously at a certain value of χ.

If the mixing free energy f_{sol} is given by eq. (3.71), the osmotic pressure Π_{sol} is given by

$$\Pi_{sol}(\phi) = \frac{k_B T}{v_c} \left(-\ln(1-\phi) - \phi - \chi\phi^2 \right) \tag{3.76}$$

The parameter χ on the right-hand side is a function of temperature T. Therefore in the following, we shall discuss the volume change of the gel when χ is changed.

Figure 3.9(a) illustrates the graphical solution of eq. (3.75). The figure shows the right-hand side and the left-hand side of eq. (3.75) as a function of ϕ for various values of χ. The solution of eq. (3.75) is given by the intersection of the two curves. Figure 3.9(b) shows the solution ϕ as a function of χ. As χ increases, i.e., as the affinity between the polymer and solvent decreases, the polymer concentration ϕ increases, and the gel shrinks.

An analytical solution of eq. (3.75) can be obtained in the case of $\phi \ll 1$. In this case, $\Pi_{sol}(\phi)$ may be approximated as

$$\Pi_{sol}(\phi) = \frac{k_B T}{v_c} \left(\frac{1}{2} - \chi \right) \phi^2 \tag{3.77}$$

and eq. (3.75) is solved for ϕ as

$$\phi = \phi_0 \left[\frac{G_0 v_c}{k_B T \phi_0^2} \frac{1}{(1/2 - \chi)} \right]^{3/5} \tag{3.78}$$

If $1/2 - \chi$ is large, ϕ is small, i.e., the gel is swollen. The gel starts to shrink quickly as χ approaches $1/2$. Notice that the critical χ value $\chi_c = 1/2$ is equal to the critical χ value in polymer solutions (see

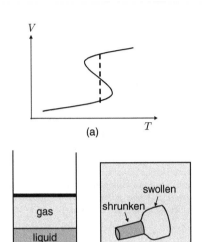

(a)

(b)   (c)

Fig. 3.10 (a) The equilibrium volume of a material showing a volume transition is plotted against temperature. (b) The coexistence state in a fluid. A gas phase and a liquid phase can coexist with negligible cost of free energy. (c) The coexistence state in a gel of cylindrical shape. When part of the shrunken gel turns into the swollen phase, both the swollen phase and the shrunken phase are deformed. As a result, there is a large cost of free energy for coexistence.

eq. (2.67)). In the case of polymer solutions, if the affinity between polymer and solvent gets worse, phase separation takes place. In the case of gels, rapid shrinking of the gel takes place.

3.4.3   Volume transition

When phase separation takes place in solutions, $\Pi_{sol}(\phi)$ has a local maximum and local minimum as shown in Fig. 2.6(b). If $\Pi_{sol}(\phi)$ has such a form, the equilibrium volume of the gel can change discontinuously as shown in Fig. 3.9(c) and (d). This phenomenon is actually observed in ionic gels (see Chapter 10), and is called a volume transition.

The volume transition of a gel is a phenomenon similar to the gas–liquid transition of a simple liquid: the volume of a liquid changes discontinuously at the boiling temperature. However, there is an essential difference between the two phenomena. The difference comes from the fact that the usual gas–liquid transition is a phenomenon taking place in a fluid, while the volume transition of a gel is a phenomenon taking place in an elastic material. The difference can be seen in the transient states occurring during the phase transition.

When a material changes from a liquid phase to a gas phase, there is an intermediate state where the two phases coexist (see Fig. 3.10). At this state, there is a cost of free energy for the coexistence of the two phases. In the case of a fluid, this cost is the interfacial energy (the energy which is proportional to the interfacial area, see Fig. 3.10(b)), which is negligibly small compared with the bulk free energy. On the other hand, in the case of a gel, the cost of the coexistence is the elastic energy of the material. For example, consider the situation that a swollen phase of a gel coexists with a shrunken phase (see Fig. 3.10(c)). Since the gel is a continuum, the phases cannot coexist keeping their shapes in the force free state; both phases must be deformed from the force free state as in Fig. 3.10(c). This cost is the elastic energy of deformation and is proportional to the volume of the material. Therefore the cost of coexistence cannot be ignored in the case of elastic materials. This affects the phase transition behaviour. For example, even if the bulk free energy of the swollen gel is equal to that of the shrunken gel at a certain temperature, the phase transition cannot take place at this temperature since for the transition to take place, a large free energy cost is needed. As a result, the temperature at which the shrunken phase turns into the swollen phase is different from the temperature at which the swollen phase turns into the shrunken phase. Therefore the volume transition of gels shows strong hysteresis.

3.4.4   Volume change by compression

Next we consider the volume change of a gel under external forces. Suppose that a weight is placed on top of a cubic gel immersed in solvent (see Fig. 3.11) at time $t = 0$. This induces compression in the vertical direction, and stretching in the horizontal direction for the gel. Let λ_z

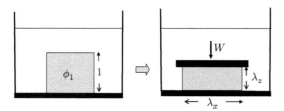

Fig. 3.11 (a) Equilibrium state of a gel in solvent. (b) Equilibrium state of a gel compressed by the weight.

and λ_x be the elongational ratios in the vertical and the horizontal directions, respectively ($\lambda_z < 1$ and $\lambda_x > 1$). Notice that λ_z and λ_x denote the elongation relative to the state of swelling equilibrium.

Immediately after the weight is placed, the gel deforms keeping its volume constant since there is no time for the solvent in the gel to move out of the gel. Therefore the horizontal elongation of the gel at time $t = 0$ is given by

$$\lambda_{x0} = \frac{1}{\sqrt{\lambda_z}} \qquad (3.79)$$

As time goes on, the solvent permeates through the gel and moves out of the gel. Accordingly the volume of the gel decreases in time. Let us consider the equilibrium state of the gel. Let ϕ_1 be the polymer volume fraction at equilibrium before the weight is placed: ϕ_1 is given by the solution of eq. (3.75). If we stretch this gel by a factor λ_z and λ_x in the vertical and horizontal directions, respectively, the elongational ratios of the deformation with respect to the reference state are $(\phi_0/\phi_1)^{1/3}\lambda_z$ and $(\phi_0/\phi_1)^{1/3}\lambda_x$. Therefore the free energy of the gel is given by

$$f_{gel} = \frac{G_0}{2}\left[\left(\frac{\phi_0}{\phi_1}\right)^{2/3}(\lambda_z^2 + 2\lambda_x^2) - 3\right] + \frac{\phi_0}{\phi}f_{sol}(\phi) \qquad (3.80)$$

The conservation of polymer volume gives $\phi = \phi_1/\lambda_z\lambda_x^2$. Therefore λ_x is expressed as

$$\lambda_x = \left(\frac{\phi_1}{\lambda_z\phi}\right)^{1/2} \qquad (3.81)$$

From eqs. (3.80) and (3.81), it follows that

$$f_{gel} = \frac{G_0}{2}\left[\left(\frac{\phi_0}{\phi_1}\right)^{2/3}\left(\lambda_z^2 + \frac{2\phi_1}{\lambda_z\phi}\right) - 3\right] + \frac{\phi_0}{\phi}f_{sol}(\phi) \qquad (3.82)$$

Given the value of λ_z, the equilibrium volume fraction ϕ is determined by the condition $\partial f_{gel}/\partial\phi = 0$. This gives the following equation for ϕ:

$$G_0\left(\frac{\phi_1}{\phi_0}\right)^{1/3}\frac{1}{\lambda_z} = \Pi_{sol}(\phi) \qquad (3.83)$$

Equation (3.83) determines the equilibrium value of ϕ for given λ_z. If $\lambda_z = 1$, eq. (3.83) reduces to eq. (3.75) and the solution is $\phi = \phi_1$. If $\lambda_z > 1$ the solution ϕ of eq. (3.83) is larger than ϕ_1 since $\partial\Pi_{sol}/\partial\phi > 0$ in a thermodynamically stable system. Therefore, the equilibrium volume of a gel decreases when it is compressed along a certain direction. This is of course what we know by experience; fluid comes out from a compressed gel.

3.5   Summary of this chapter

Elastic materials have unique equilibrium shape when all external forces are removed. This property is characterized by the non-zero value of the shear modulus. Polymer fluids (polymer melts and polymer solutions) can be made into elastic (rubbers and gels) by cross-linking the polymer chains and creating a network of polymers. The resulting materials (rubber) have small shear modulus (their shape can be changed easily), but have large bulk modulus (their volume cannot be changed). The elastic property is entirely characterized by the deformation free energy density $f(\boldsymbol{E})$.

The elasticity of rubber originates from the elasticity of the polymer chain. The shear modulus of rubber is essentially determined by the number density of subchains (or the number density of cross-links).

A gel is a mixture of a polymer network and solvents and may be regarded as an elastic polymer solution. Unlike rubber, a gel can change its volume by taking in (or out) solvents. This behaviour is again characterized by the deformation free energy density $f_{gel}(\boldsymbol{E})$. The phenomenon of phase separation in the solution is seen as a volume transition in gels.

Further reading

(1) *Introduction to Polymer Physics*, Masao Doi, Oxford University Press (1996).

(2) *The Structure and Rheology of Complex Fluids*, Ronald G. Larson, Oxford University Press (1999).

(3) *Polymer Physics*, Michael Rubinstein and Ralph H. Colby, Oxford University Press (2003).

Exercises

(3.1) Answer the following questions.

    (a) Show that the bulk modulus of an ideal gas is given by

$$K = nk_B T \qquad (3.84)$$

    where n is the number density of molecules in the system.

    (b) Estimate the bulk modulus of water assuming that eq. (3.84) is valid even in the liquid state of water. Assume that a water molecule is a sphere of diameter 0.3 [nm]. Compare this value with the actual bulk modulus of water (2 [GPa]).

    (c) Estimate the shear modulus of rubber (density 1 [g/cm^3]) which consists of subchains of molecular weight 10^4 [g/mol].

(3.2) Consider an orthogonal deformation of an isotropic elastic material. If the deformation is small, the free energy of deformation is written as a quadratic function of the elongational strain $\epsilon_\alpha = \lambda_\alpha - 1$ ($\alpha = x, y, z$) and can generally be written as

$$f(\epsilon_x, \epsilon_y, \epsilon_z) = \frac{K}{2}(\epsilon_x + \epsilon_y + \epsilon_z)^2$$
$$+ \frac{G}{3}\left[(\epsilon_x - \epsilon_y)^2 + (\epsilon_y - \epsilon_z)^2 + (\epsilon_z - \epsilon_x)^2\right] \tag{3.85}$$

where K and G are the bulk modulus and the shear modulus, respectively. If the material is stretched in the z-direction by a stress σ, the total free energy of the system is written as

$$f_{tot} = f(\epsilon_x, \epsilon_y, \epsilon_z) - \sigma\epsilon_z \tag{3.86}$$

Answer the following questions.

(a) The equilibrium values of ϵ_x, ϵ_y, and ϵ_z are determined by the condition that f_{tot} be a minimum at equilibrium. Prove that the equilibrium values are given by

$$\epsilon_z = \frac{\sigma}{E} \qquad \epsilon_x = \epsilon_y = -\nu\epsilon_z \tag{3.87}$$

with

$$E = \frac{9KG}{3K + G} \qquad \nu = \frac{3K - 2G}{2(3K + G)} \tag{3.88}$$

(b) Show that the volume of the material changes as

$$\frac{\Delta V}{V} = \frac{\sigma}{3K} \tag{3.89}$$

(3.3) Let λ_i^2 ($i = 1, 2, 3$) be the eigenvalues of the tensor $\boldsymbol{B} = \boldsymbol{E} \cdot \boldsymbol{E}^t$.

(a) Using the fact that λ_i^2 ($i = 1, 2, 3$) are the solutions of the eigenvalue equation $\det(\boldsymbol{B} - \lambda^2 \boldsymbol{I}) = 0$ ($\boldsymbol{I}$ being a unit tensor), prove the following relations

$$\lambda_1^2 + \lambda_2^2 + \lambda_3^2 = B_{\alpha\alpha} = E_{\alpha\beta}^2 \tag{3.90}$$
$$\lambda_1^2\lambda_2^2\lambda_3^2 = \det(\boldsymbol{B}) \tag{3.91}$$

(b) Prove the relation

$$\lambda_1^{-2} + \lambda_2^{-2} + \lambda_3^{-2} = (\boldsymbol{B}^{-1})_{\alpha\alpha} \tag{3.92}$$

(c) Prove that the free energy of deformation of an incompressible isotropic material can be written as a function of $\mathrm{Tr}\boldsymbol{B}$ and $\mathrm{Tr}\boldsymbol{B}^{-1}$.

(3.4) Obtain the principal elongational ratios $\lambda_1, \lambda_2, \lambda_3$ for shear deformation (3.51), and show that for small γ, they are given by

$$\lambda_1 = 1 + \gamma/2, \quad \lambda_2 = 1 - \gamma/2, \quad \lambda_3 = 1 \tag{3.93}$$

(3.5) The elastic deformation free energy density of rubber is often expressed as

$$f(E) = \frac{1}{2}C_1(\mathrm{Tr}\boldsymbol{B} - 3) + \frac{1}{2}C_2(\mathrm{Tr}\boldsymbol{B}^{-1} - 3) \tag{3.94}$$

where C_1 and C_2 are constants. Answer the following questions.

(a) Obtain the stress σ in the uniaxial elongation of such a rubber as a function of the elongational ratio λ.

(b) Obtain the shear stress σ as a function of the shear strain γ.

(c) Discuss the inflation of a balloon made of such rubber and draw the curve shown in Fig. 3.7.

(3.6) *When we inflate a tube-like rubber balloon, we often see the situation shown in Fig. 3.12: the strongly inflated part coexists with the less inflated part. Answer the following questions.

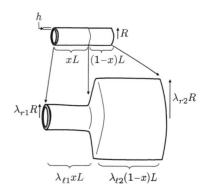

Fig. 3.12

(a) The deformation free energy density of the rubber can be written as $f(\lambda_r, \lambda_\ell)$, where λ_r and λ_ℓ are the elongational ratios in the radial and the axial directions. Let h, R, and L be the initial thickness, radius, and axial length of the tube-like balloon. Suppose that part (of fraction x) is weakly inflated, and the remaining part (of fraction $(1 - x)$) is strongly inflated. The total free energy of the system is written as

$$F_{tot} = 2\pi RLh[xf(\lambda_{r1}, \lambda_{\ell1}) + (1-x)f(\lambda_{r2}, \lambda_{\ell2})]$$
$$- \pi R^2 L\Delta P[x\lambda_{r1}^2\lambda_{\ell1} + (1-x)\lambda_{r2}^2\lambda_{\ell2} - 1] \tag{3.95}$$

where ΔP is the extra pressure in the balloon. Show that the equilibrium state is determined by the following set of equations:

$$\frac{\partial f}{\partial \lambda_{ri}} = 2p\lambda_{ri}\lambda_{\ell i} \quad \text{for} \quad i = 1, 2 \quad (3.96)$$

$$\frac{\partial f}{\partial \lambda_{\ell i}} = p\lambda_{ri}^2 \quad \text{for} \quad i = 1, 2 \quad (3.97)$$

and

$$f(\lambda_{r1}, \lambda_{\ell 1}) - f(\lambda_{r2}, \lambda_{\ell 2}) = p(\lambda_{r1}^2\lambda_{\ell 1} - \lambda_{r2}^2\lambda_{\ell 2}) \quad (3.98)$$

where

$$p = \frac{R\Delta P}{2h} \quad (3.99)$$

(b) Discuss whether such coexistence is possible or not for the deformation free energy given by eq. (3.49).

(3.7) A gel is in equilibrium in a good solvent. Suppose that the polymer which constitutes the gel is added to the outer solvent (see Fig. 3.13). Show that the gel shrinks and the equilibrium volume fraction ϕ

is determined by the equation

$$G_0 \left(\frac{\phi}{\phi_0}\right)^{1/3} = \Pi_{sol}(\phi) - \Pi_{sol}(\phi_{sol}) \quad (3.100)$$

where ϕ_{sol} is the volume fraction of polymer in the outer solution. Assume that the polymer in the solution cannot get into the gel.

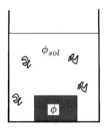

Fig. 3.13

(3.8) For the gel compressed by the weight W (see Fig. 3.11), obtain the relation between the weight W and the vertical contraction λ_z.

Surfaces and surfactants

4

In this chapter, we shall discuss the phenomena associated with material surfaces. Surfaces, or more generally, interfaces are important in soft matter by two reasons. First, the weak forces associated with surfaces, such as surface tension and inter-surface forces, play an important role in the flow and deformation of soft matter. For example, surface tension is the force governing the behaviour of liquid droplets. Second, many soft matter systems, especially colloidal dispersions, are composed of several different phases and have large interfacial area within the material. In such materials, the interfacial properties are crucially important for the bulk properties. For example, whether colloidal dispersions are stable or not is determined by the inter-surface force as we have discussed in Section 2.5.

The organization of this chapter is as follows. We start with liquid surfaces and discuss the phenomena seen in everyday life. We then discuss surfactants, the materials which effectively change surface properties. Finally, we discuss the interaction between the surfaces in close proximity. The last topics are important in the study of thin films and colloidal dispersions.

4.1 Surface tension

4.1.1 Equilibrium shape of a fluid

In the previous chapters, it has been argued that fluid does not have its own shape, and therefore the free energy of a fluid is independent of its shape. However, in daily life, we see many phenomena which indicate that fluids have certain preferred shapes. For example, a small liquid droplet placed on a plate usually takes the shape of spherical cap or pancake (see Fig. 4.1(a)). The shape changes if the plate is tilted, but it recovers the original shape when the plate is set back. This phenomenon indicates that the fluid droplet on a plate has an equilibrium shape.

The force which gives the equilibrium shape to a fluid is the force associated with the fluid surface. A surface is a special place for molecules constituting the fluid. Molecules in a fluid usually attract each other. If such a molecule sits in the bulk region (the region far from the surface), it has a negative potential energy $-z\epsilon$, where z is the coordination number and $-\epsilon$ is the van der Waals energy between neighbouring molecules (see Section 2.3.2). On the other hand, if the molecule is at the surface,

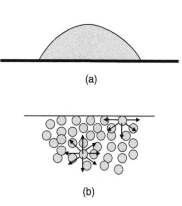

Fig. 4.1 (a) Example of the effect of surface tension: a liquid droplet on a substrate takes the shape of a spherical cap. (b) Molecular origin of surface tension.

Fig. 4.2 (a) Experimental methods to measure the surface tension. A liquid is set in the region bounded by the U-shaped supporting wire and a straight bounding wire. The force acting on the bounding wire is given by $2\gamma a$. (b) Derivation of eq. (4.2). The liquid in the dashed box is pulled to the right by the force f exerted on the bounding wire and is pulled to the left by the surface tension $2\gamma a$. The balance of these forces gives eq. (4.2). (c) Laplace pressure. To make a liquid droplet by injection of the liquid through a syringe, an excess pressure ΔP must be applied. The excess pressure at equilibrium is the Laplace pressure. For a droplet of radius r, the Laplace pressure is $\Delta P = 2\gamma/r$. (d) The excess pressure ΔP needed to blow a soap bubble is given by $\Delta P = 4\gamma/r$.

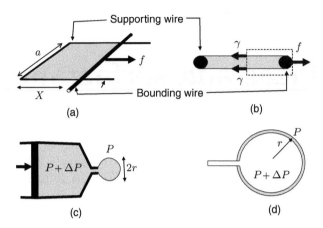

approximately half of the neighbours are lost as shown in Fig. 4.1(b), and the potential energy is about $-(z/2)\epsilon$. Therefore the molecules at the surface have a potential energy about $(z/2)\epsilon$ higher than the molecules in the bulk.

The excess energy associated with the surface is called the surface energy. (More precisely, it should be called the surface free energy, and its definition will be given later.) The surface energy is proportional to the surface area of the fluid, and the energy per unit area of the surface is called the surface energy density (or surface tension), denoted by γ.

Since the surface area of a fluid depends on the fluid shape, the equilibrium shape of a fluid is uniquely determined by the condition that the free energy of the system is a minimum. If there are no other forces acting on the fluid, the fluid takes a spherical shape, the one which minimizes the surface area for a given volume.

4.1.2   Surface tension

The surface energy of a fluid can be measured by experimental methods shown in Fig. 4.2.

In Fig. 4.2(a), a straight wire (bounding wire) is placed on a U-shaped supporting wire, and a liquid film is made in the region bounded by the two wires. The surface area of the fluid is $2ax$, where a and x are defined in the figure, and the factor of 2 appears since the fluid has two surfaces, at the top and the bottom. The surface energy of the fluid is therefore given by $G_A = 2ax\gamma$. To minimize the surface energy, the liquid film tends to shrink. To keep the surface area constant, one must apply a force f to the bounding wire. Moving the position of the bounding wire by a distance dx involves work $f dx$. This work is equal to the change of the surface free energy dG_A, and we have[1]

$$f dx = 2\gamma a dx \qquad (4.1)$$

or

$$f = 2a\gamma \qquad (4.2)$$

---

[1] Here we have assumed that γ does not depend on the surface area. This assumption is justified for surfaces at equilibrium as we shall see in Section 4.1.4. If the surface includes surfactants which are insoluble in the bulk fluid or very slow to dissolve, γ becomes a function of the surface area (see Section 4.3.5). For such a system, eq. (4.1) should be written as $f = \partial G_A/\partial x$.

Equation (4.2) indicates that γ is a measure of the force acting on the surface of a fluid. The surface area of a fluid tends to decrease, which indicates that there is a tensile force acting on the surface of the fluid, and γ is equal to the tensile force per unit length. For this reason γ is also called the surface tension. Equation (4.2) can be derived from the force balance acting on the fluid element as explained in Fig. 4.2(b).

Figure 4.2(c) shows another way of measuring the surface tension. To push out a fluid from a syringe, an excess pressure ΔP must be applied since the fluid tends to move back to the bulk to decrease the surface energy. The excess pressure ΔP at equilibrium is related to the surface energy γ by

$$\Delta P = \frac{2\gamma}{r} \tag{4.3}$$

where r is the radius of the droplet. Equation (4.3) is derived as follows. If we push the piston slightly and change the volume V of the droplet by δV, we do work $\Delta P \delta V$ to the fluid. At equilibrium, this is equal to the change of the surface energy

$$\Delta P \delta V = \gamma \delta A \tag{4.4}$$

where A is the surface area of the droplet. Since δV and δA are written as $\delta V = 4\pi r^2 \delta r$ and $\delta A = 8\pi r \delta r$, eq. (4.4) gives eq. (4.3).

The excess pressure ΔP created by the surface tension is called the Laplace pressure. It is the pressure we need to apply to create a soap bubble. In the case of a soap bubble, the excess pressure to make a bubble of radius r is given by $4\gamma/r$. The factor of 4 now comes from the fact that the soap bubble has two surfaces, the outer surface and the inner surface (see Fig. 4.2(d)).

4.1.3   Grand canonical free energy

We now define the surface free energy in a more rigorous way. For this purpose, it is convenient to use the grand canonical free energy.

The grand canonical free energy is defined for a material which occupies a volume V, and exchanges energy and molecules with surrounding materials. Let T be the temperature of the environment and μ_i be the chemical potential of the molecules of the i-th component. The grand canonical free energy is defined by

$$G(V, T, \mu_i) = F - \sum_{i=1}^{p} N_i \mu_i \tag{4.5}$$

where F is the Helmholtz free energy, and N_i is the number of molecules of the i-th component. (Notice that in this chapter, G stands for the grand canonical free energy, not the Gibbs free energy.)

For given values of T and μ_i, the free energy G is proportional to the system volume V. Therefore G can be written as

$$G(V, T, \mu_i) = V g(T, \mu_i) \tag{4.6}$$

Now it can be proven that $g(T, \mu_i)$ is equal to minus the pressure P, i.e.,

$$G(V, T, \mu_i) = -VP(T, \mu_i) \tag{4.7}$$

Equation (4.7) can be shown as follows. The total differential for the Helmholtz free energy is given by

$$dF = -SdT - PdV + \sum_i \mu_i dN_i \tag{4.8}$$

where S is the entropy of the system. From eqs. (4.5) and (4.8), it follows that

$$dG = -SdT - PdV - \sum_i N_i d\mu_i \tag{4.9}$$

On the other hand, eq. (4.6) gives

$$dG = gdV + V\frac{\partial g}{\partial T}dT + V\sum_i \frac{\partial g}{\partial \mu_i}d\mu_i \tag{4.10}$$

Comparing eq. (4.9) with eq. (4.10), we have

$$g = -P, \quad \left(\frac{\partial g}{\partial T}\right)_{\mu_i} = -\frac{S}{V}, \quad \left(\frac{\partial g}{\partial \mu_i}\right)_T = -\frac{N_i}{V} \tag{4.11}$$

The first equation in (4.11) gives eq. (4.7). The equations in (4.11) also give the following Gibbs–Duhem equation

$$VdP = SdT + \sum_i N_i d\mu_i \tag{4.12}$$

4.1.4   Interfacial free energy

Now, we define the interfacial free energy. Consider the situation shown in Fig. 4.3. Two phases I and II coexist and are in equilibrium. The temperature T and chemical potentials μ_i are common in both phases. The total grand canonical free energy of this system can be written as follows

$$G = G_I + G_{II} + G_A \tag{4.13}$$

where G_I and G_{II} represent the grand canonical free energy of the bulk phases I and II, and G_A represents the grand canonical free energy of the interface. G_I and G_{II} are proportional to the bulk volume of each phase V_I and V_{II}, and G_A is proportional to the interfacial area A, i.e.,

$$G_I = V_I \, g_I(T, \mu_i), \qquad G_{II} = V_{II} \, g_{II}(T, \mu_i) \tag{4.14}$$

and

$$G_A = A\gamma(T, \mu_i) \tag{4.15}$$

Equations (4.13)–(4.15) define the interfacial free energy. The basis of this definition is the assumption that the free energy of a system is written as a sum of bulk parts (the part proportional to the volume), and the interfacial part (the part proportional to the interfacial area).

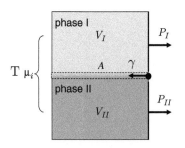

Fig. 4.3 Parameters defining the thermodynamic state of two coexisting phases I and II, where V_I, V_{II}, and A are the volume of phase I, that of phase II, and the interfacial area, respectively. Temperature T and the chemical potentials of molecules μ_i are common. The free energy of the system is uniquely determined as a function of these parameters.

As we have seen, the grand canonical free energy per unit volume g_I is equal to minus the pressure P_I, the force acting on a unit area considered in the bulk. Likewise, the grand canonical free energy per unit area of the interface γ is equal to the force f_ℓ acting on a unit length considered in the interface. The proof can be done precisely in the same way as for g_I. If the interfacial area is changed by dA, the work done to the system is $f_\ell dA$. Therefore the total differential of G_A is written as

$$dG_A = -S_A dT + f_\ell dA - \sum_i N_{Ai} d\mu_i \qquad (4.16)$$

where S_A is the entropy of the surface, and N_{Ai} is the number of molecules of component i on the surface. On the other hand, eq. (4.15) gives

$$dG_A = \gamma dA + A \frac{\partial \gamma}{\partial T} dT + A \sum_i \frac{\partial \gamma}{\partial \mu_i} d\mu_i \qquad (4.17)$$

Equations (4.16) and (4.17) give $f_\ell = \gamma$, and the following Gibbs–Duhem equation for the surface

$$A d\gamma = -S_A dT - \sum_i N_{Ai} d\mu_i \qquad (4.18)$$

4.1.5   Surface excess

The number of molecules per unit area on the surface is given by

$$\Gamma_i = \frac{N_{Ai}}{A} \qquad (4.19)$$

Γ_i is called the surface excess. It is called 'excess' since it can be positive or negative. The meaning of Γ_i is explained in Fig. 4.4. Here the profile of the number density $n_i(z)$ of molecules of species i is plotted against the coordinate taken along the axis normal to the interface. On the left side, $n_i(z)$ approaches n_{Ii}, the bulk number density in phase I, and on the right side, $n_i(z)$ approaches n_{IIi}. Near the interface, $n_i(z)$ will be different from these. Figure 4.4 shows the case that the component i is attracted to the interface. Then Γ_i is defined by[2]

$$\Gamma_i = \int_{-\infty}^{0} dz (n_i(z) - n_{Ii}) + \int_{0}^{\infty} dz (n_i(z) - n_{IIi}) \qquad (4.20)$$

If the component i likes the surface, Γ_i is positive. Surfactant molecules like the surface very much, and Γ_i can be regarded as the number of surfactant molecules adsorbed per unit area on the surface. On the other hand, ions prefer to stay in the bulk rather than at the surface, and therefore their surface excess Γ_i is negative.

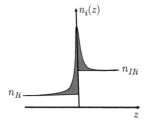

Fig. 4.4 Definition of the surface excess Γ_i. Here $n_i(z)$ represents the number density of molecules of component i at the coordinate z taken normal to the interface. The surface excess is given by the area of the shaded region in the figure, where n_{Ii} and n_{IIi} stands for the asymptotic values of $n_i(z)$ in the limit of $z \to -\infty$ and $z \to \infty$.

[2] The value of the integral in eq. (4.20) depends on the position of the interface, i.e., the position of the origin in the z-coordinate. The usual convention is to take it at the position which gives zero excess for the solvent (or the major component).

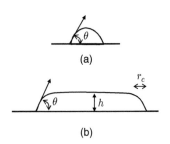

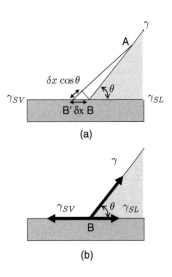

Fig. 4.5 (a) The shape of a small droplet on a substrate, and (b) the shape of a large droplet. The contact angle θ is common for two droplets. r_c denotes the capillary length.

Fig. 4.6 (a) Derivation of the Young–Dupre equation by the principle of energy minimization. Here A is a point taken on the interface sufficiently far from the contact line. (b) Derivation of the Young–Dupre equation by the force balance condition. The balance of the x-component of the surface forces acting on the point B gives the Young–Dupre equation.

4.2   Wetting

4.2.1   Thermodynamic driving force for wetting

When a liquid droplet is placed on a substrate, it spreads on the surface of the substrate. This phenomenon is called wetting. The wetting also takes place in porous materials (sponges, textiles, sands, etc.), where the liquid spreads over the internal surfaces of bulk materials.

The driving force for the wetting is the interfacial energy between the solid and the liquid. If a liquid spreads over a surface of a solid, the area which has been the interface between the solid and air is now replaced by two interfaces, the solid–liquid interface, and the liquid–air interface. In the following, we shall call the air phase the vapour phase since in a one-component system, the gas phase is filled by the vapour of the liquid.

Let γ_{SL} and γ_{SV} be the interfacial energy for the solid–liquid, and the solid–vapor interfaces. Then when the liquid forms a film of area A, the interfacial energy changes from $A\gamma_{SV}$ to $A(\gamma + \gamma_{SL})$. The difference in the coefficient of A

$$\gamma_S = \gamma_{SV} - (\gamma + \gamma_{SL}) \tag{4.21}$$

is called the spreading coefficient. If γ_S is positive, the interfacial energy decreases as A increases. In this case, the wetting proceeds and the liquid eventually wets the entire solid surface. This case is called complete wetting. On the other hand, if γ_S is negative, the liquid drop wets the solid surface partially, taking a certain equilibrium conformation on the substrate as shown in Fig. 4.5. This case is called partial wetting.

The outer rim of the wetted region is called the contact line. It is also called the triple line because at this line, the three phases (liquid, vapour, and substrate) meet.

The angle that the liquid surface makes against the substrate is called the contact angle. If the contact line can move freely on the substrate, the contact angle θ is constant and is independent of the size and the shape of the droplet (see Fig. 4.5). This can be shown as follows. Consider a wedge-like region of the fluid near the surface (see Fig. 4.6). Now suppose that the contact line is displaced from B to B$'$ on the substrate. Then the area δx of the solid–vapour interface is replaced by the solid–liquid interface, and the surface area of the liquid increases by $\delta x \cos\theta$. Therefore, the change of the interfacial energy due to the displacement of the contact line is $(\gamma_{SL} - \gamma_{SV})\delta x + \gamma \delta x \cos\theta$. At equilibrium, this must be equal to zero. Hence

$$\gamma_{SL} - \gamma_{SV} + \gamma\cos\theta = 0 \tag{4.22}$$

Equation (4.22) is called the Young–Dupre equation. According to this equation, the contact angle θ is determined by the interfacial energies only, and is independent of the droplet size and other bulk forces acting on the droplet. The Young–Dupre equation can also be derived from the force balance as explained in Fig. 4.6(b).

4.2.2   Liquid droplet on a solid surface

Capillary length

Let us now consider the equilibrium shape of a droplet placed on a substrate (see Fig. 4.5).

The equilibrium shape of the droplet is determined by minimizing the total free energy. The bulk free energy of the droplet is independent of the droplet shape, but the interfacial free energy and the potential energy due to gravity depend on the droplet shape.

Consider a liquid droplet of volume V. The characteristic length of the droplet is $r \simeq V^{1/3}$. Now the interfacial energy of the droplet is of the order of γr^2, while the potential energy of gravity is of the order of $\rho g V r \simeq \rho g r^4$, where ρ is the density of the liquid. If r is small, the latter effect is negligible, and the droplet shape is determined by the interfacial energy only. On the other hand, if r is large, the gravity is dominant. The characteristic length which distinguishes the two cases is

$$r_c = \sqrt{\frac{\gamma}{\rho g}} \qquad (4.23)$$

which is called the capillary length. The capillary length of water is 1.4 mm at room temperature.

If the size of the droplet is much less than r_c, the liquid takes the form of a spherical cap as in Fig. 4.5(a). The radius of the sphere is determined by the volume of the droplet and the contact angle θ. On the other hand, if the size of the droplet is much larger than r_c, the liquid takes a flat pancake shape with rounded edges as in Fig. 4.5(b).

Equilibrium film thickness

Let us consider the thickness of the liquid film shown in Fig. 4.5(b). If the volume of the liquid is V, and the thickness is h, the potential energy of gravity is $\rho g V h/2$. On the other hand, the total interfacial energy G_A is given by the contact area of the liquid $A = V/h$, and the spreading coefficient γ_S as $G_A = -\gamma_S V/h$. (Notice that $\gamma_S < 0$ in the present situation.) Therefore the total energy is

$$G_{tot} = \frac{1}{2}\rho g V h - \frac{\gamma_S V}{h} \qquad (4.24)$$

Minimizing this with respect to h, we have the equilibrium thickness

$$h = \sqrt{\frac{-2\gamma_S}{\rho g}} \qquad (4.25)$$

Using eq. (4.22), γ_S can be written as

$$\gamma_S = -\gamma(1 - \cos\theta) \qquad (4.26)$$

Hence, the equilibrium thickness h is written as

$$h = \sqrt{\frac{2\gamma(1 - \cos\theta)}{\rho g}} = 2r_c \sin\left(\frac{\theta}{2}\right) \qquad (4.27)$$

According to eq. (4.27), h becomes zero in the case of complete wetting. In practice, however, it has been observed that h does not go to zero even in the case of complete wetting. This is due to the effect of van der Waals forces acting between the vapour phase and the solid substrate. This will be discussed later in Section 4.4.7.

4.2.3   Liquid rise in a capillary

Another familiar phenomenon caused by surface tension is the rise of the meniscus in a capillary (see Fig. 4.7). If the meniscus rises by a height h in a capillary of radius a, the dry surface (i.e., the solid–vapor interface) of the capillary wall of area $2\pi ah$ is replaced by the solid–liquid interface. Therefore, the total energy change is given by

$$G_{tot} = 2\pi ah(\gamma_{SL} - \gamma_{SV}) + \frac{1}{2}\rho g\pi a^2 h^2 \tag{4.28}$$

Minimizing eq. (4.28) with respect to h, we have

$$h = \frac{2(\gamma_{SV} - \gamma_{SL})}{\rho ga} \tag{4.29}$$

This can be rewritten using eq. (4.22) as

$$h = \frac{2\gamma\cos\theta}{\rho ga} = \frac{2r_c^2}{a}\cos\theta \tag{4.30}$$

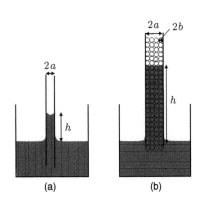

If the contact angle θ is less than $\pi/2$, the meniscus goes up ($h > 0$), and if θ is larger than $\pi/2$, the meniscus goes down ($h < 0$).

Notice that the rewriting of eq. (4.29) into eq. (4.30) is valid only for the case of partial wetting, i.e., the case of $\gamma_S < 0$. In the case of complete wetting, the capillary wall is already wetted by the fluid, and has surface energy $\gamma_{SL} + \gamma$ per unit area. As the meniscus rises, this is replaced by the bulk fluid which has surface energy γ_{SL}. Therefore the gain in the surface energy is γ per unit area. Accordingly, the height of the meniscus is given by eq. (4.30) with $\theta = 0$. Hence eq. (4.30) is actually valid both for $\gamma_S < 0$ and $\gamma_S > 0$.

Equation (4.30) indicates that the height of the meniscus increases with the decrease of the capillary radius a. This is because the surface effect is enhanced for smaller capillaries. Instead of decreasing the capillary radius, the surface effect can be enhanced by filling the capillary with small beads as in Fig. 4.7(b). If the bead radius is b, the surface area of the beads contained in the height h is approximately $\pi a^2 h/b$. Hence the capillary height is increased by a factor a/b. This is the reason why dry sands or textiles are strongly wetted (or dewetted).

Fig. 4.7 (a) Rise of the meniscus of a liquid in a capillary. (b) Rise of the meniscus in a capillary filled with glass beads of radius b.

4.2.4   Liquid droplet on liquid

The Young–Dupre equation (4.22) has been derived for a rigid substrate, i.e., a substrate which does not deform. If the substrate is a soft elastic material or a fluid, the substrate deforms when a fluid droplet is placed upon it. For example, an oil droplet placed on a water surface will take a lens-like form as shown in Fig. 4.8(a), where A, B, and C represent air, water, and oil, respectively. At equilibrium, the interfacial forces acting at the contact line must balance as shown in Fig. 4.8(b) and (c), giving

$$\vec{\gamma}_{AB} + \vec{\gamma}_{BC} + \vec{\gamma}_{CA} = 0 \qquad (4.31)$$

where $\vec{\gamma}_{XY}$ (X and Y representing one of A, B, C) stand for the interfacial force vector acting at the X–Y interface. Equation (4.31) is equivalent to the condition that the vectors $\vec{\gamma}_{AB}$, $\vec{\gamma}_{BC}$, $\vec{\gamma}_{CA}$ form a triangle. This condition is a generalization of the Young–Dupre condition and is called the Neumann condition. Using the Neumann condition, one can express the contact angles θ_A, θ_B, θ_C in terms of the interfacial energies $\gamma_{AB}, \gamma_{BC}, \gamma_{CA}$ by constructing the triangle (called the Neumann triangle) shown in Fig. 4.8(c).

The Neumann triangle cannot always be constructed. For example, it cannot be constructed for the situation $\gamma_{AB} > \gamma_{BC} + \gamma_{CA}$. In this situation, complete wetting of the C phase takes place, i.e., the liquid C forms a thin film between liquid A and liquid B.

4.3 Surfactants

4.3.1 Surfactant molecules

As discussed in Section 1.3, surfactant molecules consist of a hydrophilic (water-loving) part and a hydrophobic (water-hating, or oil-loving) part. Such molecules are not completely happy in either water or oil: in water, the contact of the hydrophobic part with water is energetically unfavourable. Likewise, in oil, the contact of the hydrophilic part with oil is energetically unfavourable. However, the surfactant molecules are completely happy at the oil–water interface because there the hydrophilic part can sit in water, and the hydrophobic part can sit in oil.

If surfactant molecules are added to a mixture of oil and water, the molecules tend to increase the interfacial area between oil and water. Therefore, in the presence of surfactant, the free energy of the system decreases with the increase of the area of the oil–water interface. In the extreme case, an oil droplet in water is broken up into smaller droplets and dispersed in water. This phenomenon is called spontaneous emulsification.

4.3.2 Surface tension of surfactant solutions

If surfactant is added to water, the surfactant molecules assemble at the surface because the hydrophobic part of the surfactant molecule is happier to sit in air than in water. As this happens, the energetic cost of the surface decreases, i.e., the surface tension of water with surfactants is smaller than that of pure water.

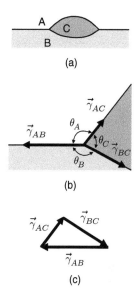

Fig. 4.8 (a) Oil droplet placed on the surface of water. (b) The contact angles around the triple line of three phases A, B, and C. (c) The Neumann triangle representing the force balance at the triple line.

The reduction of the surface tension by surfactants can be observed experimentally using the setup shown in Fig. 4.9. A trough containing water is separated into two chambers by a flexible membrane fixed at the bottom and connected to a floating bar. If surfactant is added to the left chamber, the surface tension in the left chamber decreases, and therefore the bar moves to the right. The force acting on the bar per unit length is called the surface pressure, and is given by

$$\Pi_A = \gamma_0 - \gamma \tag{4.32}$$

where γ_0 denotes the surface tension of the original liquid.

Another way of looking at the phenomenon is the following. Surfactant molecules assemble at the surface, and tend to expand their area bounded by the bar. This mechanism is analogous to that of osmotic pressure: the solute molecules confined by a semi-permeable membrane tend to expand and exert osmotic pressure. Likewise, the surfactant molecules confined by the bar create surface pressure.

The above example indicates that if surface tension is not constant along the surface, the gradient of the surface tension exerts a net force for the fluid elements at the surface, and induces a macroscopic flow. This effect is called the Marangoni effect.

The Marangoni effect is seen in daily life. If a drop of liquid detergent is placed on a water surface, any dust particles floating on the water are pushed away from the point at which the detergent is placed. In this example, the Marangoni effect is caused by the gradient of surfactant concentration. The Marangoni effect can also be caused by a temperature gradient. The surface tension is a function of temperature (surface tension usually decreases with an increase of temperature), so if there is a temperature gradient at the surface, the fluid will flow from the high-temperature region to the low-temperature region.

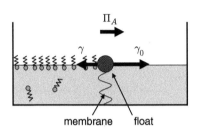

Fig. 4.9 Experimental setup to measure the change of surface tension due to the addition of surfactant.

4.3.3   Surface adsorption and surface tension

The surface tension of a fluid decreases if we add materials which prefer the surface rather than the bulk. This effect can be proven using the Gibbs–Duhem equation for the surface. From eq. (4.18) it follows that

$$\left(\frac{\partial \gamma}{\partial \mu_i}\right)_T = -\frac{N_{Ai}}{A} = -\Gamma_i \tag{4.33}$$

For simplicity of notation, we consider the situation that the solution is a two-component system made of solvent and surfactant and omit the subscript i. We therefore write eq. (4.33) as

$$\left(\frac{\partial \gamma}{\partial \mu}\right)_T = -\frac{N_A}{A} = -\Gamma \tag{4.34}$$

where $\Gamma = N_A/A$ is the number of surfactant molecules adsorbed at the surface of unit area.

The chemical potential μ of the surfactant is a function of the surfactant concentration. Let n be the number density of surfactant molecules in the bulk. In a dilute solution of surfactant, μ can be written as

$$\mu(T, n) = \mu_0(T) + k_B T \ln n \tag{4.35}$$

In this case, the change of surface tension caused by the change of surfactant concentration $(\partial\gamma/\partial n)_T$ can be calculated as

$$\left(\frac{\partial\gamma}{\partial n}\right)_T = \left(\frac{\partial\gamma}{\partial\mu}\right)_T \left(\frac{\partial\mu}{\partial n}\right)_T = -k_B T \frac{\Gamma}{n} \tag{4.36}$$

Since Γ is positive for surfactants, eq. (4.36) indicates that the surface tension of the liquid decreases with the increase of surfactant concentration.

Notice that eq. (4.36) is valid for any additives. Therefore if we add materials which have negative surface excess ($\Gamma < 0$), the surface tension increases with the increase of the additives.

The relation between n (the number density of additives in the bulk), and Γ (the number density of the additive adsorbed at the surface) is called an adsorption equation. A typical adsorption equation is the Langmuir equation given by

$$\Gamma = \frac{n\Gamma_s}{n_s + n} \tag{4.37}$$

where Γ_s and n_s are constants. Equation (4.37) describes the behaviour shown in Fig. 4.10: Γ increases linearly with n for $n < n_s$, and then starts to saturate for $n > n_s$, approaching Γ_s as $n \to \infty$.

Equations (4.36) and (4.37) give the following differential equation

$$\frac{\partial\gamma}{\partial n} = -k_B T \frac{\Gamma_s}{n_s + n} \tag{4.38}$$

The solution of this equation is

$$\gamma = \gamma_0 - \Gamma_s k_B T \ln\left(1 + \frac{n}{n_s}\right) \tag{4.39}$$

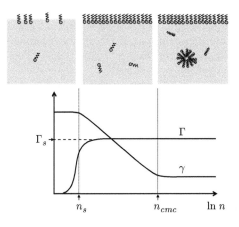

Fig. 4.10 The surface tension γ and the surface excess Γ of a surfactant are plotted against the number density of surfactant in the bulk. n_s is the concentration around which the surface is saturated with surfactant, and n_{cmc} is the concentration above which micelles are formed.

The surface tension γ is also shown in Fig. 4.10. As the surfactant concentration increases, γ first decreases quickly, where $\gamma_0 - \gamma$ is proportional to n, but soon γ starts to decrease slowly, where $\gamma_0 - \gamma$ is proportional to $\ln n$. This is because Γ soon saturates.

According to eq. (4.39), the surface tension γ keeps decreasing as a function of surfactant concentration n. In reality, the decrease of the surface tension stops at a certain concentration n_{cmc} called the critical micelle concentration, as shown in Fig. 4.10. Beyond this concentration, the surface tension does not decrease further even if we add more surfactant. This is because any added surfactant is used to form micelles beyond the concentration n_{cmc}. This subject is discussed in the following section.

4.3.4   Micelles

At concentration higher than n_{cmc}, surfactant molecules assemble to form a structure called micelles. Various kinds of micelles are known, as shown in Fig. 1.5. Once micelles are formed, any additional surfactant molecules added to the system are taken into micelles, and therefore the surface tension does not decrease any more. This can be shown analytically as follows.

We consider the simple case of spherical micelles formed in water. In this case, the outer shell of the micelle is made of hydrophilic groups, and the inner core is made of hydrophobic groups. This structure requires an optimum number of surfactant molecules contained in a micelle: if the number is too small, some hydrophobic groups are exposed to water, and if the number is too large, some hydrophilic groups cannot contact with water. Therefore spherical micelles have an optimal size. Let us assume that all micelles have the same optimal size and consist of m surfactant molecules: i.e., we assume that a surfactant molecule can exist either as a single molecule (called the unimer) or as a member in a micelle of size m. Let n_1 and n_m be the number of unimers and micelles in unit volume. Then

$$n_1 + mn_m = n \tag{4.40}$$

where n is the total number of surfactant molecules in the solution (per unit volume). The chemical potential of the unimer and that of the micelle are respectively given by

$$\mu_1 = \mu_1^0 + k_B T \ln n_1, \qquad \mu_m = \mu_m^0 + k_B T \ln n_m \tag{4.41}$$

Since m unimers can turn into a micelle or vice versa, the following equation holds at equilibrium,

$$m\mu_1 = \mu_m \tag{4.42}$$

From eqs. (4.41)–(4.42), we have

$$n_m = n_1^m \exp[(m\mu_1^0 - \mu_m^0)/k_B T] = n_c \left(\frac{n_1}{n_c}\right)^m \tag{4.43}$$

where n_c is defined by

$$\exp[(m\mu_1^0 - \mu_m^0)/k_B T] = n_c^{-(m-1)} \qquad (4.44)$$

Equation (4.40) is then written as

$$\frac{n_1}{n_c} + m\left(\frac{n_1}{n_c}\right)^m = \frac{n}{n_c} \qquad (4.45)$$

Since m is a large number (typically $30 \sim 100$), the second term on the left-hand side of eq. (4.45) is a very steep function of n_1/n_c: if $n_1/n_c < 1$ it is negligibly small compared with the first term, and if $n_1/n_c > 1$, it becomes much larger than the first. Therefore the solution of eq. (4.45) is obtained as

$$n_1 = \begin{cases} n & n < n_c \\ n_c \left(\frac{n}{mn_c}\right)^{1/m} & n > n_c \end{cases} \qquad (4.46)$$

Therefore if $n < n_c$, all surfactants exist as a unimer. On the other hand, if $n > n_c$, n_1 changes little with n since the exponent $1/m$ is very small. Physically, the reason why n_1 changes little is that added surfactant molecules are taken into the micelles, and do not remain as unimers. One can see from eqs. (4.43) and (4.46) that the number density of micelles n_m is essentially zero for $n < n_c$ and is equal to n/m for $n > n_c$. Therefore n_c is identical to n_{cmc}, the concentration at which micelles start to appear. Since the chemical potential μ_1 of the unimer changes little with n for $n > n_{cmc}$, the surface tension γ ceases to decrease with further addition of surfactants for $n > n_{cmc}$.

Statistical mechanical theory has been developed for other types of micelles (cylindrical, lamellar) shown in Fig. 4.10, and the size distribution and n_{cmc} have been discussed for such micelles.

4.3.5 Gibbs monolayer and Langmuir monolayer

In the above discussion, we have assumed that the surfactant molecules at the surface can dissolve in the liquid, and, conversely, the surfactant in the liquid can adsorb at the surface. In this case, the number density Γ of surfactant molecules on the surface is determined by the condition that the chemical potential of surfactant on the surface is equal to that in the bulk.

If the surfactant molecule has a big hydrophobic group (if the number of alkyl groups is larger than, say, 15), they cannot dissolve in water any more. In this case, Γ is determined by the amount of surfactant added to the system, i.e., $\Gamma = N/A$, where N is the number of surfactant molecules added to the system, and A is the surface area.

The insoluble surfactant layer formed at surfaces is called a Langmuir monolayer. In contrast, the surfactant layer we have been discussing is called a Gibbs monolayer. In the Gibbs monolayer, the surface pressure Π_A (or the surface tension γ) is studied as a function of the surfactant concentration n in the bulk, while in the Langmuir monolayer Π_A is studied as a function of the surface area A.

The relation between the surface pressure Π_A and the surface area A in the Langmuir monolayer is very much like the relation between the pressure P and the volume V of a simple fluid. In this case, the Π_A–A relation can be described by a two-dimensional model of the fluid. At large A, the system is in a gas phase, and Π_A is inversely proportional to A. As the surface area A decreases, the system becomes a condensed phase (liquid phase or solid phase), and Π_A starts to increase rapidly.

The Langmuir monolayer can be transferred to a solid substrate and can be made into multi-layer materials with molecules standing normal to the layer. The Langmuir monolayer formed on solid surfaces has various applications since it changes the surface properties (mechanical, electrical, and optical, etc). To make the monolayer strongly attached to solid surfaces, chemical reactions are often used. Such a monolayer is called a self-assembling monolayer (SAM).

4.4 Inter-surface potential

4.4.1 Interaction between surfaces

We now consider the interaction between surfaces. If two surfaces of materials, I and II, come close enough to each other as shown in Fig. 4.11, the interaction between the surfaces becomes non-negligible and the free energy of the system becomes a function of the distance h between the surfaces. The interaction between the surfaces can be characterized by the inter-surface potential $W(h)$ which is the work needed to bring two parallel plates made of material I and II from infinity to a separation distance h. Equivalently, $W(h)$ may be defined as the difference of the free energy for the two situations, one in which the plates are placed parallel to each other with a gap distance h, and the other in which they are placed indefinitely far apart.

For a large surface, $W(h)$ is proportional to the area A of the surface. The potential per unit area $w(h) = W(h)/A$ is called the inter-surface potential density.

4.4.2 Inter-surface potential and the interaction of colloidal particles

The inter-surface potential is important in the interaction between large colloidal particles. As discussed in Section 2.5, the interaction force between two colloidal particles acts only when they are close to each other. Therefore, when the forces are acting, the two surfaces of the particles are nearly parallel to each other. In such a situation the interaction energy $U(h)$ acting between the particles can be calculated from $w(h)$.

Consider two spherical particles of radius R placed with the gap distance h (see Fig. 4.12). The distance $\tilde{h}(\rho)$ in Fig. 4.12 represents the

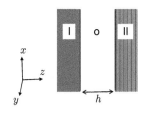

Fig. 4.11 The inter-surface potential for two interfaces I/0 and II/0.

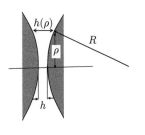

Fig. 4.12 Coordinate used in Derjaguin's approximation.

surface-to-surface distance along a parallel line separated by a distance ρ from the line connecting the centres of the particles. For $\rho \ll R$, $\tilde{h}(\rho)$ is approximated as

$$\tilde{h}(\rho) = h + 2(R - \sqrt{R^2 - \rho^2}) \approx h + \frac{\rho^2}{R} \qquad (4.47)$$

Consider a ring-like region on the surface of the particle bounded by two cylinders of radius ρ and $\rho + d\rho$. If $\rho \ll R$, the area of the ring is $2\pi\rho d\rho$, and the interaction energy is $w(\tilde{h})2\pi\rho d\rho$. Integrating this with respect to ρ, we have the interaction potential between the particles:

$$U(h) = \int_0^R w(\tilde{h})2\pi\rho d\rho \qquad (4.48)$$

By using a new variable $x = h + \rho^2/R$, the right-hand side is rewritten as

$$U(h) = \pi R \int_h^\infty w(x)dx \qquad (4.49)$$

where the upper bound of the integral has been replaced by ∞. If the spheres have different radii, say R_1 and R_2, $U(h)$ is given by

$$U(h) = 2\pi \frac{R_1 R_2}{R_1 + R_2} \int_h^\infty w(x)dx \qquad (4.50)$$

Equations (4.49) and (4.50) are called Derjaguin's approximation, and are valid if $w(h)$ becomes negligibly small at $h \simeq R$.

We now discuss the interaction potential between the colloidal particles using this approximation.

4.4.3   Van der Waals forces

The most common interaction between surfaces is the van der Waals interaction. This has the same origin as the van der Waals force acting between atoms (or molecules), but a rigorous theoretical calculation becomes rather sophisticated (see *Intermolecular and Surface Forces* by J.N. Israelachrili listed in the Further reading). Here we use a simple and approximate treatment.

The van der Waals potential acting between two atoms separated by a distance r is given by

$$u_{atom}(\boldsymbol{r}) = -C \left(\frac{a_0}{r}\right)^6 \qquad (4.51)$$

where a_0 is the atomic radius, and C is a constant, which is of the order of $k_B T$. If we assume that the materials I and II in Fig. 4.11 are made of atoms which are interacting via the potential (4.51), the total interaction energy between the materials is given by

$$W(h) = -\int_{\boldsymbol{r}_1 \in V_I} d\boldsymbol{r}_1 \int_{\boldsymbol{r}_2 \in V_{II}} d\boldsymbol{r}_2\, n^2 \frac{C a_0^6}{|\boldsymbol{r}_1 - \boldsymbol{r}_2|^6} \qquad (4.52)$$

where n is the number density of atoms in the wall materials, and the symbol $r_1 \in V_I$ means that the integral for r_1 is done for the region occupied by the material I. If we do the integral over r_2 first, the result is independent of x_1 and y_1. Hence $W(h)$ is proportional to the interfacial area A and the energy per unit area is given by

$$w(h) = -\int_{z_1 \in V_I} dz_1 \int_{r_2 \in V_{II}} dr_2 \frac{n^2 C a_0^6}{|r_1 - r_2|^6} \tag{4.53}$$

By dimensional analysis, it is easy to show that the right-hand side of eq. (4.53) is proportional to h^{-2}. The integral can be calculated analytically, and it is customary to write the result in the following form:

$$w(h) = -\frac{A_H}{12\pi h^2} \tag{4.54}$$

where A_H is defined by $n^2 C a_0^6 \pi^2$ and is called the Hamaker constant.[3]

The interaction potential acting between two spherical particles of radius R is calculated from eqs. (4.49) and (4.54) as

$$U(h) = -\frac{A_H}{12}\frac{R}{h} \tag{4.56}$$

The Hamaker constant is of the order of $k_B T$. Therefore for $R = 0.5\,\mu\text{m}$ and $h = 0.5\,\text{nm}$, the interaction potential is about $100 k_B T$. With such a large attractive energy, the colloidal particles will aggregate, and subsequently precipitate (or form a gel). In order to have a stable colloidal dispersion, the surfaces of the colloidal particles must be modified to include repulsive forces.

Two strategies are taken to create such repulsive forces. One is to make surfaces charged as in Fig. 4.13(a), and the other is to cover surfaces with polymers as in Fig. 4.13(b). The interaction between such surfaces will be discussed in the following sections.

4.4.4   Charged surfaces

If ionic groups of the same sign of charges are attached to the walls, the walls repel each other due to the Coulombic repulsion. Though this statement is true in vacuum, it is not always correct in solutions because there are free ions floating around the walls in the solutions. In fact, evaluation of the forces in solutions requires careful consideration. Detailed discussion on this force is given in Section 10.3.6. Here, we summarize the main result.

If charged walls are inserted in an electrolyte solution, counter-ions (ions which have charges opposite to the wall) accumulate around the walls and screen the wall charge. As a result, the effect of the wall charge decays exponentially with the distance x from the wall. For example, the electric potential $\psi(x)$ decays as

$$\psi \propto e^{-\kappa x} \tag{4.57}$$

[3] Equation (4.53) is written as

$$w(h) = -\int_{-\infty}^{0} dz_1 \int_{h}^{\infty} dz_2 \int_{0}^{\infty} d\rho\, 2\pi$$
$$\rho \frac{n^2 C a_0^6}{[(z_1 - z_2)^2 + \rho^2]^3} \tag{4.55}$$

The integral gives eq. (4.54).

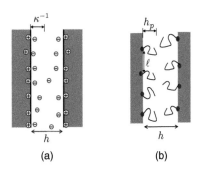

Fig. 4.13 Surface modifications to create repulsive forces between the surfaces. (a) Attaching charges on the surfaces. (b) Attaching polymers on the surfaces.

The length $1/\kappa$ is called the Debye length. For an electrolyte solution consisting of monovalent ions having charges e_0 and $-e_0$, κ is given by

$$\kappa = \left(\frac{2n_s e_0^2}{\epsilon k_B T}\right)^{1/2} \tag{4.58}$$

where n_s is the number density of ions in the solution, and ϵ is the dielectric constant of the solvent (usually water).

The repulsive force between the charged walls is strong if the ion concentration n_s is small, but decreases quickly with the increase of n_s. An example of the inter-particle potential between charged particles is shown in Fig. 2.10(b). When the ion concentration is small, the repulsive interaction due to the charged surface dominates, and the particles do not aggregate. If salts are added, the van der Waals interaction becomes dominant, and the particles start to aggregate.

4.4.5 Polymer grafted surfaces

Stabilization of colloidal dispersions by surface charge does not work at high salt concentration, or in organic solvents. Another way of stabilizing colloidal particles is to cover particles with polymers. If we attach polymers which have good affinity to the solvent, the polymers form a thick layer (see Fig. 4.13(b)) which prevents the particles from direct contact, and hinders the aggregation. The attached polymer is called grafted polymer.

Let Γ_p be the number of polymer chains bound to a unit area of the surface. The mean distance between the neighbouring binding sites is given by $\ell = \Gamma_p^{-1/2}$. If ℓ is much less than the mean size of a polymer coil R_g, the polymers form a dense uniform layer on the surface. Within the layer, the polymer chains are stretched normal to the surface. Such a layer is called a polymer brush.

The thickness of the polymer brush is determined by the same physics which determines the swelling of a gel, i.e., the competition between the mixing energy and the elastic energy of the polymer chain. If the polymer brush has thickness h_p, a polymer chain occupies the volume Nv_c in a volume $\ell^2 h_p$, where v_c is the volume of the polymer segment, and N is the number of segments in the polymer chain. Hence the volume fraction ϕ of polymer within the brush is given by

$$\phi = \frac{Nv_c}{\ell^2 h_p} = v_c \frac{\Gamma_p N}{h_p} \tag{4.59}$$

The free energy per unit area of the polymer grafted surface is written as

$$u(h_p) = \Gamma_p \frac{3k_B T}{2Nb^2} h_p^2 + h_p f_{mix}(\phi) \tag{4.60}$$

The first term stands for the elastic energy of polymer chains, and the second term is the mixing energy of polymer and solvent. If we use the lattice model, the mixing free energy density is given by

$$f_{mix}(\phi) = \frac{k_B T}{v_c} \left[(1 - \phi) \ln(1 - \phi) + \chi \phi (1 - \phi) \right]$$

$$\approx \frac{k_B T}{v_c} \left[(\chi - 1)\phi + \left(\frac{1}{2} - \chi \right) \phi^2 \right] \tag{4.61}$$

where we have expanded $f_{mix}(\phi)$ with respect to ϕ.

Minimizing eq. (4.60) with respect to h_p, we have the equilibrium thickness of the polymer brush

$$h_p^{eq} = N \Gamma_p^{1/3} \left[\frac{v_c b^2}{3} \left(\frac{1}{2} - \chi \right) \right]^{1/3} \tag{4.62}$$

The equilibrium thickness of the polymer brush is proportional to N. Therefore the polymer chains in the brush are strongly stretched relative to the natural state where the size is about $\sqrt{N}b$. The thickness of the brush layer increases with the increase of the grafting density Γ_p.

Equation (4.62) represents the thickness of one brush layer. Now if two polymer grafted surfaces are separated by a distance less than $2h_p^{eq}$, the polymer brush is compressed, and the inter-surface energy increases. Since the thickness of the polymer brush is now given by $h/2$, the inter-surface potential is given by

$$w(h) = 2[u(h/2) - u(h_p^{eq})] \qquad \text{for} \quad h < 2h_p \tag{4.63}$$

The right-hand side can be expanded with respect to $h - 2h_p^{eq}$, and the result is written as

$$w(h) = \frac{1}{2} k_{brush} (h - 2h_p^{eq})^2 \tag{4.64}$$

with

$$k_{brush} = \left. \frac{\partial^2 w(h)}{\partial h^2} \right|_{h = 2h_p^{eq}} = \frac{9 k_B T}{N b^2} \Gamma_p \tag{4.65}$$

Again the repulsive force increases with the increase of the graft density Γ_p.

4.4.6   Effect of non-grafted polymers

The grafted polymer provides a repulsive contribution to the inter-surface potential, and stabilizes the colloidal dispersions. If the polymer is not grafted, the situation is completely different. If a homopolymer (a polymer made with the same monomers) is added to colloidal dispersions, it always provides an attractive contribution to the inter-surface potential irrespective of whether the polymer is attracted to or repelled by the surfaces. The reasons are explained below.

Polymer attracted to the surface

If the interaction between the polymer and the surface is attractive, the polymer is adsorbed on the surface forming an adsorbed layer. Unlike the polymer brush, the adsorbed layer gives an attractive interaction between the surfaces. The reason is that now individual polymers can simultaneously adsorb both surfaces and can make bridges between the surfaces[4] (see Fig. 4.14(a)). As the surfaces come close to each other, more bridges are formed and the free energy of the system decreases. Therefore the inter-surface potential becomes attractive.

Polymer repelled by the surface

In the opposite situation that the polymers are repelled by the surfaces, one might expect that the polymer provides a repulsive contribution to the inter-surface forces. In fact, such a polymer gives an attractive contribution. This surprising effect is called the Asakura–Oosawa effect, or the depletion effect.

To see the origin of the attractive force, let us assume that a polymer molecule occupies a spherical region of radius R_g, and that the centre of the molecule is not allowed to be within the distance R_g from the surface (see Fig. 4.14(b)).

Now suppose that the distance h between the surfaces becomes less than $2R_g$. In this situation, the polymer molecule cannot enter the gap region between the surfaces. This is equivalent to the situation that the gap region between the surfaces is sealed by a semi-permeable membrane which prevents polymer molecules from entering into the region. Since polymer molecules cannot enter the gap region, the pressure in the gap region is less than the pressure outside by the osmotic pressure. In a dilute solution, the osmotic pressure of the solution is given by $\Pi = n_p k_B T$, where n_p is the number density of polymer molecules in the outer solution. Therefore, the surfaces are pushed inward with force Π per unit area, i.e., there is an attractive force between the particles. Since the attractive force arises from the fact that the polymer molecules are depleted at the surface, the force is called the depletion force.

The inter-surface potential for the depletion force can be written as

$$w_{depletion}(h) = \begin{cases} 0 & h > 2R_g \\ n_p k_B T(h - 2R_g) & h < 2R_g \end{cases} \qquad (4.66)$$

The depletion force can be derived by a slightly different argument. Due to the repulsive interaction between polymer and surface, there appears a region depleted of polymers near the surface. The volume of the depleted region decreases as h becomes less than $2R_g$, and the resulting gain of the entropy gives the depletion force. Equation (4.66) can be derived by this consideration (see problem (4.6)).

4.4.7 Disjoining pressure

The inter-surface potential is also important in thin films. If the thickness of the material 0 in Fig. 4.11 becomes very small, the surface free energy should be written as

$$G_A = A(\gamma_I + \gamma_{II} + w(h)) \qquad (4.67)$$

[4] Here it is assumed that the adsorbed layer is at equilibrium. The adsorbed layer may contribute to the repulsive force in the non-equilibrium state where the layer acts like a bumper.

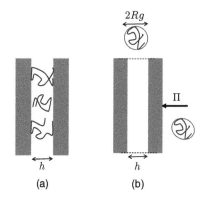

Fig. 4.14 (a) If the polymer is attracted to the surface, it creates bridges between the surfaces, and gives an attractive force between the surfaces. (b) If the polymer is repelled by the surface, it is excluded from the gap region between the surfaces. This effect also gives an attractive contribution to the inter-surface force.

where γ_I and γ_{II} are the interfacial energies for the I/0 interface and the II/0 interface. G_A has an extra contribution from the inter-surface potential $w(h)$, which is negligible for large h, but can be important for small h. In the limit of $h \to 0$, G_A/A becomes equal to $\gamma_{I,II}$, the interfacial energy between I and II. Hence

$$w(0) = \gamma_{I,II} - \gamma_I - \gamma_{II} \tag{4.68}$$

which is equal to the spreading coefficient γ_S.

The van der Waals force acting between the surfaces shown in Fig. 4.11 depends on three materials: material I, material II, and the material 0 which occupies the space between the two. If materials I and II are the same, the van der Waals force is always attractive. On the other hand, if materials I and II are different, the van der Waals force can be repulsive.

To a good approximation, the Hamaker constant A_H is proportional to $(\alpha_I - \alpha_0)(\alpha_{II} - \alpha_0)$, where α_i $(i = \text{I}, \text{II}, 0)$ is the polarizability of the material i. If $\alpha_I = \alpha_{II}$, A_H is always positive, and the van der Waals force is attractive. On the other hand, if $\alpha_I > \alpha_0 > \alpha_{II}$, the van der Waals force is repulsive. In a thin liquid film on a solid substrate, the materials I, 0, and II correspond to solid, liquid film, and air, respectively. In this situation, the condition $\alpha_I > \alpha_0 > \alpha_{II}$ is fulfilled, and the van der Waals force becomes repulsive, i.e., it acts to increase the thickness of the liquid film. The repulsive van der Waals force acting in a thin liquid film is called the disjoining pressure.

The disjoining pressure P_d is given by the force derived from the van der Waals potential, eq. (4.54)

$$P_d = -\frac{\partial w}{\partial h} = -\frac{A_H}{6\pi h^3} \tag{4.69}$$

where A_H is negative.

The disjoining pressure becomes important in thin liquid films. According to the reasoning in Section 4.2.2, if the spreading coefficient γ_S is positive, the liquid continues to spread on the surface, and the final film thickness becomes zero. In reality, the spreading of the liquid (i.e., the thinning of the liquid film) stops at a certain thickness due to the disjoining pressure.

Consider a liquid film of volume V wetting the area A on the surface of a substrate. The thickness of the film is $h = V/A$. The equilibrium thickness is determined by minimizing the following free energy

$$F = -\gamma_S A + w(h)A = V \left[-\frac{\gamma_S}{h} + \frac{w(h)}{h} \right] \tag{4.70}$$

This gives the following equation for the equilibrium thickness h_e

$$\gamma_S = w(h_e) - h_e w'(h_e) \tag{4.71}$$

h_e is usually few tens of nm. The disjoining pressure is negligible if the film is much thicker than this value.

4.5 Summary of this chapter

The surface tension of a liquid is defined as the free energy per unit area of the liquid surface. It is also defined as the force acting across a line per unit length taken on the surface. The equilibrium shape of a liquid, say the shape of a liquid droplet on a table, is obtained by minimizing the total free energy accounting for the surface tension.

Surfactants consist of a hydrophobic part and a hydrophilic part. When added to a liquid, they assemble at the surface and decrease the surface tension. The decrease of the surface tension, however, stops when the surfactant concentration in the liquid exceeds the critical micellar concentration c_{cmc}. Above c_{cmc}, the added surfactants stay in the bulk forming micelles, and do not affect the surface property.

If two surfaces come close to each other, the interaction between the surfaces becomes important. The extra energy per unit area due to the interaction is called the inter-surface potential. The inter-surface potential is important in the dispersion of colloidal particles. Colloidal particles usually attract each other by the van der Walls force. To stabilize the dispersion, repulsive forces have to be added. This is done by attaching charges or polymers on the surface. On the other hand, if homopolymers are added to the solution, they create attractive forces and cause coagulation of the particles. The inter-surface potential acting in a thin liquid film on a substrate is repulsive and is called the disjoining pressure.

Further reading

(1) *An Introduction to Interfaces and Colloids: The Bridge to Nanoscience*, John C. Berg, World Scientific (2009).

(2) *Capillarity and Wetting Phenomena*, Pierre-Gilles De Gennes, Francoise Brochard-Wyart, David Quere, Springer (2012).

(3) *Intermolecular and Surface Forces*, Jacob N. Israelachvili, 3rd edition, Academic Press (2010).

Exercises

(4.1) Two glass plates with one edge in contact at an angle θ ($\theta \ll 1$) are inserted vertically into a water pool (see Fig. 4.15). The water rises in the gap between the plates due to the capillary force. Show that the profile of the wetting front can be written as $y = C/x$ and express C in terms of the material parameters and θ. Assume complete wetting for water on the glass plate.

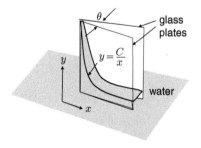

Fig. 4.15

(4.2) A water droplet of volume V is sandwiched between two glass plates separated by a distance h (see Fig. 4.16). Calculate the capillary force acting on the plates. Assume that the radius of the wet region r is much larger than h and that the contact angle of the water on the glass is θ.

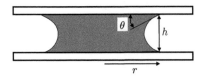

Fig. 4.16

(4.3) An oil droplet (A) is placed on a water pool (B) (see Fig. 4.17). Calculate the equilibrium thickness of the droplet $h = h_1 + h_2$ for a large droplet. Assume that the surface tensions of oil and water, and the interfacial tension of the oil–water interface are γ_A, γ_B, and γ_{AB}, respectively, and that the densities of oil and water are ρ_A and ρ_B.

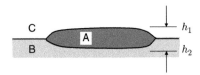

Fig. 4.17

(4.4) Answer the following questions. Assume that the surface tension of water is $\gamma = 70\,\mathrm{mN/m}$.

(a) Calculate the internal pressure of a water droplet of radius $1\,\mu\mathrm{m}$ suspended in air at atmospheric pressure.

(b) Show that if a pure fluid is assumed to be incompressible, the chemical potential of a fluid molecule can be written as

$$\mu(T, P) = \mu_0(T) + vP, \qquad (4.72)$$

where v is the specific volume of the molecule, and P is the pressure.

(c) The vapour pressure of bulk water is 0.02 atm at temperature 20 °C. Calculate the vapour pressure of a water droplet of radius $1\,\mu\mathrm{m}$. The molar volume of water is $18\,\mathrm{cm}^3/\mathrm{mol}$.

(4.5) The chemical potential of a rod-like micelle which consists of m surfactant molecules is written as

$$\mu_m = \mu_m^0 + k_B T \ln n_m, \qquad \text{with} \quad \mu_m^0 = a + mb \qquad (4.73)$$

where a and b are constants independent of m. Answer the following questions.

(a) Show that at equilibrium the number density of micelles of size m can be written as

$$n_m = n_c \left(\frac{n_1}{n_c} \right)^m \qquad (4.74)$$

Obtain an expression for n_c.

(b) n_m satisfies the equation

$$\sum_{m=1}^{\infty} m n_m = n \qquad (4.75)$$

where n is the number density of surfactants in the solution. Express n_1 as a function n.

(c) Express the surface tension as a function of surfactant density n, and discuss the meaning of n_c.

(4.6) The depletion force discussed in Section 4.4.6 can be derived as follows. Consider a gas of rigid spherical molecules of radius a confined in a volume V. If two rigid spheres of radius R are introduced to the system with gap distance h as in Fig. 4.18(a), each molecule is excluded from the dotted region in Fig. 4.18. Let $\Delta V(h)$ be the volume of the dotted region. Answer the following questions.

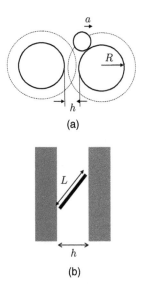

Fig. 4.18

(a) By considering the free energy of the particle, show that the average force acting on the sphere is given by

$$F = -nVk_BT\frac{\partial \ln(V - \Delta V(h))}{\partial h}$$

$$= nk_BT\frac{\partial \Delta V(h)}{\partial h} \qquad (4.76)$$

where n is the number density of the molecule.

(b) Calculate F as a function of h.

(c) Show that the above result agrees with eq. (4.66) in the limit of $R \gg h$.

(4.7) *Calculate the inter-surface force created by the depletion effect of rod-like molecules (see Fig. 4.18(b)).

Liquid crystals

isotropic liquid

(a)

nematic liquid

(b)

Fig. 5.1 Isotropic state (a) and nematic state (b) of a liquid made of ellipsoidal molecules.

A liquid crystal is a state of matter which has an intermediate order between liquid and crystal. While fluid in nature, liquid crystals possess order in molecular orientation. Due to this ordering, the molecular orientation of liquid crystals is easily controlled by relatively weak external forces such as electric fields. In usual liquids, a very large electric field (of the order of MV/mm) is needed to orient the molecules which are vigorously moving by thermal motion. Once the material is in the liquid crystalline state, the molecular orientation can be controlled by field strengths of a few tenths of a V/mm. This aspect of liquid crystals is extensively used in display devices.

A liquid crystal is an example that shows that the collective nature of soft matter is created by phase transitions. The molecules constituting liquid crystals are not large (their size is ca. 1 nm), but in the liquid crystalline state, these molecules arrange themselves in order, and move collectively.

In this chapter, we shall discuss how the interaction of molecules creates spontaneous ordering in materials. This phenomenon is generally called an order–disorder transition. The transition from normal liquid to liquid crystal is an instructive example of order–disorder transition.

5.1 Nematic liquid crystals

5.1.1 Birefringent liquid

There are many types of liquid crystalline order, but here we shall focus on a specific type, i.e., the nematics. Nematics are the simplest type of liquid crystals, yet it is the most widely used type.

The nematic phase is formed by rod-like molecules (see Fig. 5.1). At high temperature, the orientation of the molecules is completely random, and the system is an isotropic liquid (Fig. 5.1(a)). With decreasing temperature, the molecules start to orient in a common direction as shown in Fig. 5.1(b) and form the nematic phase. The nematic phase is an anisotropic liquid: its optical, electrical, and magnetic properties are anisotropic like a crystal, yet the system is still in a liquid state and can flow.

Apart from a few exceptions, nematics are uniaxial, i.e., the system has a rotational symmetry around a certain axis. This axis is called the director, and is specified by a unit vector n (see Fig. 5.1(b)).

The most apparent characteristic of the nematic phase is birefringence, i.e., anisotropy in the transmission of light. Consider the setup shown in Fig. 5.2. A nematic sample is sandwiched between two polarizing films (polarizer and analyser), with their polarization axes set orthogonal to each other, and incident light is directed normal to the film. If the sample is an isotropic liquid, the intensity of the transmitted light is zero since the polarization axes of the two films are orthogonal. If the sample is nematic, however, a certain amount of the incident light is transmitted. The intensity of the transmitted light is a function of the angle θ that the director $\boldsymbol{n}$ makes with the polarizing axis; the transmission is zero when θ is equal to 0 or $\pi/2$, and takes a maximum when $\theta = \pi/4$.[1] Since the director $\boldsymbol{n}$ can be changed by an electric field, the transmission of light can be easily controlled. This is the principle of how nematics work in display devices.

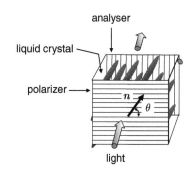

Fig. 5.2 Principle of the optical cell of liquid crystal devices. A nematic liquid crystal is sandwiched between crossed polarizers. The intensity of the transmitted light depends on the angle θ that the nematic director $\boldsymbol{n}$ makes with the direction of the polarizer.

[1] The intensity of the transmitted light is proportional to $\sin^2(2\theta)$.

5.1.2   Orientational distribution function

To distinguish the nematic phase from the isotropic phase, let us consider the distribution of the molecular orientation. Let $\boldsymbol{u}$ be a unit vector in the direction of the long axis of the molecule (see Fig. 5.1(a)). In the isotropic phase, $\boldsymbol{u}$ will be uniformly distributed on a unit sphere $|\boldsymbol{u}| = 1$. In the nematic phase, the distribution of $\boldsymbol{u}$ becomes non-uniform on the sphere. Let $\psi(\boldsymbol{u})$ be the distribution function of $\boldsymbol{u}$ on the unit sphere. It is normalized as

$$\int d\boldsymbol{u} \; \psi(\boldsymbol{u}) = 1 \qquad (5.1)$$

where $d\boldsymbol{u}$ stands for the surface element of the sphere $|\boldsymbol{u}| = 1$, and the integral is done over the entire surface.

In the isotropic phase, $\psi(\boldsymbol{u})$ is constant, independent of $\boldsymbol{u}$, and therefore

$$\psi(\boldsymbol{u}) = \frac{1}{4\pi} \qquad (5.2)$$

On the other hand, in the nematic phase, $\psi(\boldsymbol{u})$ will be oriented toward the direction of the director $\boldsymbol{n}$, and $\psi(\boldsymbol{u})$ becomes anisotropic.

5.1.3   Order parameter for nematics

Let us consider a parameter which characterizes the anisotropy of $\psi(\boldsymbol{u})$ in the nematic phase. An obvious candidate for such a parameter is the first moment, $\langle u_\alpha \rangle$ $(\alpha = x, y, z)$, where $\langle \cdots \rangle$ stands for the average for the distribution function $\psi(\boldsymbol{u})$, i.e.,

$$\langle \cdots \rangle = \int d\boldsymbol{u} \cdots \psi(\boldsymbol{u}) \qquad (5.3)$$

However, for the symmetric ellipsoidal molecules shown in Fig. 5.1(a), $\langle u_\alpha \rangle$ is always equal to zero since the states specified by $\boldsymbol{u}$ and $-\boldsymbol{u}$ are identical, and therefore $\psi(\boldsymbol{u})$ has an inversion symmetry $\psi(\boldsymbol{u}) = \psi(-\boldsymbol{u})$.

[2] This is seen as follows. In the isotropic state, $\langle u_x^2 \rangle = \langle u_y^2 \rangle = \langle u_z^2 \rangle$, and $\langle u_x u_y \rangle = \langle u_y u_z \rangle = \langle u_z u_x \rangle = 0$. Therefore $\langle u_\alpha u_\beta \rangle$ is written as $\langle u_\alpha u_\beta \rangle = A\delta_{\alpha\beta}$. Since $u_x^2 + u_y^2 + u_z^2 = 1$, A is equal to 1/3.

Let us therefore consider the second moment $\langle u_\alpha u_\beta \rangle$. In the isotropic state,[2]

$$\langle u_\alpha u_\beta \rangle = \frac{1}{3}\delta_{\alpha\beta} \tag{5.4}$$

On the other hand, if $\boldsymbol{u}$ is completely aligned along $\boldsymbol{n}$,

$$\langle u_\alpha u_\beta \rangle = n_\alpha n_\beta \tag{5.5}$$

Thus we consider the parameter

$$Q_{\alpha\beta} = \left\langle u_\alpha u_\beta - \frac{1}{3}\delta_{\alpha\beta} \right\rangle \tag{5.6}$$

$Q_{\alpha\beta}$ represents the orientational order of the molecules in the nematic phase and is called the order parameter. It is zero in the isotropic phase, and becomes non-zero in the nematic phase.

If the distribution of $\boldsymbol{u}$ has a uniaxial symmetry around the axis of $\boldsymbol{n}$, $Q_{\alpha\beta}$ can be written as[3]

[3] The second rank tensor for a system which has uniaxial symmetry around $\boldsymbol{n}$ is always written as $An_\alpha n_\beta + B\delta_{\alpha\beta}$. Since $Q_{\alpha\alpha} = 0$, we have eq. (5.7).

$$Q_{\alpha\beta} = S\left(n_\alpha n_\beta - \frac{1}{3}\delta_{\alpha\beta}\right) \tag{5.7}$$

where S is a parameter representing how perfectly the molecules are aligned along $\boldsymbol{n}$. If the alignment is perfect, S is equal to 1. If there is no alignment, S is equal to 0. Thus S also represents the degree of the order in the nematic phase, and is called the scalar order parameter. To avoid confusion with S, $Q_{\alpha\beta}$ is often called the tensor order parameter.

The tensor order parameter $Q_{\alpha\beta}$ includes two pieces of information. One is how strongly the molecules are aligned; this is represented by the scalar order parameter S. The other is the direction in which the molecules are aligned; this is represented by the director $\boldsymbol{n}$. By eq. (5.6),

$$\left\langle (\boldsymbol{u} \cdot \boldsymbol{n})^2 - \frac{1}{3} \right\rangle = n_\alpha n_\beta Q_{\alpha\beta} \tag{5.8}$$

The right-hand side is equal to $(2/3)S$ by eq. (5.7). Hence S is expressed as

$$S = \frac{3}{2}\left\langle (\boldsymbol{u} \cdot \boldsymbol{n})^2 - \frac{1}{3} \right\rangle \tag{5.9}$$

5.2   Mean field theory for the isotropic–nematic transition

5.2.1   Free energy functional for the orientational distribution function

Let us now study the phase transition from the isotropic state to the nematic state. Theories for phase transitions usually become approximate, involving some kind of mean field approximation. In the case of the isotropic–nematic transition of rod-like molecules, Onsager showed that rigorous theory can be developed in the limit of large aspect ratio. This beautiful theory is for the phase transition in lyotropic liquid crystals (liquid crystals formed in solutions by a change of concentration). It is discussed at the end of this chapter. Here, we present another classical theory due to Maier and Saupe. It is a mean field theory for thermotropic liquid crystals (liquid crystals formed by a change of temperature).

The basic assumption of the Maier–Saupe theory is that nematic forming molecules have an interaction potential which forces them to align in the same direction. Let $w(\boldsymbol{u}, \boldsymbol{u}')$ be the interaction potential acting between two neighbouring molecules each pointing along the $\boldsymbol{u}$ and $\boldsymbol{u}'$ directions.[4] The interaction potential $w(\boldsymbol{u}, \boldsymbol{u}')$ has the property that it decreases with decrease of the angle Θ between $\boldsymbol{u}$ and $\boldsymbol{u}'$, and becomes smallest when two molecules align in the same direction. For molecules which do not have polarity (such as ellipsoidal or rod-like molecules), the interaction potential should not change under the transformation $\boldsymbol{u} \rightarrow -\boldsymbol{u}$. The simplest form of such a potential is

$$w(\boldsymbol{u}, \boldsymbol{u}') = -\tilde{U}(\boldsymbol{u} \cdot \boldsymbol{u}')^2 \qquad (5.10)$$

where $\tilde{U}$ is a positive constant.

Now if the system consists of N such molecules, and if their orientational distribution is given by $\psi(\boldsymbol{u})$, the average energy of the system is given by

$$E[\psi] = \frac{zN}{2} \int d\boldsymbol{u} \int d\boldsymbol{u}' \; w(\boldsymbol{u}, \boldsymbol{u}') \psi(\boldsymbol{u}) \psi(\boldsymbol{u}') \qquad (5.11)$$

where z is the mean number of neighbouring molecules (i.e., the coordination number introduced in Section 2.3.2).

The interaction potential tends to align the molecules in the same direction, but this is opposed by thermal motion which tends to randomize the molecular orientation. For a given orientational distribution $\psi(\boldsymbol{u})$, the orientational entropy is given by[5]

$$S[\psi] = -Nk_B \int d\boldsymbol{u} \; \psi(\boldsymbol{u}) \ln \psi(\boldsymbol{u}) \qquad (5.13)$$

and therefore the free energy of the system is given by

$$
\begin{aligned}
F[\psi] &= E[\psi] - TS[\psi] \\
&= N\left[k_B T \int d\boldsymbol{u} \; \psi(\boldsymbol{u}) \ln \psi(\boldsymbol{u}) - \frac{U}{2} \int d\boldsymbol{u} \int d\boldsymbol{u}' \; (\boldsymbol{u} \cdot \boldsymbol{u}')^2 \psi(\boldsymbol{u}) \psi(\boldsymbol{u}') \right]
\end{aligned}
$$
$$(5.14)$$

where $U = z\tilde{U}$.

The orientational distribution at equilibrium is determined by the condition that eq. (5.14) be a minimum with respect to $\psi(\boldsymbol{u})$, i.e., by the condition

$$\frac{\delta}{\delta \psi}\left[F[\psi] - \lambda \int d\boldsymbol{u}\psi(\boldsymbol{u}) \right] = 0 \qquad (5.15)$$

[4] The interaction potential generally depends not only on the orientation vectors $\boldsymbol{u}$ and $\boldsymbol{u}'$, but also on the vector $\boldsymbol{r}$ which joins the centres of the two molecules. Here $w(\boldsymbol{u}, \boldsymbol{u}')$ represents the effective potential in which the average for $\boldsymbol{r}$ has already been taken.

[5] This equation is derived as follows. Let us imagine that the surface of the unit sphere $|\boldsymbol{u}| = 1$ is divided into M small cells of area $du = 4\pi/M$, each pointing to the direction $\boldsymbol{u}_i$ ($i = 1, 2, \ldots, M$). If their orientational distribution is $\psi(\boldsymbol{u})$, the number of molecules in region i is $N_i = N\psi(\boldsymbol{u}_i)du$ (N being the total number of molecules in the system). The number of ways of putting the molecules in such a state is $W = N!/(N_1!N_2! \cdots N_M!)$. Therefore the entropy is given by

$$
\begin{aligned}
S &= k_B \ln W \\
&= k_B\left[N(\ln N - 1) - \sum_i N_i(\ln N_i - 1) \right] \\
&= -Nk_B \sum_i (N_i/N) \ln(N_i/N)
\end{aligned}
$$
$$(5.12)$$

where we have used $N = \sum_i N_i$. Equation (5.12) gives eq. (5.13).

where $\delta \cdots / \delta \psi$ stands for the functional derivative with respect to ψ.[6] The second term on the left-hand side of eq. (5.15) is introduced to account for the normalization condition for ψ (eq. (5.1)).

For $F[\psi]$ given by eq. (5.14), eq. (5.15) gives

$$k_B T \left[\ln \psi(\boldsymbol{u}) + 1\right] - U \int d\boldsymbol{u}' \, (\boldsymbol{u} \cdot \boldsymbol{u}')^2 \psi(\boldsymbol{u}') - \lambda = 0 \qquad (5.16)$$

This gives

$$\psi(\boldsymbol{u}) = C \exp[-\beta w_{mf}(\boldsymbol{u})] \qquad (5.17)$$

where $\beta = 1/k_B T$, C is a normalization constant, and $w_{mf}(\boldsymbol{u})$ is given by

$$w_{mf}(\boldsymbol{u}) = -U \int d\boldsymbol{u}' \, (\boldsymbol{u} \cdot \boldsymbol{u}')^2 \psi(\boldsymbol{u}') \qquad (5.18)$$

Equation (5.17) indicates that at equilibrium, the distribution of $\boldsymbol{u}$ is given by the Boltzmann distribution in the mean field potential $w_{mf}(\boldsymbol{u})$. Equations (5.17) and (5.18) form an integral equation for $\psi(\boldsymbol{u})$. This integral equation can be solved rigorously as we shall see in the next section.

5.2.2   Self-consistent equation

The mean field potential in eq. (5.18) is written as

$$w_{mf}(\boldsymbol{u}) = -U u_\alpha u_\beta \langle u'_\alpha u'_\beta \rangle = -U u_\alpha u_\beta \langle u_\alpha u_\beta \rangle \qquad (5.19)$$

where we have used the fact that the distribution of $\boldsymbol{u}'$ is the same as that of $\boldsymbol{u}$. The average in eq. (5.19) can be expressed by the scalar order parameter S. Let us take the z-axis in the direction of $\boldsymbol{n}$. Then eq. (5.9) is written as

$$S = \frac{3}{2} \left\langle u_z^2 - \frac{1}{3} \right\rangle \qquad (5.20)$$

Since the distribution of $\boldsymbol{u}$ has uniaxial symmetry around the z-axis, $\langle u_\alpha u_\beta \rangle$ is zero for $\alpha \neq \beta$, and the other components are calculated from eq. (5.20) as

$$\langle u_z^2 \rangle = \frac{1}{3}(2S + 1) \qquad (5.21)$$

$$\langle u_x^2 \rangle = \langle u_y^2 \rangle = \frac{1}{2} \left(1 - \langle u_z^2 \rangle\right) = \frac{1}{3}(-S + 1) \qquad (5.22)$$

Hence the mean field potential (5.19) is calculated as

$$\begin{aligned}
w_{mf}(\boldsymbol{u}) &= -U \left[u_x^2 \langle u_x^2 \rangle + u_y^2 \langle u_y^2 \rangle + u_z^2 \langle u_z^2 \rangle \right] \\
&= -U \left[\frac{1}{3}(-S + 1)\left(u_x^2 + u_y^2\right) + \frac{1}{3}(2S + 1)u_z^2 \right] \\
&= -U \left[\frac{1}{3}(-S + 1)\left(1 - u_z^2\right) + \frac{1}{3}(2S + 1)u_z^2 \right] \\
&= -U S u_z^2 + \text{constant} \qquad (5.23)
\end{aligned}$$

Therefore the equilibrium distribution function is given by

$$\psi(\boldsymbol{u}) = Ce^{\beta U S u_z^2} \tag{5.24}$$

Equations (5.20) and (5.24) give the following self-consistent equation for S

$$S = \frac{\int d\boldsymbol{u} \frac{3}{2}\left(u_z^2 - \frac{1}{3}\right)e^{\beta U S u_z^2}}{\int d\boldsymbol{u} e^{\beta U S u_z^2}} \tag{5.25}$$

We introduce the parameter $x = \beta U S$. Equation (5.25) is then written as

$$\frac{k_B T}{U}x = I(x) \tag{5.26}$$

where

$$I(x) = \frac{\int_0^1 dt \, \frac{3}{2}\left(t^2 - \frac{1}{3}\right)e^{xt^2}}{\int_0^1 dt \, e^{xt^2}} \tag{5.27}$$

Equation (5.26) can be solved by a graphical method: the solution is given by the intersection between the line $y = (k_B T/U)x$ and the curve $y = I(x)$. This is done in Fig. 5.3(a). At high temperature, there is only one solution at $x = 0$ which corresponds to the isotropic phase ($S = 0$). With decreasing temperature, two non-zero solutions appear below the temperature T_{c1} defined in Fig. 5.3(a).

With decreasing the temperature further, one solution increases, while the other solution decreases, crossing zero at the temperature T_{c2}. Figure 5.3(b) summarizes this behaviour of the solution S plotted against the temperature T.

The above analysis indicates that the nematic state can appear as the temperature is decreased, but does not indicate where the actual transition takes place. This is discussed in the next section.

5.2.3   Free energy function for the order parameter

In order to identify the transition temperature, we consider the free energy $F(S;T)$ of the system for a given value of the order parameter S. $F(S;T)$ represents the free energy of a system in which the order parameter is hypothetically constrained at S. If the system consists of N molecules each pointing the direction $\boldsymbol{u}_i$ ($i = 1, 2, \ldots, N$), $F(S;T)$ is defined as the free energy of the system under the constraint

$$\frac{1}{N}\sum_i \frac{3}{2}\left(u_{iz}^2 - \frac{1}{3}\right) = S \tag{5.28}$$

In general, the free energy under certain constraints is called the restricted free energy. $F(S;T)$ is an example of the restricted free energy. A general discussion on the restricted free energy is given in Appendix B.

If the free energy $F(S;T)$ is known, the equilibrium value of S is determined by the condition that $F(S;T)$ becomes a minimum at equilibrium, i.e.,

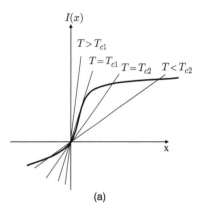

(a)

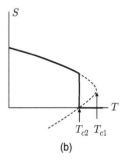

(b)

Fig. 5.3 (a) Graphical solution of eq. (5.26). T_{c1} corresponds to the temperature at which the line $y = (k_B T/U)x$ touches the curve $y = I(x)$ at certain positive x. T_{c2} corresponds to the temperature at which the line $y = (k_B T/U)x$ touches the curve $y = I(x)$ at $x = 0$. (b) The scalar order parameter S is plotted against temperature. The solid line indicates the trajectory that the system follows when the temperature is decreased continuously.

$$\frac{\partial F}{\partial S} = 0 \tag{5.29}$$

In the mean field approximation, $F(S;T)$ can be calculated by minimizing the free energy functional (5.14) with respect to ψ under the constraint

$$\int d\boldsymbol{u} \ \psi(\boldsymbol{u}) \frac{3}{2} \left(u_z^2 - \frac{1}{3} \right) = S \tag{5.30}$$

Equation (5.25) is also derived from eq. (5.29) by this method (see problem (5.3)).

Alternatively, if the temperature dependence of the solution of eq. (5.29) is known, the qualitative form of $F(S;T)$ can be inferred from the fact that the solution corresponds to the extremum of the function $F(S;T)$. This is shown in Fig. 5.4. For example, since there is only one solution at $S = 0$ for $T > T_{c1}$, $F(S;T)$ should have only one minimum at $S = 0$ for $T > T_{c1}$. This corresponds to the curve (i) in Fig. 5.4. Below T_{c1}, the equation $\partial F/\partial S = 0$ has three solutions. This corresponds to two local minima and one local maximum. The solution $S = 0$ corresponds to the isotropic state, and the other solutions correspond to the nematic state. Therefore, below T_{c1}, $F(S;T)$ behaves as shown by curve (iii) in Fig. 5.4. At temperature T_{c2}, the local minimum at $S = 0$ now becomes a local maximum, i.e., the isotropic state becomes an unstable state. Therefore $F(S;T)$ behaves as shown by the curve (vii) in Fig. 5.4. The free energy of the isotropic state becomes equal to that of the nematic state at a certain temperature T_e between T_{c1} and T_{c2}. The temperature T_e corresponds to the equilibrium transition temperature.

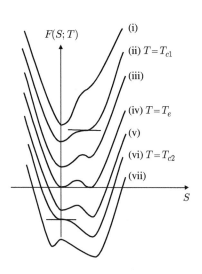

$F(S;T)$

(i)

(ii) $T = T_{c1}$

(iii)

(iv) $T = T_e$

(v)

(vi) $T = T_{c2}$

(vii)

S

Fig. 5.4 Temperature dependence of the free energy curve $F(S;T)$ which gives the curve $S(T)$ shown in Fig. 5.3(b).

5.3   Landau–de Gennes theory

5.3.1   Expression for the free energy near the transition point

Phase transitions which involve changes in the ordering of molecules (or atoms) are generally called order–disorder transitions. The order–disorder transition is characterized by the order parameter which is equal to zero in the disordered state, and is non-zero in the ordered state. The isotropic–nematic transition is an example of an order–disorder transition. Another example of an order–disorder transition is the magnetic transition. The magnetic state of ferromagnets changes from an ordered state to a disordered state at a certain temperature T_c. Below T_c, the magnetic moments of atoms are ordered, and there is a spontaneous magnetic moment M. Above T_c, the magnetic moments of atoms become random, and M vanishes. Therefore the magnetic moment M can be regarded as the order parameter in the magnetic transition.

Though the nematic transition looks similar to the magnetic transition, there is an essential difference. The difference exists in the behaviour of the order parameter near the phase transition. In the case of the magnetic transition, the magnetic moment M changes continuously as a function of temperature as shown in Fig. 5.5(a). The magnetic moment is zero above T_c, and is non-zero below T_c, but it changes continuously at T_c. On the other hand, in the case of the nematic transition, the order parameter S changes discontinuously at T_c.

An order–disorder transition in which the order parameter changes discontinuously at the transition point is called a discontinuous transition or first-order transition. On the other hand, a phase transition in which the order parameter changes continuously is called a continuous transition or second-order transition.

Landau has shown that the distinction between these transitions has its root in the nature of the ordering that takes place at the transition. To see this, let us consider a simple case where the order–disorder transition is characterized by a single order parameter x. We assume that the system is in the disordered phase above T_c. Therefore the equilibrium value of x is zero above T_c, and becomes non-zero below T_c. Let $F(x;T)$ be the free energy (more precisely the restricted free energy) at temperature T for a given value of the order parameter x. For small x, $F(x;T)$ can be expressed as a power series in x:

$$F(x;T) = a_0(T) + a_1(T)x + a_2(T)x^2 + a_3(T)x^3 + a_4(T)x^4 + \cdots \quad (5.31)$$

Now since the disordered state becomes unstable at temperature T_c, the state of $x = 0$ changes from a local minimum of $F(x;T)$ to a local maximum. For this to happen, $a_1(T)$ must be identically zero, and $a_2(T)$ must have the form $a_2(T) = A(T - T_c)$, where A is a positive constant.

Since the important point is that the sign of $a_2(T)$ changes at T_c, we may assume that a_3 and a_4 are constants independent of temperature. Therefore

$$F(x;T) = A(T - T_c)x^2 + a_3 x^3 + a_4 x^4 + \cdots \quad (5.32)$$

where the constant term a_0 has been ignored as it does not affect the behaviour of the transition.

Now in the case of a magnetic transition, the free energy must be an even function of M since the state of $-M$ is obtained by rotating the system by 180 degrees, and (if there is no magnetic field) it must have the same free energy as the state of M. Therefore the coefficient a_3 must be zero. In this case, the temperature dependence of the free energy $F(M;T)$ must be as shown in Fig. 5.5(b). Above T_c, the minimum of $F(M;T)$ is at $M = 0$, and below T_c, the minima of the free energy are at $M = \pm\sqrt{A(T_c - T)/2a_4}$. The order parameter M changes continuously at T_c, and the transition is continuous.

On the other hand, in the case of a nematic transition, the state of $-S$ is different from the state of S. In the state of $S > 0$, the molecules align in the z-direction, while in the state of $-S < 0$, the molecules align normal to the z-axis. Hence a_3 is not equal to zero. In this case, the change of the free energy function $F(S;T)$ becomes as shown in Fig. 5.4. Accordingly, the transition becomes discontinuous.

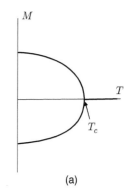

(a)

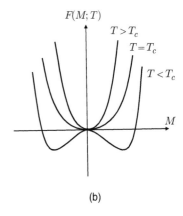

(b)

Fig. 5.5 (a) Temperature dependence of the magnetic moment M of a ferromagnet. (b) Temperature dependence of the free energy curve $F(M;T)$ which gives the curve shown in (a).

The above argument can be made clearer if we go back to the definition of the order parameter. The order parameter for the nematic phase is a tensor $\boldsymbol{Q}$, while the free energy is a scalar. If the free energy is expressed as a power series of the tensor $\boldsymbol{Q}$, the coefficients have to satisfy certain constraints. For example, one can immediately see that there is no linear term in $F(\boldsymbol{Q};T)$ since the only scalar constructed from a symmetric tensor $\boldsymbol{Q}$ is $\mathrm{Tr}\,\boldsymbol{Q}$, but $\mathrm{Tr}\,\boldsymbol{Q}$ is zero by the definition of eq. (5.6). By similar reasoning, one can show that $F(\boldsymbol{Q};T)$ must have the following form

$$F(\boldsymbol{Q};T) = a_2(T)\mathrm{Tr}(\boldsymbol{Q}^2) + a_3\mathrm{Tr}(\boldsymbol{Q}^3) + a_4\mathrm{Tr}(\boldsymbol{Q}^4) + \cdots \qquad (5.33)$$

In this case, there is a third-order term in $\boldsymbol{Q}$. If eq. (5.7) is used for $\boldsymbol{Q}$, eq. (5.33) becomes

$$F(\boldsymbol{Q};T) = \frac{2}{3}a_2(T)S^2 + \frac{2}{9}a_3 S^3 + \frac{2}{9}a_4 S^4 + \cdots \qquad (5.34)$$

The free energy depends on S, but does not depend on $\boldsymbol{n}$. This must be so since the free energy should depend on how strongly the molecules are aligned along $\boldsymbol{n}$, but should not depend on the direction in which $\boldsymbol{n}$ is pointing.

From eq. (5.34), $F(\boldsymbol{Q};T)$ may be expressed as

$$F(\boldsymbol{Q};T) = \frac{1}{2}A(T - T_c)S^2 - \frac{1}{3}BS^3 + \frac{1}{4}CS^4 \qquad (5.35)$$

where A, B, C are positive constants independent of temperature.[7] The free energy of eq. (5.33) or (5.35) for nematic forming materials is called the Landau–de Gennes free energy. The Landau–de Gennes free energy represents the essential feature of the isotropic–nematic transition, and is often used as a model free energy for the transition.

The coefficients A, B, C can be calculated using mean field theory, and the result is (see Appendix B)

$$A \simeq Nk_B, \quad B \simeq Nk_BT_c, \quad C \simeq Nk_BT_c \qquad (5.36)$$

5.3.2 Alignment of nematic molecules by a magnetic field

The Landau–de Gennes free energy is convenient to see the characteristic behaviour of the system near the transition temperature. As an example, let us consider the effect of a magnetic field on the molecular orientation.

When a magnetic field $\boldsymbol{H}$ is applied to a nematic forming molecule, a magnetic moment is induced in the molecule. The induced magnetic moment $\boldsymbol{m}$ depends on the angle that the molecular axis $\boldsymbol{u}$ makes with the magnetic field $\boldsymbol{H}$ (see Fig. 5.6). Let $\alpha_\parallel$ (and $\alpha_\perp$) be the magnetic susceptibility of the molecule when the magnetic field $\boldsymbol{H}$ is applied parallel (and perpendicular) to the molecular axis $\boldsymbol{u}$. If a magnetic field $\boldsymbol{H}$ is applied to a molecule pointing in the direction $\boldsymbol{u}$, the molecule will have a magnetic moment $\boldsymbol{m}_\parallel = \alpha_\parallel(\boldsymbol{H} \cdot \boldsymbol{u})\boldsymbol{u}$ along $\boldsymbol{u}$, and a magnetic moment $\boldsymbol{m}_\perp = \alpha_\perp[\boldsymbol{H} - (\boldsymbol{H} \cdot \boldsymbol{u})\boldsymbol{u}]$ in the direction normal to $\boldsymbol{u}$ (see Fig. 5.6). Therefore the molecule feels a potential energy

[7] The temperature T_c here corresponds to the temperature T_{c2} in the previous section.

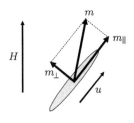

Fig. 5.6 The magnetic moment induced in an ellipsoidal molecule by a magnetic field $\boldsymbol{H}$. The magnetic field $\boldsymbol{H}$ can be decomposed into components parallel and perpendicular to the molecular axis $\boldsymbol{u}$: $\boldsymbol{H}_\parallel = (\boldsymbol{H} \cdot \boldsymbol{u})\boldsymbol{u}$ and $\boldsymbol{H}_\perp = \boldsymbol{H} - \boldsymbol{H}_\parallel$. Each component induces the magnetic moment $\boldsymbol{m}_\parallel = \alpha_\parallel\boldsymbol{H}_\parallel$ and $\boldsymbol{m}_\perp = \alpha_\perp\boldsymbol{H}_\perp$ in the molecule. The magnetic moment induced in the molecule is the sum of the two: $\boldsymbol{m} = \boldsymbol{m}_\parallel + \boldsymbol{m}_\perp$.

$$w_H(\boldsymbol{u}) = -\frac{1}{2}\alpha_\parallel(\boldsymbol{H}\cdot\boldsymbol{u})^2 - \frac{1}{2}\alpha_\perp[\boldsymbol{H} - (\boldsymbol{H}\cdot\boldsymbol{u})\boldsymbol{u}]^2$$

$$= -\frac{1}{2}\alpha_\parallel(\boldsymbol{H}\cdot\boldsymbol{u})^2 + \frac{1}{2}\alpha_\perp(\boldsymbol{H}\cdot\boldsymbol{u})^2 + \text{terms independent of } \boldsymbol{u}$$

$$= -\frac{1}{2}\alpha_d(\boldsymbol{H}\cdot\boldsymbol{u})^2 + \text{terms independent of } \boldsymbol{u} \qquad (5.37)$$

where $\alpha_d = \alpha_\parallel - \alpha_\perp$. If $\alpha_d > 0$, the magnetic field tends to align the molecule in the direction of $\boldsymbol{H}$, while if $\alpha_d < 0$, the magnetic field tends to rotate the molecule to the direction perpendicular to $\boldsymbol{H}$. In what follows, we assume $\alpha_d > 0$.

If there are N molecules in the system, the total potential energy due to the magnetic field is given by

$$F_H = -\frac{N}{2}\alpha_d\langle(\boldsymbol{H}\cdot\boldsymbol{u})^2\rangle \qquad (5.38)$$

Using the definition of $\boldsymbol{Q}$ (eq. (5.6)), this can be written as

$$F_H = -\frac{N}{2}\alpha_d\boldsymbol{H}\cdot\boldsymbol{Q}\cdot\boldsymbol{H} = -\frac{N}{2}\alpha_d S(\boldsymbol{H}\cdot\boldsymbol{n})^2 \qquad (5.39)$$

where we have used eq. (5.7) and ignored the terms which do not depend on $\boldsymbol{Q}$. Therefore the total free energy of the system is

$$F(\boldsymbol{Q};T) = \frac{1}{2}A(T - T_c)S^2 - \frac{1}{3}BS^3 + \frac{1}{4}CS^4 - \frac{SN}{2}\alpha_d(\boldsymbol{H}\cdot\boldsymbol{n})^2 \quad (5.40)$$

The order parameter $\boldsymbol{Q}$ in the presence of the magnetic field is given by S and $\boldsymbol{n}$ which minimize eq. (5.40). The response of $\boldsymbol{Q}$ to the magnetic field is quite different in the disordered phase and in the ordered phase.

Response in the disordered phase

In the disordered phase, the order parameter S is zero when there is no magnetic field. When a magnetic field is applied, S becomes non-zero, but is small. Therefore we may approximate eq. (5.40) as

$$F(\boldsymbol{Q};T) = \frac{1}{2}A(T - T_c)S^2 - \frac{SN}{2}\alpha_d\boldsymbol{H}^2 \qquad (5.41)$$

where we have used the fact that $\boldsymbol{n}$ is parallel to $\boldsymbol{H}$. Minimization of this with respect to S gives

$$S = \frac{N\alpha_d\boldsymbol{H}^2}{2A(T - T_c)} \qquad (5.42)$$

Equation (5.42) indicates that as we approach T_c, the order parameter S increases, and diverges at $T = T_c$.

The divergence of S at T_c is due to the appearance of some locally ordered region in the disordered phase. In the disordered phase, there is no macroscopic order, but the tendency that neighbouring molecules align in the same direction increases as T approaches T_c. Accordingly, near T_c, the system is divided into regions within which the molecules align in the same direction. The size of such ordered regions increases as T approaches T_c, and therefore S diverges at T_c.

The phenomenon that the response to external fields in the disordered phase diverges at T_c is seen quite generally in order–disorder transitions and is called a critical phenomenon.[8] In a continuous transition, the divergence is always seen since the system remains in the disordered phase as long as T is larger than T_c. In a discontinuous transition, the divergence may not be seen since the transition to the ordered phase can take place before the factor $1/(T - T_c)$ becomes dominant. In the isotropic–nematic transition, the temperatures T_{c1}, T_e, and T_{c2} are close to each other, and the divergence has been observed.

[8] Critical phenomena are also seen near the critical point in the gas–liquid transition and in the phase separation of solutions.

Response in the ordered phase

In the ordered phase, S has a non-zero value S_{eq} in the absence of a magnetic field. If a magnetic field is applied, S changes slightly from S_{eq}, but the deviation $S - S_{eq}$ is very small (being of the order of $\alpha_d H^2 / k_B T_c$). Therefore the magnetic field has little effect on the scalar order parameter S. On the other hand, the magnetic field has a crucial effect on the tensor order parameter $\boldsymbol{Q}$ since the director $\boldsymbol{n}$ is strongly affected by the magnetic field.

In the disordered phase, the rotation of molecules is essentially independent of each other. Therefore if one wants to align the molecules, one needs to apply a very large magnetic field which satisfies $\alpha_d H^2 > k_B T$. On the other hand, in the ordered phase, N molecules move together as a single entity. Therefore if one applies a magnetic field which satisfies $S_{eq} N \alpha_d H^2 > k_B T$, one can rotate all molecules in the system. Since N is a macroscopic number ($N \simeq 10^{20}$), the necessary magnetic field is very small. This is another example of the principle discussed in Section 1.5 that molecular collectivity produces a very large response to external fields.

5.4   Effect of a spatial gradient on the nematic order

5.4.1   Free energy functional for a non-uniformly ordered state

So far, we have been assuming that the order parameter does not depend on position, i.e., the system is spatially uniform. Let us now consider the situation that the order parameter $\boldsymbol{Q}$ varies with position. This situation has practical importance since the order parameter of nematics contained in a cell is strongly influenced by the constraints imposed by the cell walls.

To discuss the equilibrium of a non-uniformly ordered state, we consider the free energy functional $F_{tot}[\boldsymbol{Q}(\boldsymbol{r})]$ that is representative of the system in which the order parameter at position $\boldsymbol{r}$ is $\boldsymbol{Q}(\boldsymbol{r})$.

If there are no external fields, and no boundaries, the equilibrium state should be the uniform state in which the order parameter $Q(r)$ is independent of position r. If the order parameter varies with position, the free energy of the system must be larger than that of the uniform state. Therefore $F_{tot}[Q(r)]$ should be written as follows

$$F_{tot} = \int dr[f(Q(r)) + f_{el}(Q, \nabla Q)] \qquad (5.43)$$

The first term $f(Q)$ is the free energy density for a uniform system, and is essentially the same as that given by (5.40). (A small letter f is used to emphasize that it is a free energy per unit volume.) The second term $f_{el}(Q, \nabla Q)$ is the excess free energy due to the spatial gradient of Q. If ∇Q is small, f_{el} can be expanded as a power series of ∇Q. Since the free energy is a minimum in the uniform state (the state of $\nabla Q = 0$), the lowest term must be written as

$$f_{el}(Q, \nabla Q) = \frac{1}{2} K_{\alpha\beta\gamma, \alpha'\beta'\gamma'} \nabla_\alpha Q_{\beta\gamma} \nabla_{\alpha'} Q_{\beta'\gamma'} \qquad (5.44)$$

Since this is an expansion with respect to ∇Q, the coefficient $K_{\alpha\beta\gamma, \alpha'\beta'\gamma'}$ does not depend on ∇Q, but can depend on Q. It represents the component of a positive definite symmetric tensor. We shall now discuss the explicit form of $f_{el}(Q, \nabla Q)$ for the isotropic and nematic states.

5.4.2   Effect of the gradient terms in the disordered phase

In the isotropic state, we may assume that $K_{\alpha\beta\gamma, \alpha'\beta'\gamma'}$ is independent of Q since Q is small in the isotropic state. Hence f_{el} is a scalar constructed by a quadratic form of ∇Q. Using the properties $Q_{\alpha\beta} = Q_{\beta\alpha}$, and $Q_{\alpha\alpha} = 0$, we can show that f_{el} is written in the following form:[9]

$$f_{el} = \frac{1}{2} K_1 \nabla_\alpha Q_{\beta\gamma} \nabla_\alpha Q_{\beta\gamma} + \frac{1}{2} K_2 \nabla_\alpha Q_{\alpha\gamma} \nabla_\beta Q_{\beta\gamma} \qquad (5.45)$$

where K_1 and K_2 are positive constants. Thus the free energy functional in the isotropic state is written as

$$F_{tot} = \int dr \left[\frac{1}{2} A(T - T_c) S^2 + \frac{1}{2} K_1 \nabla_\alpha Q_{\beta\gamma} \nabla_\alpha Q_{\beta\gamma} + \frac{1}{2} K_2 \nabla_\alpha Q_{\alpha\gamma} \nabla_\beta Q_{\beta\gamma} \right] \qquad (5.46)$$

where we have ignored higher order terms in Q.

As an application of eq. (5.46), let us consider the local ordering induced by a wall of solid substrate. The molecules near the wall feel the potential of the wall, and their orientational distribution is not isotropic even in the isotropic state. Consider the situation shown in Fig. 5.7(a), where the molecules tend to align in the direction normal to the wall. If we take the x, y, z coordinates as in Fig. 5.7(a), the order parameter $Q_{\alpha\beta}(x)$ can be written as follows

$$Q_{xx} = \frac{2}{3} S, \quad Q_{yy} = Q_{zz} = -\frac{1}{3} S, \quad Q_{xy} = Q_{yz} = Q_{zx} = 0 \qquad (5.47)$$

[9] The scalar $\nabla_\alpha Q_{\beta\gamma} \nabla_\beta Q_{\alpha\gamma}$ is not included in eq. (5.45) since it is transformed to the second term on the right-hand side of eq. (5.45) by integration by parts.

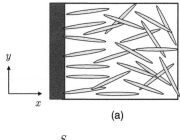

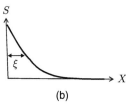

Fig. 5.7 (a) Wall induced ordering of nematic forming molecules in the isotropic state. The molecules at the wall are assumed to be oriented in the x-direction by the wall. (b) Order parameter plotted against the distance from the wall. ξ represents the correlation length.

Substituting this for $Q_{\alpha\beta}(x)$ in eq. (5.46), we have

$$F_{tot} = \int d\boldsymbol{x} \left[\frac{1}{2} A(T-T_c)S^2 + \frac{1}{3}K_1 \left(\frac{dS}{dx}\right)^2 + \frac{2}{9}K_2 \left(\frac{dS}{dx}\right)^2 \right]$$

$$= \frac{1}{2} A(T-T_c) \int d\boldsymbol{x} \left[S^2 + \xi^2 \left(\frac{dS}{dx}\right)^2 \right] \tag{5.48}$$

where ξ is defined by

$$\xi = \sqrt{\frac{2(3K_1 + 2K_2)}{9A(T-T_c)}} \tag{5.49}$$

For eq. (5.48), the condition $\delta F_{tot}/\delta S = 0$ gives

$$\xi^2 \frac{d^2 S}{dx^2} = S \tag{5.50}$$

Therefore

$$S = S_0 e^{-x/\xi} \tag{5.51}$$

where S_0 is the value of S at the wall.

Equation (5.51) indicates that the wall affects the molecular orientation up to the distance ξ from the wall. Notice that the wall effect is intrinsically short ranged: only the molecules close to the wall feel the wall potential. However, the effect of the wall propagates into the bulk due to the tendency of the local ordering of the molecules. The length ξ represents how far this effect persists, and is called the correlation length. The correlation length ξ can be regarded as the size of the locally ordered region in the disordered phase. As indicated by eq. (5.49), ξ diverges as T approaches T_c. Therefore, near the transition temperature, a large number of molecules move collectively. This is the reason why the response of the order parameter diverges at the transition temperature T_c (see eq. (5.42)).

5.4.3   Effect of the gradient terms in the ordered phase

As discussed in Section 5.3.2, when an external field is applied to the ordered phase, the scalar order parameter S changes little, while the director $\boldsymbol{n}$ changes drastically. Therefore we may assume that in the ordered phase the tensor order parameter $\boldsymbol{Q}$ can be written as

$$\boldsymbol{Q}(\boldsymbol{r}) = S_{eq} \left[\boldsymbol{n}(\boldsymbol{r})\boldsymbol{n}(\boldsymbol{r}) - \frac{1}{3}\boldsymbol{I} \right] \tag{5.52}$$

where S_{eq} is the equilibrium value of S in the absence of an external field.

If eq. (5.52) is used for $\boldsymbol{Q}$ in eq. (5.43), $f(\boldsymbol{Q})$ becomes constant, and may be dropped in the subsequent calculation. On the other hand, f_{el}

can be written as a quadratic form of $\boldsymbol{\nabla}\boldsymbol{n}$. The coefficient $K_{\alpha\beta\gamma,\alpha'\beta'\gamma'}$ may depend on $\boldsymbol{n}$. Repeating the same argument as in the previous section, we can show that f_{el} can be written in the following form

$$f_{el} = \frac{1}{2}K_1(\boldsymbol{\nabla}\cdot\boldsymbol{n})^2 + \frac{1}{2}K_2(\boldsymbol{n}\cdot\boldsymbol{\nabla}\times\boldsymbol{n})^2 + \frac{1}{2}K_3(\boldsymbol{n}\times\boldsymbol{\nabla}\times\boldsymbol{n})^2 \quad (5.53)$$

where K_1, K_2, K_3 are constants having dimension of [J/m], and are called the Frank elastic constants. Each constant represents the resistance of the nematics to the spatial variation of $\boldsymbol{n}$ shown in Fig. 5.8. The constants K_1, K_2, and K_3 are called the splay, twist, and bend constants, respectively.

If there is a magnetic field $\boldsymbol{H}$, the following term should be added to eq. (5.53):

$$f_H = -\frac{1}{2}\Delta\chi(\boldsymbol{H}\cdot\boldsymbol{n})^2 + \text{terms independent of } \boldsymbol{n} \quad (5.54)$$

where $\Delta\chi = \alpha_d n_p S_{eq}$, and n_p is the number of nematic molecules per unit volume.

The wall effect is accounted for by the boundary condition for $\boldsymbol{n}$. By using a proper surface treatment, it is possible to let $\boldsymbol{n}$ align along a certain direction in the wall plane, or along the direction normal to the plane.

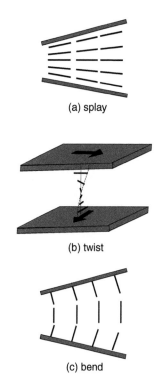

(a) splay

(b) twist

(c) bend

Fig. 5.8 Examples of spatial variation of director $\boldsymbol{n}$ which gives the Frank elastic energy represented by eq. (5.53): (a) splay (for which $\boldsymbol{\nabla}\cdot\boldsymbol{n}\neq 0$); (b) twist ($\boldsymbol{\nabla}\times\boldsymbol{n}\neq 0$); (c) bend ($\boldsymbol{n}\times(\boldsymbol{\nabla}\times\boldsymbol{n})\neq 0$).

5.4.4   Fredericks transition

As an application of the above theory, let us consider the problem illustrated in Fig. 5.9(a). A nematic liquid crystal is sandwiched between two plates. We take the z-axis normal to the plate. At the wall of the plate, we assume that the director is in the x-direction. Suppose that we apply a magnetic field H in the z-direction. Then the molecules tend to orient along the z-direction, but this is not compatible with the boundary conditions at $x = 0$ and $x = L$. What will be the equilibrium configuration in this situation?

Fig. 5.9 (a) Model of an optical switch cell utilizing the Fredericks transition. Light is transmitted across the cell shown in Fig. 5.2, where the direction of the polarizer makes 45 degrees with the x-axis in the x–y plane. The cell consists of two polarizing plates and a nematic layer. The director $\boldsymbol{n}$ is fixed in the x-direction at the cell wall. Upon the inception of the magnetic field, the nematics tend to align in the z-direction. (b) If H is less than the critical field H_c, the nematic remains aligned in the x-direction, and the light is transmitted. If H is larger than H_c, the nematics start to align in the z-direction, and the light is not transmitted. The intensity of the transmitted light changes steeply at H_c.

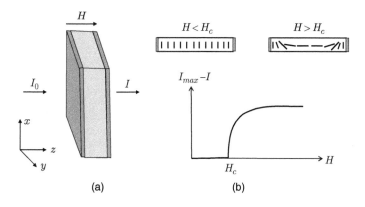

(a)

(b)

By the symmetry of the problem, we may assume that n is in x–z plane, and only depends on the z-coordinate. Hence the x, y, z components of n are given by

$$n_x(z) = \cos\theta(z), \qquad n_y(z) = 0, \qquad n_z(z) = \sin\theta(z) \qquad (5.55)$$

In this case, $\nabla \cdot n$ and $\nabla \times n$ are given by

$$\nabla \cdot n = \frac{\partial n_z}{\partial z} = \cos\theta \frac{d\theta}{dz} \qquad (5.56)$$

$$\nabla \times n = \left(0, \frac{\partial n_x}{\partial z}, 0\right) = \left(0, -\sin\theta \frac{d\theta}{dz}, 0\right) \qquad (5.57)$$

Using these relations, the free energy is given by

$$F_{tot} = \int dz (f_{el} + f_H)$$

$$= \int dz \left[\frac{1}{2}K_1 \cos^2\theta \left(\frac{d\theta}{dz}\right)^2 + \frac{1}{2}K_3 \sin^2\theta \left(\frac{d\theta}{dz}\right)^2 - \frac{1}{2}\Delta\chi H^2 \sin^2\theta\right]$$

$$(5.58)$$

$\theta(z)$ has to satisfy the following boundary condition

$$\theta(0) = \theta(L) = 0 \qquad (5.59)$$

The equilibrium solution is given by the $\theta(z)$ which minimizes eq. (5.58) under the condition (5.59). If the magnetic field H is weak, the solution is given by $\theta(z) = 0$. As the magnetic field increases, the solution becomes unstable, and a new solution appears. Therefore the structure of the problem is the same as that in the order–disorder transition. To pursue this resemblance, we assume that the solution has the following functional form

$$\theta(z) = \theta_0 \sin\left(\frac{\pi z}{L}\right) \qquad (5.60)$$

and express the free energy F_{tot} as a function of θ_0. If θ_0 is small, F_{tot} can be calculated analytically, and is given by

$$F_{tot} = \frac{1}{4}\left[\frac{K_1\pi^2}{L} - \Delta\chi H^2 L\right]\theta_0^2 = \frac{1}{4}\Delta\chi L(H_c^2 - H^2)\theta_0^2 \qquad (5.61)$$

where

$$H_c = \sqrt{\frac{K_1\pi^2}{\Delta\chi L^2}} \qquad (5.62)$$

Therefore if $H < H_c$, the solution $\theta_0 = 0$ is stable, and if $H > H_c$ the solution is unstable. Notice that the transition taking place at H_c is continuous, since the free energy is an even function of θ_0. Such a transition is called the Fredericks transition.

The Fredericks transition is important in the application of nematics to displays since the light transmission across the nematic layer changes sharply at the critical field.

5.5 Onsager's theory for the isotropic–nematic transition of rod-like particles

Onsager predicted that a solution of rod-like particles forms a nematic phase above a certain concentration.[10] Let L and D be the length and diameter of the particle. The aspect ratio L/D is assumed to be very large. In the limit of $L/D \to \infty$, Onsager's theory for the phase transition becomes rigorous.

Now consider two rod-like particles 1 and 2, each pointing in directions $\boldsymbol{u}$ and $\boldsymbol{u}'$, respectively. These particles cannot overlap each other. Therefore if the position of rod 1 is fixed, the centre of mass of rod 2 cannot enter into a certain region shown in Fig. 5.10. The volume of this region is called the excluded volume. The excluded volume is a function of the angle Θ that $\boldsymbol{u}'$ makes with $\boldsymbol{u}$. As indicated in Fig. 5.10, this volume is given by

$$v_{ex}(\boldsymbol{u}, \boldsymbol{u}') = 2DL^2 \sin \Theta = 2DL^2 |\boldsymbol{u} \times \boldsymbol{u}'| \qquad (5.63)$$

where we have ignored the smaller terms which are of the order of D/L less than that in eq. (5.63). Equation (5.63) indicates that the excluded volume decreases with decrease of Θ. Therefore if rod 2 moves around rod 1 keeping its direction $\boldsymbol{u}'$ constant, the region allowed for rod 2 increases as Θ decreases. Therefore the pair with small Θ is entropically more favourable than the pair with large Θ, and the rod-like particles tend to align in the same direction. This is why rod-like particles form a nematic phase at high concentration.

Let us consider a solution consisting of N rod-like particles in volume V. Let us consider the probability $\psi(\boldsymbol{u})$ that a particular particle, say particle 1, is pointing in the direction $\boldsymbol{u}$. Due to the principle of equal weight, $\psi(\boldsymbol{u})$ is proportional to the probability that all other particles $2, 3, \ldots, N$ do not overlap particle 1. Since the probability that particle j does not overlap with particle 1 is equal to $1 - v_{ex}(\boldsymbol{u}, \boldsymbol{u}_j)/V$, $\psi(\boldsymbol{u})$ is given by

$$\psi(\boldsymbol{u}) \propto \Pi_{j=2}^{N} \left[1 - \frac{v_{ex}(\boldsymbol{u}, \boldsymbol{u}_j)}{V} \right] = \exp\left[-\sum_{j=2}^{N} \frac{v_{ex}(\boldsymbol{u}, \boldsymbol{u}_j)}{V} \right] \qquad (5.64)$$

The summation in the exponent can be written as

$$\sum_{j=2}^{N} \frac{v_{ex}(\boldsymbol{u}, \boldsymbol{u}_j)}{V} = n \int d\boldsymbol{u}' v_{ex}(\boldsymbol{u}, \boldsymbol{u}') \psi(\boldsymbol{u}') \qquad (5.65)$$

[10] In the theory explained here, the solvent is regarded as a continuum background to solute particles and is not explicitly considered.

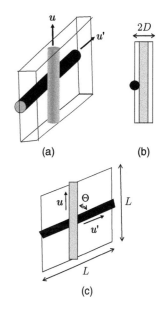

Fig. 5.10 Excluded volume of rod-like particles. (a) When a particle pointing along $\boldsymbol{u}$ is fixed, the centre of mass of the other particle pointing along $\boldsymbol{u}'$ cannot enter into the parallelepiped region shown in this figure. (b) The parallelepiped viewed from the direction of $\boldsymbol{u}'$. (c) The parallelepiped viewed from the direction of $\boldsymbol{u} \times \boldsymbol{u}'$. The parallelepiped has the base of a rhombus with area $L^2 \sin \Theta$, and thickness $2D$.

where $n = N/V$ is the number density of the particles, and we have used the fact that the orientational distribution of other particles is also given by $\psi(\boldsymbol{u})$. Equations (5.64) and (5.65) give the following self-consistent equation for $\psi(\boldsymbol{u})$:

$$\psi(\boldsymbol{u}) = C \exp\left[-n \int d\boldsymbol{u}' v_{ex}(\boldsymbol{u}, \boldsymbol{u}')\psi(\boldsymbol{u}')\right] \tag{5.66}$$

Equation (5.66) has the same structure as eq. (5.17). The interaction potential $w(\boldsymbol{u}, \boldsymbol{u}')$ is now replaced by

$$w_{eff}(\boldsymbol{u}, \boldsymbol{u}') = nk_BTv_{ex}(\boldsymbol{u}, \boldsymbol{u}') = 2nDL^2k_BT|\boldsymbol{u} \times \boldsymbol{u}'| \tag{5.67}$$

The potential $w_{eff}(\boldsymbol{u}, \boldsymbol{u}')$ takes a minimum when $\boldsymbol{u}$ and $\boldsymbol{u}'$ are parallel (or antiparallel) to each other. The strength of the interaction is now characterized by nDL^2. Therefore if nDL^2 exceeds a certain critical value, the isotropic state becomes unstable, and the nematic state appears.

Numerical solution of eq. (5.66) indicates that the isotropic state becomes unstable when $nDL^2 > 5.1$. The volume fraction at this density is given by

$$\phi_{c2} = \frac{5.1\pi}{4}\frac{D}{L} \simeq 4\frac{D}{L} \tag{5.68}$$

Above this concentration, the isotropic state cannot be stable, and the system turns into the nematic state.

In deriving eq. (5.66), we have ignored the correlation in the molecular configuration of particles $2, \ldots, N$. This approximation is justified for particles having large aspect ratio since the volume fraction ϕ_{c2} becomes very small for large aspect ratio. In the limit of large aspect ratio $L/D \to \infty$, ϕ_{c2} goes to zero, and Onsager's theory becomes an exact theory.

In the case of lyotropic liquid crystals, the transition from the isotropic state to the nematic state is accompanied by phase separation, i.e., when the nematic phase appears, the solution separates into two solutions, an isotropic solution with concentration ϕ_A, and a nematic solution with concentration ϕ_B. Both ϕ_A and ϕ_B are of the same order as ϕ_{c2}, i.e., $\phi_A \simeq \phi_B \simeq D/L$.

5.6 Summary of this chapter

A liquid crystal is an anisotropic fluid. In nematic liquid crystals, molecules spontaneously align along a certain direction denoted by a unit vector $\boldsymbol{n}$ called the director. The degree of molecular order is characterized by the scalar order parameter S or the tensor order parameter $\boldsymbol{Q}$. The tensor order parameter includes information on $\boldsymbol{n}$ and S.

The transition from the isotropic state to the nematic state can be discussed by molecular mean field theories (such as the Maier–Saupe theory, or the Onsager theory) which considers the equilibrium distribution of molecular orientation in the mean field. Alternatively, the transition can be discussed by a phenomenological theory (such as the Landau–de Gennes theory) which consider the free energy of the system for given order parameter S or $\boldsymbol{Q}$. The phenomenological theory describes the behaviour of the system near the transition temperature.

External fields (say a magnetic field H) orient the molecules along its direction (if $\alpha_d > 0$) and change the value of the order parameter $\boldsymbol{Q}$. Well above the transition temperature T_c, the change in the order parameter is very small. As the temperature T approaches T_c, the molecules tend to move collectively. This gives a large response to external fields. In particular, the scalar order parameter changes as $1/(T - T_c)$ above T_c. Below T_c, the scalar order parameter S changes little, but the director $\boldsymbol{n}$ changes significantly. This behaviour can be discussed by the Lanau–de Gennes theory.

Further reading

(1) *Principles of Condensed Matter Physics*, Paul M. Chaikin and Tom C. Lubensky, Cambridge University Press (1997).

(2) *Liquid Crystals*, 2nd edition, Sivaramakrishna Chandrasekher, Cambridge University Press (1992).

(3) *The Physics of Liquid Crystals*, 2nd edition, Pierre-Gilles de Gennes and Jacques Prost, Oxford University Press (1995).

Exercises

(5.1) Suppose that the unit vector $\boldsymbol{u}$ is isotropically distributed on the sphere $|\boldsymbol{u}| = 1$. Let $\langle \cdots \rangle_0$ be the average for this distribution

$$\langle \cdots \rangle_0 = \frac{1}{4\pi} \int d\boldsymbol{u} \cdots \qquad (5.69)$$

Answer the following questions.

(a) Derive the following equations

$$\langle u_z^2 \rangle_0 = \frac{1}{3}, \quad \langle u_z^4 \rangle_0 = \frac{1}{5},$$

$$\langle u_z^{2m} \rangle_0 = \frac{1}{2m + 1}$$

(b) Derive the following equations

$$\langle u_\alpha u_\beta \rangle_0 = \frac{1}{3} \delta_{\alpha\beta}$$

$$\langle u_\alpha u_\beta u_\mu u_\nu \rangle_0 = \frac{1}{15}(\delta_{\alpha\beta}\delta_{\mu\nu} + \delta_{\alpha\mu}\delta_{\beta\nu} + \delta_{\alpha\nu}\delta_{\beta\mu})$$

(5.2) The mean field theory predicts that the scalar order parameter S at equilibrium is given by the solution of eq. (5.26) with $S = x/\beta U$. Answer the following questions.

(a) For small x, express the function $I(x)$ as a power series in x up to order x^3.

(b) Obtain the temperature T_{c2} at which the isotropic phase becomes unstable.

(5.3) Minimize the free energy functional $F[\psi]$ of eq. (5.14) under the constraint that the scalar order parameter is fixed at S

$$S = \int d\boldsymbol{u} \frac{3}{2}\left(u_z^2 - \frac{1}{3}\right)\psi(\boldsymbol{u}) \qquad (5.70)$$

and show that the equilibrium distribution function is given by eqs. (5.24) and (5.25).

(5.4) Suppose that the free energy of the fluid which consists of polar molecules is written in the following form

$$F[\psi] = N\left[k_B T \int d\boldsymbol{u} \, \psi(\boldsymbol{u}) \ln \psi(\boldsymbol{u}) \right.$$
$$\left. - \frac{U}{2} \int d\boldsymbol{u} \int d\boldsymbol{u}' \, (\boldsymbol{u} \cdot \boldsymbol{u}') \psi(\boldsymbol{u}) \psi(\boldsymbol{u}') \right]$$
(5.71)

Notice that the interaction energy between the molecules parallel to $\boldsymbol{u}$ and $\boldsymbol{u}'$ is proportional to $\boldsymbol{u} \cdot \boldsymbol{u}'$ (not $(\boldsymbol{u} \cdot \boldsymbol{u}')^2$). Answer the following questions.

(a) Show that the distribution function which minimizes eq. (5.71) can be written as

$$\psi(\boldsymbol{u}) = C \exp(\beta U \boldsymbol{P} \cdot \boldsymbol{u}) \qquad (5.72)$$

where $\boldsymbol{P}$ is defined by

$$\boldsymbol{P} = \langle \boldsymbol{u} \rangle \qquad (5.73)$$

(b) Show that eqs. (5.72) and (5.73) give the following equation for $x = \beta U |\boldsymbol{P}|$:

$$\frac{k_B T}{U} x = \coth x - \frac{1}{x} \qquad (5.74)$$

(c) Show that eq. (5.74) has one solution $x = 0$ for $T > T_c$ and three solutions for $T < T_c$, where $T_c = U/3k_B$, and discuss the phase transition of this system.

(5.5) The entropy S given by (5.13) can be expressed as a function of the order parameter. Suppose that $\psi(\boldsymbol{u})$ satisfies the constraint

$$P = \int d\boldsymbol{u} \, u_z \psi(\boldsymbol{u}) \qquad (5.75)$$

The maximum value of the entropy $S[\psi(\boldsymbol{u})]$ under this constraint can be written as a function of P. This defines the entropy $S(P)$ as a function of the order parameter P. Answer the following questions.

(a) Obtain $S(P)$ by maximizing the functional

$$\tilde{S} = -Nk_B\left[\int d\boldsymbol{u} \, \psi(\boldsymbol{u}) \ln \psi(\boldsymbol{u}) \right.$$
$$\left. - \lambda \int d\boldsymbol{u} \, \psi(\boldsymbol{u})(u_z - P) \right]$$
(5.76)

and show that it is given by ψ which satisfies

$$\psi(\boldsymbol{u}) = C e^{\lambda u_z} \qquad (5.77)$$

where C is a normalization constant, and λ is determined by eqs. (5.75) and (5.77).

(b) For small P show that λ is given by

$$\lambda = 3P \qquad (5.78)$$

and that the entropy $S(P)$ is given by

$$S(P) = -\frac{3}{2} k_B P^2 \qquad (5.79)$$

(c) The above procedure allows us to write the free energy as a function of the order parameter. For the free energy given by eq. (5.71), show that the free energy $F(P)$ is given by

$$F(P; T) = \frac{3}{2} k_B T P^2 - \frac{U}{2} P^2 \qquad (5.80)$$

Obtain the critical temperature T_c.

Brownian motion
and thermal fluctuations

In the previous chapters, we have discussed soft matter at equilibrium. In the present and following chapters, we shall consider soft matter in non-equilibrium states. We shall discuss how soft matter in non-equilibrium states approaches equilibrium, and how the system evolves in time when external conditions are changed.

In this chapter we shall discuss time-dependent phenomena taking place at equilibrium, or near equilibrium.

An equilibrium state is a stationary state, and therefore macroscopic physical quantities, such as pressure or concentration, should not depend on time. However, if we measure these quantities for small systems, their values fluctuate in time. For example (i) the pressure acting on the wall of a cubic box of size $(10\,\mathrm{nm})^3$ fluctuates in time, and (ii) the number of solute molecules in a region of size $(10\,\mathrm{nm})^3$ fluctuates in time. These fluctuations are caused by the random thermal motion of molecules constituting the system. The objective of this chapter is to discuss the time dependence of such fluctuations.

The fluctuations are most clearly seen as Brownian motion: small particles suspended in a fluid move randomly. Einstein showed that this motion is a manifestation of the thermal motion of molecules, and therefore it links our macroscopic world to the microscopic world of molecules. His theory of Brownian motion relates the microscopic fluctuations at equilibrium to the macroscopic response of the system to external forces. This theorem, called the fluctuation–dissipation theorem, leads us to an important principle called the Onsager principle which gives a common base in the dynamics of soft matter as we shall see in subsequent chapters.

6.1 Random motion of small particles

6.1.1 Time correlation functions

Small particles (size less than $1\,\mu\mathrm{m}$) suspended in a fluid undergo random irregular motions: spherical particles change their position randomly as in Fig. 6.1(a), and rod-like particles change their position and orientation randomly as in Fig. 6.1(b). Such random motion is called Brownian motion.

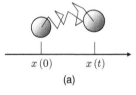

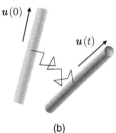

Fig. 6.1 (a) Brownian motion of a spherical particle. (b) Brownian motion of a rod-like particle.

Brownian motion is a manifestation of the thermal motion of molecules. According to statistical mechanics, any particle of mass m, placed in an environment at temperature T, moves randomly with velocity of the order of $\sqrt{k_B T/m}$. For a macroscopic particle, having size 1 cm and mass 1 g, this velocity is of the order of 1 nm/s, and is hardly noticeable. On the other hand, for a colloidal particle having size 1 μm and mass 10^{-12} g, the velocity becomes about 1 mm/s and becomes observable by microscopy. Particles undergoing Brownian motion are called Brownian particles.

If we record the values of certain physical quantities x_i $(i = 1, 2, \ldots)$ of a Brownian particle, say particle position or orientational angle, we will have a sequence of data $x_i(t)$ which varies randomly in time as shown in Fig. 6.2(a). If we do such a measurement for many systems at equilibrium, we can define the average $\langle \cdots \rangle$.[1] For a Brownian particle at equilibrium, the average $\langle x_i(t) \rangle$ is independent of time, and is written as $\bar{x}_i$

$$\bar{x}_i = \langle x_i(t) \rangle \tag{6.1}$$

Let $\Delta x_i(t)$ be the deviation of $x_i(t)$ from the average

$$\Delta x_i(t) = x_i(t) - \bar{x}_i \tag{6.2}$$

The correlation of $\Delta x_i(t)$ and $\Delta x_j(t)$ at different times t_1 and t_2, $\langle \Delta x_i(t_1) \Delta x_j(t_2) \rangle$, is called a time correlation function. A time correlation function for the same physical quantity $\langle \Delta x_i(t_1) \Delta x_i(t_2) \rangle$ is called the autocorrelation function, and that for different quantities $\langle \Delta x_i(t_1) \Delta x_j(t_2) \rangle$ $(i \neq j)$ is called the cross-correlation function.

At equilibrium, the time correlation function $\langle \Delta x_i(t_1) \Delta x_j(t_2) \rangle$ does not depend on the origin of time, and depends on the difference of time, $t_1 - t_2$, only. Therefore the time correlation function has the following property:

$$\langle \Delta x_i(t_1) \Delta x_j(t_2) \rangle = \langle \Delta x_i(t_1 - t_2) \Delta x_j(0) \rangle = \langle \Delta x_i(0) \Delta x_j(t_2 - t_1) \rangle \tag{6.3}$$

This property is called time translational invariance.

By eq. (6.3), it is shown that the autocorrelation function is an even function of time:

$$\langle \Delta x_i(t) \Delta x_i(0) \rangle = \langle \Delta x_i(0) \Delta x_i(-t) \rangle = \langle \Delta x_i(-t) \Delta x_i(0) \rangle \tag{6.4}$$

6.1.2   Symmetry properties of the time correlation functions

Onsager has shown that if x_i $(i = 1, 2, \ldots)$ are the quantities representing the particle configuration (position and orientation) or their functions (e.g., the dipole moment), the cross-correlation function also becomes an even function of time, i.e.,

$$\langle \Delta x_i(t) \Delta x_j(0) \rangle = \langle \Delta x_i(-t) \Delta x_j(0) \rangle \tag{6.5}$$

This simple relation derives from a very fundamental law of physics: the time reversal symmetry in the fluctuations at equilibrium.

[1] Here the average $\langle \cdots \rangle$ is defined as the ensemble average, i.e., the average for many equivalent systems. In practice, the average $\langle \cdots \rangle$ is obtained by the average for many experiments for a particular system. In the latter case, the origin of the time is defined for each experiment.

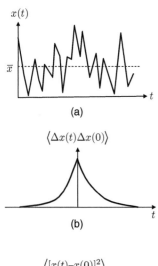

(a)

(b)

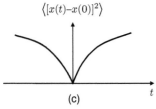

(c)

Fig. 6.2 (a) Example of the time dependence of a stochastic quantity $x(t)$. (b) Time dependence of the correlation function $\langle \Delta x(t) \Delta x(0) \rangle$ for the deviation from the average $\Delta x(t) = x(t) - \bar{x}$. (c) Time dependence of the mean square deviation $\langle (x(t) - x(0))^2 \rangle$.

Suppose that a video of Brownian motion is taken and shown to us in reverse mode. The time reversal symmetry says that we cannot tell whether the video is shown in reverse mode or in normal mode as long as the video is taken at equilibrium. This is because the event shown in reverse mode is physically possible and takes place with the same probability as the event shown in normal mode. This is a result of the following principle of statistical mechanics.

From the molecular viewpoint, Brownian motion is a deterministic process. It looks like a stochastic process if we focus on the motion of the Brownian particle only, but if we look at the motion of the Brownian particle and the surrounding fluid molecules, the process is deterministic, governed by Hamilton's equations of motion of mechanics. Let $q_1, q_2, \ldots, q_f$ and $p_1, p_2, \ldots, p_f$ be the set of coordinates and momenta of all particles constituting the system. We shall denote the whole set of variables by the symbol Γ, i.e., $\Gamma = (q_1, q_2, \ldots, q_f, p_1, p_2, \ldots, p_f)$, and denote the Hamiltonian of the system by $H(\Gamma)$. In the reverse mode, the state Γ changes to $(q_1, q_2, \ldots, q_f, -p_1, -p_2, \ldots, -p_f)$ and the time t changes to $-t$. We shall denote the state $(q_1, q_2, \ldots, q_f, -p_1, -p_2, \ldots, -p_f)$ as $-\Gamma$. Since the Hamiltonian has the property that $H(\Gamma) = H(-\Gamma)$, the states Γ and $-\Gamma$ are equally possible at equilibrium, i.e., the probability of finding the system in state Γ is equal to the probability of finding the system in state $-\Gamma$. Also the time evolution of the normal mode and that of the reverse mode are both allowed by Hamilton's equations of motion, i.e., if the state Γ_1 evolves to the state Γ_2 in time t, the state $-\Gamma_2$ evolves to $-\Gamma_1$ in time t. This is the reason for eq. (6.5).

Any observable physical quantity, say x_i, is expressed as a function of Γ such as $\hat{x}_i(\Gamma)$. When Γ is replaced by $-\Gamma$, the function $\hat{x}_i(\Gamma)$ usually remains unchanged or changes sign. Therefore we can define

$$\hat{x}_i(-\Gamma) = \epsilon_i \hat{x}_i(\Gamma) \tag{6.6}$$

where ϵ_i takes the value 1 or -1: ϵ_i is equal to 1 for quantities such as particle position or energy, and -1 for quantities such as velocity or angular momentum. Then the relation (6.5) is generally written as

$$\langle \Delta x_i(t) \Delta x_j(0) \rangle = \epsilon_i \epsilon_j \langle \Delta x_i(-t) \Delta x_j(0) \rangle \tag{6.7}$$

In the following discussion, we shall mostly consider the case of $\epsilon_i = 1$, and use eq. (6.5).

Using time translational invariance, relation (6.5) can be written as

$$\begin{aligned} \langle \Delta x_i(t) \Delta x_j(0) \rangle &= \langle \Delta x_i(-t) \Delta x_j(0) \rangle \\ &= \langle \Delta x_j(0) \Delta x_i(-t) \rangle = \langle \Delta x_j(t) \Delta x_i(0) \rangle \end{aligned} \tag{6.8}$$

i.e., the time correlation function $\langle \Delta x_i(t) \Delta x_j(0) \rangle$ is symmetric with respect to the exchange of i and j.

For certain physical quantities, such as the position of a free Brownian particle, it is not convenient to consider the time correlation function

$\langle \Delta x_i(t) \Delta x_j(0) \rangle$ since the averages $\bar{x}_i$ and $\bar{x}_j$ are not defined. For such quantities, it is convenient to consider the correlation of the displacement $x_i(t) - x_i(0)$, i.e., $\langle (x_i(t) - x_i(0))(x_j(t) - x_j(0)) \rangle$. The two correlation functions are related to each other if $\langle \Delta x(0)^2 \rangle$ is finite. This is shown as follows.

$$
\begin{aligned}
\langle (x_i(t) - x_i(0))(x_j(t) - x_j(0)) \rangle &= \langle (\Delta x_i(t) - \Delta x_i(0))(\Delta x_j(t) - \Delta x_j(0)) \rangle \\
&= \langle \Delta x_i(t) \Delta x_j(t) \rangle - \langle \Delta x_i(t) \Delta x_j(0) \rangle \\
&\quad - \langle \Delta x_i(0) \Delta x_j(t) \rangle + \langle \Delta x_i(0) \Delta x_j(0) \rangle
\end{aligned}
$$

$$(6.9)$$

By use of eqs. (6.3) and.(6.5), this can be written as

$$
\langle (x_i(t) - x_i(0))(x_j(t) - x_j(0)) \rangle = 2[\langle \Delta x_i(0) \Delta x_j(0) \rangle - \langle \Delta x_i(t) \Delta x_j(0) \rangle]
$$

$$(6.10)$$

Hence the dynamics of the fluctuations can be characterized either by the time correlation functions of the deviation $\Delta x_i(t) = x_i(t) - \bar{x}_i$ or the correlation of the displacement $x_i(t) - x_i(0)$.

6.1.3   Velocity correlation functions

Let $\dot{x}_i(t)$ be the time derivative of $x_i(t)$, i.e., $\dot{x}_i(t) = dx_i(t)/dt$. We shall call $\dot{x}_i(t)$ the 'velocity' of $x_i(t)$. The time correlation function of the velocity $\dot{x}_i(t)$ is obtained from the time correlation function of $x_i(t)$.

$$
\begin{aligned}
\langle \dot{x}_i(t) \dot{x}_j(t') \rangle &= \frac{\partial^2}{\partial t \partial t'} \langle \Delta x_i(t) \Delta x_j(t') \rangle \\
&= \frac{\partial^2}{\partial t \partial t'} \langle \Delta x_i(t - t') \Delta x_j(0) \rangle \\
&= -\frac{\partial^2}{\partial t^2} \langle \Delta x_i(t - t') \Delta x_j(0) \rangle
\end{aligned}
$$

$$(6.11)$$

Hence

$$
\begin{aligned}
\langle \dot{x}_i(t) \dot{x}_j(0) \rangle &= -\frac{\partial^2}{\partial t^2} \langle \Delta x_i(t) \Delta x_j(0) \rangle \\
&= \frac{1}{2} \frac{\partial^2}{\partial t^2} \langle (x_i(t) - x_i(0))(x_j(t) - x_j(0)) \rangle
\end{aligned}
$$

$$(6.12)$$

where eq. (6.10) has been used. Therefore the velocity correlation is obtained from the displacement correlation. Conversely, the displacement correlation is obtained from the velocity correlation by

$$
\langle (x_i(t) - x_i(0))(x_j(t) - x_j(0)) \rangle = 2 \int_0^t dt_1 \int_0^{t_1} dt_2 \langle \dot{x}_i(t_2) \dot{x}_j(0) \rangle \quad (6.13)
$$

6.2   Brownian motion of a free particle

6.2.1   Langevin equation for particle velocity

Let us now consider the Brownian motion of a point particle in free space. Let $x(t)$ be the x-coordinate of the particle position at time t. If the particle is macroscopic, the time evolution of $x(t)$ is described by the usual equation of motion of mechanics.

When a particle moves in a viscous fluid with velocity $v = \dot{x}$, it experiences a drag F_f from the fluid. Stokes has shown that for small velocity, F_f is proportional to the particle velocity v:

$$F_f = -\zeta v \tag{6.14}$$

The constant ζ is called the friction constant. For a spherical particle of radius a, ζ is given by

$$\zeta = 6\pi\eta a \tag{6.15}$$

where η is the viscosity of the fluid. Equation (6.15) is derived by hydrodynamics (see Appendix A). Therefore, for a macroscopic free particle, the equation of motion becomes

$$m\frac{dv}{dt} = -\zeta v \tag{6.16}$$

where m is the mass of the particle. According to eq. (6.16), the particle velocity decreases to zero as

$$v(t) = v_0 e^{-t/\tau_v} \tag{6.17}$$

where

$$\tau_v = m/\zeta \tag{6.18}$$

Equation (6.16) is not appropriate to describe the motion of small particles undergoing Brownian motion. According to eq. (6.16), the particle eventually stops moving, while actual Brownian particles keep moving. To describe Brownian motion, we have to take into account the fluctuations of the forces acting on the particle.

The fluid surrounding the particle is made of small molecules, which collide with the particle, giving randomly fluctuating forces. This effect can be described by the following model equation

$$m\frac{dv}{dt} = -\zeta v + F_r(t) \tag{6.19}$$

where $F_r(t)$ stands for the fluctuating force exerted on the particle by the fluid.

Equation (6.19) is a differential equation which involves a stochastic variable $F_r(t)$. Such an equation is called the Langevin equation. To define the Langevin equation, we need to specify the statistical property of the stochastic variable $F_r(t)$. This is done in the following section.

6.2.2   Time correlation of random forces

To specify the statistical property of the random force $F_r(t)$, let us consider the average $\langle F_r(t) \rangle$ and the time correlation $\langle F_r(t)F_r(t') \rangle$. It is natural to assume that the average of the random force is zero, i.e.,

$$\langle F_r(t) \rangle = 0 \tag{6.20}$$

Equation (6.20) is a necessary condition for eq. (6.19) to be valid; if it is not satisfied, eq. (6.19) gives the incorrect result that the average velocity $\langle v(t) \rangle$ of a free particle is non-zero.

Next, let us consider the time correlation of the random force. The random force $F_r(t)$ arises from the random collision of fluid molecules on the Brownian particle. Therefore the correlation of the random forces is expected to decay very quickly; the correlation time is of the order of a molecular collision time, i.e., psec. Therefore, on macroscopic time-scales, we may assume the following form for the time correlation function

$$\langle F_r(t_1)F_r(t_2) \rangle = A\delta(t_1 - t_2) \tag{6.21}$$

where A is a constant to be determined.

To determine A, we calculate $\langle v(t)^2 \rangle$ by eq. (6.19), and use the condition that it must be equal to $k_B T/m$ at equilibrium. Equation (6.19) is solved for $v(t)$ as

$$v(t) = \frac{1}{m}\int_{-\infty}^{t} dt_1 e^{-(t-t_1)/\tau_v}F_r(t_1) \tag{6.22}$$

Therefore $\langle v(t)^2 \rangle$ is calculated as

$$\langle v(t)^2 \rangle = \frac{1}{m^2}\left\langle \int_{-\infty}^{t} dt_1 e^{-(t-t_1)/\tau_v}F_r(t_1) \int_{-\infty}^{t} dt_2 e^{-(t-t_2)/\tau_v}F_r(t_2) \right\rangle$$
$$= \frac{1}{m^2}\int_{-\infty}^{t} dt_1 \int_{-\infty}^{t} dt_2 e^{-(t-t_1)/\tau_v}e^{-(t-t_2)/\tau_v}\langle F_r(t_1)F_r(t_2) \rangle \tag{6.23}$$

The last integral in eq. (6.23) can be calculated by use of eq. (6.21) as

$$\langle v(t)^2 \rangle = \frac{A}{m^2}\int_{-\infty}^{t} dt_1 e^{-2(t-t_1)/\tau_v} = \frac{A\tau_v}{2m^2} \tag{6.24}$$

Therefore the condition $\langle v(t)^2 \rangle = k_B T/m$ gives

$$A = \frac{2mk_B T}{\tau_v} = 2\zeta k_B T \tag{6.25}$$

Hence eq. (6.21) can be written as

$$\langle F_r(t_1)F_r(t_2) \rangle = 2\zeta k_B T\delta(t_1 - t_2) \tag{6.26}$$

Equation(6.26) indicates that the fluctuations of the forces exerted on the particle by the fluid are expressed by ζ, the parameter which represents the dissipative force of the fluid exerted on the particle. Such a relation exists generally and is called the fluctuation–dissipation relation. This will be discussed further in a later part of this chapter.

Notice that the time correlation of $F_r(t)$ is independent of the particle mass m. This is reasonable since $F_r(t)$ is determined by fluid molecules alone and thus its statistical properties should be independent of the mass of the particle.

Equations (6.20) and (6.26) are not sufficient to specify the statistical properties of the random force completely. For a complete specification, we have to make further assumptions. Since the random force is a sum of the forces exerted by many fluid molecules near the particle, its distribution is expected to be Gaussian.[2] Therefore we may assume that $F_r(t)$ is a Gaussian random variable whose mean and variances are given by eqs. (6.20) and (6.26). This completes the specification of $F_r(t)$ and defines the Langevin equation.

[2] This is the result of the central limit theorem discussed in Section 3.2.2.

6.2.3   Einstein relation

Next, let us calculate the time correlation of the particle velocity $\langle v(t)v(t') \rangle$. Using eqs. (6.22) and (6.26), we have

$$\langle v(t)v(t') \rangle = \frac{1}{m^2} \int_{-\infty}^{t} dt_1 \int_{-\infty}^{t'} dt_2 e^{-(t-t_1)/\tau_v} e^{-(t'-t_2)/\tau_v} 2\zeta k_B T \delta(t_1 - t_2)$$

(6.27)

The integral has to be done separately depending on whether t is larger than t' or not. If $t > t'$, the integral for t_2 is done first, and the result is

$$\langle v(t)v(t') \rangle = \frac{1}{m^2} \int_{-\infty}^{t} dt_1 e^{-(t-t_1)/\tau_v} e^{-(t'-t_1)/\tau_v} 2\zeta k_B T = \frac{k_B T}{m} e^{-(t-t')/\tau_v}$$

(6.28)

If $t < t'$, the integral for t_1 is done first, and we eventually have the following result;

$$\langle v(t)v(t') \rangle = \frac{k_B T}{m} e^{-|t-t'|/\tau_v}$$

(6.29)

Notice that eq. (6.29) satisfies the general condition that the autocorrelation function is an even function of time.

The mean square displacement $\langle (x(t) - x(0))^2 \rangle$ can be calculated by eq. (6.13):

$$\langle (x(t) - x(0))^2 \rangle = 2 \int_0^t dt_1 \int_0^{t_1} dt_2 \langle v(t_2)v(0) \rangle$$

(6.30)

The right-hand side can be calculated easily using eq. (6.29). To simplify the calculation, we consider the case that the velocity correlation time τ_v is negligibly small in the time-scale under consideration, i.e., $\tau_v \ll t$. In this case, the characteristic value of t_1 in eq. (6.30) is much larger than τ_v. Therefore, the integral for t_2 in eq. (6.30) can be approximated as

$$\int_0^{t_1} dt_2 \langle v(t_2)v(0)\rangle \approx \int_0^\infty dt_2 \langle v(t_2)v(0)\rangle \tag{6.31}$$

The right-hand side is a constant, and is written as

$$D = \int_0^\infty dt \langle v(t)v(0)\rangle \tag{6.32}$$

From eqs. (6.30) and (6.31), the mean square displacement is calculated as

$$\langle (x(t) - x(0))^2 \rangle = 2D|t| \qquad \text{for} \quad |t| \gg \tau_v \tag{6.33}$$

The constant D is called the self-diffusion constant.[3]

From eqs. (6.29) and (6.32), we have

$$D = \frac{k_B T}{m} \tau_v \tag{6.34}$$

or by eq. (6.18)

$$D = \frac{k_B T}{\zeta} \tag{6.35}$$

Equation (6.35) is called the Einstein relation. The Einstein relation is another example of the fluctuation–dissipation relation. In this case, the fluctuation of particle position, which is represented by D, is related to the friction constant ζ. The general form of the Einstein relation is given in Section 6.5.3.

In the above argument, we have made the assumption that $t \gg \tau_v$. This assumption is usually satisfied in soft matter. The velocity correlation time τ_v is very short compared with the time-scale of soft matter. For example, for a typical colloidal particle of radius $a = 0.1\,\mu\text{m}$, τ_v is about 10 ns, while the characteristic relaxation time of colloidal particles is of the order of ms. Therefore, in the following discussion, we assume that τ_v is infinitesimally small. In this limit, the velocity correlation function can be written as

$$\langle v(t)v(t')\rangle = 2D\delta(t - t') \tag{6.36}$$

The coefficient of the delta function must be equal to $2D$ by the relation (6.32).[4]

[3] The self-diffusion constant is different from the usual diffusion constant which appears in the diffusion equation for particle concentration. The latter, called the collective diffusion constant, agrees with the self-diffusion constant in dilute solutions, but is generally different from the self-diffusion constant. This will be discussed in more detail in Chapter 7.

[4] Since the delta function is an even function, we have

$$\int_0^\infty dt\,\delta(t) = (1/2) \int_{-\infty}^\infty dt\,\delta(t) = 1/2.$$

6.3   Brownian motion in a potential field

6.3.1   Langevin equation for the particle coordinate

So far, we have been considering the Brownian motion of a free particle. Let us now consider the Brownian motion of a particle in a potential field $U(x)$. Since the particle is subject to the potential force $-\partial U/\partial x$, the Langevin equation is now written as

$$m\ddot{x} = -\zeta\dot{x} - \frac{\partial U}{\partial x} + F_r(t) \tag{6.37}$$

We may assume that the random force is given by the same Gaussian variable for the free particle. This is because the random force is determined by the thermal motion of fluid molecules and will not be affected by the potential $U(x)$ acting on the particle.

As shown above, the time $\tau_v = m/\zeta$ is very small compared with the characteristic time of the Brownian particle. Therefore we shall consider the limit $\tau_v \to 0$. In this limit, we may ignore the inertia term $m\ddot{x}$ and write eq. (6.37) as

$$-\zeta\dot{x} - \frac{\partial U}{\partial x} + F_r(t) = 0 \qquad (6.38)$$

or

$$\frac{dx}{dt} = -\frac{1}{\zeta}\frac{\partial U}{\partial x} + v_r(t) \qquad (6.39)$$

where $v_r(t) = F_r(t)/\zeta$ is the random velocity. The time correlation function for $v_r(t)$ is calculated as

$$\langle v_r(t)v_r(t')\rangle = \frac{1}{\zeta^2}\langle F_r(t)F_r(t')\rangle = \frac{2k_BT}{\zeta}\delta(t-t') = 2D\delta(t-t') \quad (6.40)$$

where eqs. (6.26) and (6.35) have been used. Thus $v_r(t)$ in eq. (6.39) is a Gaussian random variable characterized by

$$\langle v_r(t)\rangle = 0, \qquad \langle v_r(t)v_r(t')\rangle = 2D\delta(t-t') \qquad (6.41)$$

The correlation of the fluctuating velocity $v_r(t)$ is equal to the correlation of the velocity $v(t)$ of the free particles (see eq. (6.36)). This has to be so since the fluctuating velocity should be independent of the potential $U(x)$, and eq. (6.39) has to be valid for the case $U = 0$.

In the following, we shall use the Langevin equation defined by eqs. (6.39) and (6.41).

6.3.2   Brownian motion in a harmonic potential

As an application of the Langevin equation, let us discuss the Brownian motion of a particle bound by a harmonic potential

$$U(x) = \frac{1}{2}kx^2 \qquad (6.42)$$

The Langevin equation (6.39) now becomes

$$\frac{dx}{dt} = -\frac{k}{\zeta}x + v_r(t) = -\frac{x}{\tau} + v_r(t) \qquad (6.43)$$

where

$$\tau = \frac{\zeta}{k} \qquad (6.44)$$

Since the Langevin equation (6.43) for $x(t)$ has the same structure as that for velocity $v(t)$ (see eq. (6.19)), the time correlation function $\langle x(t)x(0)\rangle$ can be obtained in the same way as for $\langle v(t)v(0)\rangle$. Here we show a slightly different derivation for later purposes.

Let x_0 be the position of the particle at time $t = 0$:

$$x_0 = \int_{-\infty}^{0} dt_1 e^{-(t-t_1)/\tau} v_r(t_1) \tag{6.45}$$

Then eq. (6.43) is solved as

$$x(t) = x_0 e^{-t/\tau} + \int_{0}^{t} dt_1 e^{-(t-t_1)/\tau} v_r(t_1) \tag{6.46}$$

Multiplying both sides of eq. (6.46) by x_0 and taking the average, we have

$$\langle x(t)x(0)\rangle = \langle x_0^2\rangle e^{-t/\tau} + \int_{0}^{t} dt_1 e^{-(t-t_1)/\tau} \langle x_0 v_r(t_1)\rangle \tag{6.47}$$

The second term on the right-hand side is equal to zero due to eqs. (6.41) and (6.45). The distribution of x_0 is given by the Boltzmann distribution, and is proportional to $\exp(-kx_0^2/2k_BT)$. Therefore

$$\langle x_0^2\rangle = \frac{k_BT}{k} \tag{6.48}$$

Equations (6.47) and (6.48) give

$$\langle x(t)x(0)\rangle = \frac{k_BT}{k} e^{-|t|/\tau} \tag{6.49}$$

Here we have replaced t by $|t|$ to account for the case of $t < 0$.

From eq. (6.49), the mean square displacement of the particle is calculated as

$$\langle (x(t) - x(0))^2\rangle = 2\left[\langle x(0)^2\rangle - \langle x(t)x(0)\rangle\right] = \frac{2k_BT}{k}\left(1 - e^{-|t|/\tau}\right) \tag{6.50}$$

The velocity correlation function $\langle \dot{x}(t)\dot{x}(0)\rangle$ is calculated using the formula (6.12), and the result is[5]

$$\langle \dot{x}(t)\dot{x}(0)\rangle = 2D\delta(t) - \frac{D}{\tau}e^{-|t|/\tau} \tag{6.52}$$

The mean square displacement $\langle (x(t)-x(0))^2\rangle$ and the velocity correlation function $\langle \dot{x}(t)\dot{x}(0)\rangle$ are shown in Fig. 6.3(a) and (b), respectively. The velocity correlation function $\langle \dot{x}(t)\dot{x}(0)\rangle$ has a sharp peak at $t = 0$ (represented by the delta function), followed by a slowly decaying negative correlation. The sharp peak arises from the noise term in the Langevin equation, and the slowly decaying part is a result of the restoring force of the harmonic spring. It is important to note that the short-time behaviour of the time correlation functions $\langle (x(t) - x(0))^2\rangle$ and $\langle \dot{x}(t)\dot{x}(0)\rangle$ are the same as that for the free Brownian particle. For short time, eqs. (6.50) and (6.52) are approximated as

$$\langle (x(t) - x(0))^2\rangle = 2D|t|, \qquad \langle \dot{x}(t)\dot{x}(0)\rangle = 2D\delta(t) \qquad \text{for} \quad t \ll \tau \tag{6.53}$$

[5] To get this result, it is convenient to write $e^{-|t|/\tau}$ using the step function $\Theta(t)$ which is equal to 1 for $t > 0$, and 0 for $t < 0$ as

$$e^{-|t|/\tau} = e^{t/\tau}\Theta(-t) + e^{-t/\tau}\Theta(t) \tag{6.51}$$

The derivative of this expression can be calculated using the formula $d\Theta(t)dt = \delta(t)$.

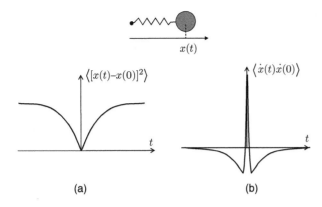

Fig. 6.3 (a) Time correlation function for the position $x(t)$ of a Brownian particle bound by a harmonic spring. (b) Time correlation function for the velocity $\dot{x}(t)$ of the same particle. Here the very narrow peak at the centre is given by the delta function $2D\delta(t)$ or, more precisely by $(k_BT/m)e^{-|t|/\tau_v}$.

These equations are equal to eqs. (6.33) and (6.36), respectively. This again demonstrates that the short-time behaviour is dominated by the stochastic term $v_r(t)$ in eq. (6.39), and is independent of the potential $U(x)$.

6.4   Brownian motion of particles of general shape

6.4.1   Langevin equation for particle configuration

Let us now consider the Brownian motion of a particle of general shape (say the rod-like particle shown in Fig. 6.1(b)), or a collection of such particles. The configuration of such systems is described by a set of generalized coordinates $x = (x_1, x_2, \ldots, x_f)$. For example, the configuration of a rigid particle is described by six coordinates, three representing the position, and three representing the orientation.

Suppose that such particles are moving in a potential field $U(x)$. The Langevin equation for this situation is obtained by generalizing the argument in the previous section. The Langevin equation (6.38) represents the balance of three forces, the potential force $F_p = -\partial U/\partial x$, the frictional force $F_f = -\zeta\dot{x}$, and the random force $F_r(t)$.

In the general case, the potential force is defined by the change of the potential $U(x)$ when the particle is moved by dx_i:

$$dU = -\sum_i F_{pi}dx_i \qquad (6.54)$$

therefore

$$F_{pi} = -\frac{\partial U}{\partial x_i} \qquad (6.55)$$

F_{pi} is called the generalized potential force conjugate to x_i. Notice that F_{pi} represents various 'forces' depending on x_i. If x_i represents the position vector of the particle, F_{pi} represents the usual force acting on the particle. On the other hand, if x_i represents the angular coordinate of the particle, F_{pi} represents the torque.

Likewise, the frictional force F_{fi} is generally defined by the irreversible work W_{irr} done to the fluid per unit time when the particle is moving with velocity $\dot{x} = (\dot{x}_1, \dot{x}_2, \ldots)$:

$$W_{irr} = \sum_i F_{fi}\dot{x}_i \tag{6.56}$$

As shown in Appendix D.1, F_{fi} is a linear function of $\dot{x}$, and is written as

$$F_{fi} = -\sum_j \zeta_{ij}(x)\dot{x}_j \tag{6.57}$$

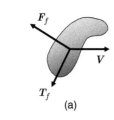

where $\zeta_{ij}(x)$ is the generalized friction coefficient Notice that the frictional force $\boldsymbol{F}_{fi}$ conjugate to x_i generally depends on the velocity of the other coordinates $\dot{x}_j$. This can be seen in the example shown in Fig. 6.4(a). When a bean-like particle moves in the x-direction, the frictional force $\boldsymbol{F}_f$ will not be parallel to the velocity vector $\boldsymbol{V}$: the motion in the x-direction induces a frictional force in the y- and z-directions. Also the translational motion of such a particle induces a frictional torque $\boldsymbol{T}_f$. These frictional forces and torques can be calculated by hydrodynamics, and they are written in the form of eq. (6.57). A detailed discussion of how to calculate F_{fi} is given in Appendix D.1.

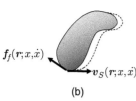

The Langevin equation for the generalized coordinates of the particle is therefore given by

$$-\sum_j \zeta_{ij}\dot{x}_j - \frac{\partial U}{\partial x_i} + F_{ri}(t) = 0 \tag{6.58}$$

Fig. 6.4 (a) The frictional force $\boldsymbol{F}_f$ and the torque $\boldsymbol{T}_f$ exerted on a particle moving with velocity $\boldsymbol{V}$ in a viscous fluid. (b) When the generalized coordinates of a particle change at rate $\dot{x}$, the point located at $\boldsymbol{r}$ on the surface of the particle moves with a certain velocity $\boldsymbol{v}_S(\boldsymbol{r}; x, \dot{x})$. The hydrodynamic force $\boldsymbol{f}_f(\boldsymbol{r}; x, \dot{x})$ acting on this point (per unit area) is calculated by solving the Stokes equation, and the total force $\boldsymbol{F}_f$ and torque $\boldsymbol{T}_f$ are calculated from $\boldsymbol{f}_f(\boldsymbol{r}; x, \dot{x})$.

The last term on the left-hand side of eq. (6.58) represents the random force. As in the case of a point particle, the random force $F_{ri}(t)$ is shown to satisfy the following equation

$$\langle F_{ri}(t)\rangle = 0, \qquad \langle F_{ri}(t)F_{rj}(t')\rangle = 2\zeta_{ij}k_BT\delta(t-t') \tag{6.59}$$

This is a generalization of eq. (6.26). A derivation of these equations is given in Appendix D.3.

6.4.2 Reciprocal relation

The friction coefficient $\zeta_{ij}(x)$ defined above has the following properties:

(i) $\zeta_{ij}(x)$ is symmetric

$$\zeta_{ij}(x) = \zeta_{ji}(x) \tag{6.60}$$

(ii) The matrix of $\zeta_{ij}(x)$ is positive definite, i.e., for any $\dot{x}_i$,

$$\sum_{i,j} \zeta_{ij}(x)\dot{x}_i\dot{x}_j \geq 0 \tag{6.61}$$

Equation (6.60) is known as the Lorentz reciprocal relation in hydrodynamics. The reciprocal relation is not a trivial relation. For example, consider the situation shown in Fig. 6.5. When a particle moves in the x-direction with velocity V_x, the hydrodynamic drag generally involves a y-component F_{fy} (Fig. 6.5(a)), and when the particle moves in the y-direction with velocity V_y, the hydrodynamic drag involves an x-component F_{fx} (Fig. 6.5(b)). The reciprocal relation indicates that F_{fy}/V_x is equal to F_{fx}/V_y, which is not trivial. A more striking example is seen for a particle of helical shape. For such a particle, the translational velocity V creates a frictional torque T_f (Fig. 6.5(c)), and the rotational velocity ω creates a frictional force F_f. The reciprocal relation gives the identity $T_f/V = F_f/\omega$, which is again very non-trivial.

The reciprocal relation can be proven by hydrodynamics using the definition of the friction matrix (see Appendix D.1).

Onsager has shown that the reciprocal relation can be proven without invoking hydrodynamics.[6] The essential point of his proof is the time reversal symmetry of the time correlation function at equilibrium (eq. (6.5)). Since the friction coefficient ζ_{ij} depends on the particle configuration x only, and is independent of the potential $U(x)$, we may consider the special case that the particles are subject to a harmonic potential

$$U(x) = \frac{1}{2}\sum_i k_i(x_i - x_i^0)^2 \qquad (6.62)$$

and that the particle configuration cannot deviate significantly from the equilibrium configuration $x^0 = (x_1^0, x_2^0, \ldots)$. In this case, eq. (6.58) can be solved for the deviation $\xi_i(t) = x_i(t) - x_i^0$. Then the requirement that the time correlation function $\langle \xi_i(t)\xi_j(0)\rangle$ is symmetric with respect to the exchange of i and j gives the reciprocal relation (6.60). A detailed explanation is given in Appendix D.3.

[6] Onsager actually derived the reciprocal relation for the kinetic coefficients in irreversible thermodynamics, but the essence of his proof is the same as that given here. Here his proof is presented in a form adapted to Brownian particles.

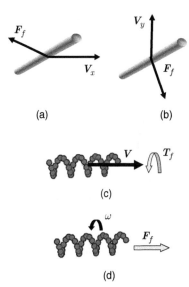

Fig. 6.5 Hydrodynamic reciprocal relation. See the text for explanation.

6.5   Fluctuation–dissipation theorem

6.5.1   Fluctuations and material parameters

We have seen that the time correlation of random forces at equilibrium is expressed in terms of the friction constant (eqs. (6.26) and (6.59)). The random force is the fluctuating part of the force that the fluid exerts on the particle when the system is at equilibrium. The friction constant represents the mean force that the fluid exerts on the particle when the particle is moving. In general the fluctuations taking place at equilibrium can be related to the response of the system when external parameters are changed. Such a relation is generally called a fluctuation–dissipation relation, and the existence of such a relation is called the fluctuation–dissipation theorem.

The fluctuation–dissipation relation is a general result in statistical mechanics and can be proven starting from Hamilton's equations of motion in molecular systems. In fact eqs. (6.26) and (6.35) are special cases of such a general relation. In this section, we explain the general form of the fluctuation–dissipation relation, and discuss its relevance in the theory of Brownian motion.

6.5.2 Time correlation of random forces

Statistical mechanics regards any system as a dynamical system consisting of molecules which obey Hamilton's equations of motion. If we know the Hamiltonian of the system, the time evolution of the system can be calculated by solving Hamilton's equations of motion.

The Hamiltonian of the system can be written as $H(\Gamma; x)$, where Γ represents the set of coordinates specifying the positions and momenta of all molecules in the system, and $x = (x_1, x_2, \ldots, x_n)$ represents external parameters, the parameters which appear in the Hamiltonian (thus affecting the motion of the molecules, but can be controlled externally).

The fluctuation–dissipation theorem can be stated for this general definition of x. However, to make the relevance of the theorem clear, we explain the theorem for the system of Brownian particles.

We consider a big particle (or particles) placed in a fluid, and regard Γ as the coordinates specifying the positions and the momenta of the fluid molecules, and $x = (x_1, x_2, \ldots, x_f)$ as the parameters specifying the configuration of the Brownian particles (see Fig. 6.6). Notice that we are considering the system made of fluid molecules only, and the configurational parameters x of the Brownian particles are regarded as external parameters in the Hamiltonian for fluid molecules.

Now the forces acting on the Brownian particles are given by

$$\hat{F}_i(\Gamma; x) = -\frac{\partial H(\Gamma; x)}{\partial x_i} \tag{6.63}$$

$\hat{F}_i(\Gamma; x)$ represents the instantaneous force acting on the particle by fluid molecules which are in the microscopic state Γ.[7] The macroscopic force at time t is the average of $\hat{F}_i(\Gamma; x)$ for the distribution function $\psi(\Gamma; x, t)$

$$\langle F_i(t) \rangle = \int d\Gamma \, \hat{F}_i(\Gamma; x) \psi(\Gamma; x, t) \tag{6.64}$$

If the fluid molecules are at equilibrium for a given value of x, the average force is calculated by

$$\langle F_i \rangle_{eq, x} = \int d\Gamma \, \hat{F}_i(\Gamma; x) \psi_{eq}(\Gamma; x) \tag{6.65}$$

where ψ_{eq} is the equilibrium distribution function for the fluid molecules. This is given by the canonical distribution:

$$\psi_{eq}(\Gamma; x) = \frac{e^{-\beta H(\Gamma; x)}}{\int d\Gamma \, e^{-\beta H(\Gamma; x)}} \tag{6.66}$$

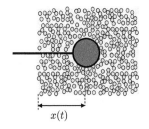

$x(t)$

Fig. 6.6 If the position of a large particle placed in a fluid can be controlled externally, the position coordinate x of the particle can be regarded as an external parameter for fluid molecules.

[7] For example, in the example shown in Fig. 6.6, the Hamiltonian involves the interaction potential $U_B(\mathbf{R}, \{\mathbf{r}_i\})$ between the Brownian particle (located at $\mathbf{R}$), and fluid molecules (located at $\{\mathbf{r}_i\} = (\mathbf{r}_1, \mathbf{r}_2, \ldots)$). In this case $-\partial H/\partial \mathbf{R} = -\partial U_B/\partial \mathbf{R}$ stands for the force exerted on the Brownian particle by the fluid molecules.

It is easy to show that $\langle F_i \rangle_{eq,x}$ is given by the partial derivative of the free energy

$$A(x) = -k_B T \ln \int d\Gamma\, e^{-\beta H(\Gamma : x)} \tag{6.67}$$

as[8]

$$\langle F_i \rangle_{eq,x} = -\frac{\partial A(x)}{\partial x_i} \tag{6.68}$$

Now consider that at time $t = 0$, the system was at equilibrium for given values of the external parameters $x(0)$, and that $x(t)$ changes slowly with velocity $\dot{x}_i(t)$ for $t > 0$. The average force $\langle F_i(t) \rangle_{x+\delta x}$ acting on the particle at time t can be calculated by solving the time evolution equation for the distribution function $\psi(\Gamma; x, t)$ (i.e., the Liouville equation in statistical mechanics). If the deviation $\delta x_i(t) = x_i(t) - x_i(0)$ is small, the result can be written in the following form (see Appendix E):

$$\langle F_i(t) \rangle_{x+\delta x} = -\frac{\partial A(x)}{\partial x_i} - \int_0^t dt' \sum_j \alpha_{ij}(x; t-t')\dot{x}_j(t') \tag{6.69}$$

The first term on the right-hand side represents the equilibrium force F_{eq} when x_i is fixed at $x_i(0)$, and the second term represents the first-order correction due to the time dependence of $x_i(t)$. It can be proven that the function $\alpha_{ij}(x; t)$ is related to the time correlation function of the fluctuating force $\hat{F}_{ri}(t)$ at the initial equilibrium state where $x(t)$ is fixed at the initial value $x(0)$, i.e.,

$$\alpha_{ij}(x; t) = \frac{1}{k_B T}\langle F_{ri}(t)F_{rj}(0) \rangle_{eq,x} \tag{6.70}$$

and

$$F_{ri}(t) = \hat{F}_i(\Gamma_t; x) - \langle F_i \rangle_{eq,x} = \hat{F}_i(\Gamma_t; x) + \frac{\partial A(x)}{\partial x_i} \tag{6.71}$$

where Γ_t is the value of the microscopic state variable Γ at time t.

Notice that $F_{ri}(t)$ changes with time since Γ_t changes with time.[9]

Equation (6.70) is a generalization of eq. (6.26). Suppose that a particle placed in a viscous fluid starts to move with constant velocity v at $t = 0$ (see Fig. 6.7(a)). Then eq. (6.69) indicates that the average force acting on the particle is given by

$$\langle F(t) \rangle_{x+\delta x} = -\int_0^t dt'\alpha(x; t-t')v = -v\int_0^t dt'\alpha(x; t') \tag{6.72}$$

(Notice that the free energy of the fluid is independent of x, and therefore $\partial A/\partial x$ is equal to zero.) On the other hand, the phenomenological friction law indicates that the viscous force F_f acting on the particle is given by $-\zeta v$. Comparing this with eq. (6.72), we have

$$\int_0^t dt'\alpha(x; t') = \zeta(x) \qquad \text{for} \quad t > 0 \tag{6.73}$$

[8] An example of $\langle F_i \rangle_{eq,x}$ is the buoyancy force acting on the particle.

[9] According to the derivation shown in Appendix E, x in eqs. (6.71) and (6.75) must be $x(0)$, i.e., the external parameter at $t = 0$. However since the change of $x(t) - x(0)$ is small, x can be interpreted as $x(t)$, the current external parameter at time t.

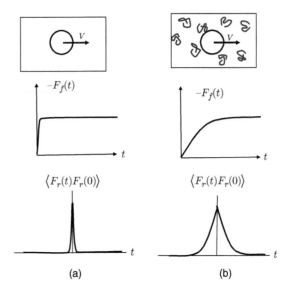

Fig. 6.7 (a) If a particle starts to move with constant velocity v in a viscous medium, the frictional force immediately reaches the steady state value $-\zeta v$. Therefore the time correlation function of the random force has the form of a delta function. (b) If the particle moves in a polymer solution, the frictional force will exhibit transient behaviour due to the viscoelasticity of the polymer solution. Accordingly, the time correlation function of the random force in a polymer solution will have a finite correlation time.

For this to be true for any time t, $\alpha(t)$ must be written as

$$\alpha(x;t) = 2\zeta(x)\delta(t) \tag{6.74}$$

Therefore eq. (6.70) gives the relation

$$\langle F_r(t)F_r(0)\rangle_{eq,x} = 2\zeta(x)k_BT\delta(t) \tag{6.75}$$

which is equivalent to eq. (6.26)

If the fluid surrounding the particle is a viscoelastic fluid, such as a polymer solution, the frictional force $F_f(t)$ behaves as shown in Fig. 6.7(b), i.e., $F_f(t)$ approaches the steady state value with some delay time. In this case, $\alpha(t)$ also has a delay time. Equation (6.70) then indicates that the fluctuating force acting on the particle has a finite correlation time as shown in Fig. 6.7(b).

6.5.3   Generalized Einstein relation

We now discuss a special case of the fluctuation–dissipation relation. This gives a generalization of the Einstein relation, and relates the time correlation function of fluctuations to the system response to external fields.

To derive such a relation, we now regard the Brownian particles as a part of the system. Therefore Γ now represents the positions and momenta of fluid molecules and Brownian particles. We assume that external fields are applied to the system for $t > 0$, and that the Hamiltonian of the system for $t > 0$ is given by

$$H(\Gamma; h) = H_0(\Gamma) - \sum_i h_i(t)\hat{M}_i(\Gamma) \tag{6.76}$$

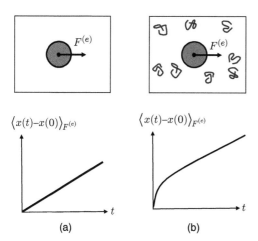

$\langle x(t)-x(0)\rangle_{F^{(e)}}$

$\langle x(t)-x(0)\rangle_{F^{(e)}}$

(a) (b)

Fig. 6.8 (a) Top: a constant external force $F^{(e)}$ begins to be applied to the particle placed in a Newtonian fluid. Bottom: the particle displacement caused by the force. (b) The same situation as in the left, but the particle is now moving in a polymer solution. Due to the viscoelasticity of polymer solutions, transient behaviour is seen in the particle displacement.

Here we have changed the notation; the external parameters are now denoted by $h_i(t)$, and the corresponding conjugate quantities are denoted by $\hat{M}_i(\Gamma)$:

$$\hat{M}_i(\Gamma) = -\frac{\partial H}{\partial h_i} \tag{6.77}$$

This is to make the meaning of $h_i(t)$ and $\hat{M}_i(\Gamma)$ clear. Usually, $h_i(t)$ represents some external force, and $\hat{M}_i(\Gamma)$ represents the quantity conjugate to h_i. For example, if an external force $F^{(e)}$ is applied to a particle (see Fig. 6.8), the potential energy of the particle will include a term $-xF^{(e)}$, where x is the x-coordinate of the particle. In this case h_i and $\hat{M}_i(\Gamma)$ correspond to $F^{(e)}$ and x, respectively. Another example is that of a magnetic field H applied to a magnetic particle having a permanent magnetic moment μ_m (see Fig. 6.9). The magnetic field creates a potential energy $-\mu_m H \cos\theta$, with θ being the angle between the magnetic moment and the field. In this case, h_i and $\hat{M}_i(\Gamma)$ correspond to H and $\mu_m \cos\theta$, respectively.

As in the previous section, we consider the situation that the system was at equilibrium at $t = 0$ with zero external fields, and that step fields h_i are applied for $t \geq \epsilon$ (ϵ being a small positive time), i.e.,

$$h_i(t) = h_i\Theta(t - \epsilon) \quad \text{or} \quad \dot{h}_i(t) = h_i\delta(t - \epsilon) \tag{6.78}$$

We define the response function $\chi_{ij}(t)$ for this situation

$$\langle M_i(t) - M_i(0)\rangle_h = \sum_j \chi_{ij}(t)h_j \tag{6.79}$$

Then it can be proven (see Appendix E) that the response function $\chi_{ij}(t)$ is related to the correlation function of the displacement:

$$\chi_{ij}(t) = \frac{1}{2k_BT}\langle [M_i(t) - M_i(0)][M_j(t) - M_j(0)]\rangle_{eq,0} \tag{6.80}$$

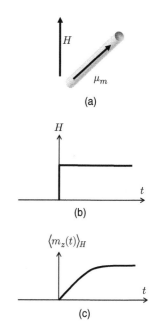

H

μ_m

(a)

H

(b)

$\langle m_z(t)\rangle_H$

(c)

Fig. 6.9 (a) A rod-like particle having a permanent magnetic moment μ_m is placed in a magnetic field. When a constant magnetic field is applied for $t > 0$ as in (b), the average magnetic moment changes as shown in (c).

Equations (6.79) and (6.80) give a generalization of the Einstein relation. Suppose that a constant force $F^{(e)}$ is applied to a Brownian particle in a solution at time $t = 0$. Then the particle starts to move with constant speed $v = F^{(e)}/\zeta$. Therefore the average displacement of the particle $\langle x(t) - x(0) \rangle_{F^{(e)}}$ is given by $F^{(e)}t/\zeta$ and $\chi(t) = t/\zeta$ (see Fig. 6.8(a)). Equation (6.80) therefore indicates

$$\frac{t}{\zeta} = \frac{1}{2k_B T} \left\langle (x(t) - x(0))^2 \right\rangle_{eq,0} \tag{6.81}$$

On the other hand, the average on the right-hand side of eq. (6.81) is expressed by the self-diffusion constant D as

$$\left\langle (x(t) - x(0))^2 \right\rangle_{eq,0} = 2Dt \tag{6.82}$$

Equations (6.81) and (6.82) give the Einstein relation (6.35).

If the particle is moving in a polymer solution, the response function will be as shown in Fig. 6.8(b): the particle first moves quickly and then slows down. Accordingly, the mean square displacement of the particle will behave as shown in the bottom of Fig. 6.8(b). Therefore eq. (6.80) is a generalization of the Einstein relation.

A similar relation holds for the response of rod-like particles (Fig. 6.9). Analysis of this case will be given in the next chapter.

6.6   Summary of this chapter

Brownian motion of small particles in a fluid is a manifestation of fluctuations of physical quantities viewed on a molecular scale. The time correlation functions of fluctuations at equilibrium have the time reversal symmetry described by eq. (6.7).

The Brownian motion of particles can be described by the Langevin equation (6.58) which involves fluctuating forces $F_{ri}(t)$. The time correlation functions of the fluctuating forces are related to the friction matrix ζ_{ij}. Such relation is called a fluctuation–dissipation relation.

The fluctuation–dissipation relation exists generally between the time correlation functions of any physical quantity F_i at equilibrium and the change of the average of F_i when the parameter conjugate to F_i is changed externally. This theorem is called the fluctuation–dissipation theorem. A special case of the fluctuation–dissipation theorem is the Einstein relation which relates the diffusion constant (the quantity that represents the fluctuation of particle position) and the friction constant (the quantity which represents the change of the particle position when an external force is applied).

An important consequence of the fluctuation–dissipation theorem is that the friction matrix ζ_{ij} generally becomes a symmetric matrix. This fact is the basis of the variational principle which will be discussed in subsequent chapters.

Further reading

(1) *Low Reynolds Number Hydrodynamics*, John Happel and Howard Brenner, Springer (1983).

(2) *Reciprocal relations in irreversible processes I, II*, Onsager L. Phys. Rev. **37**, 405–426 (1931), **38** 2265–2279 (1931).

Exercises

(6.1) Consider a colloidal particle of radius $0.1\,\mu m$, density $2\,g\,cm^{-3}$ moving in water (density $1\,g\,cm^{-3}$ viscosity $1\,mPa\cdot s$). Answer the following questions.

(a) Calculate the sedimentation velocity V_s of the particle.

(b) Calculate the diffusion constant D of the particle

(c) Calculate the average distance $V_s t$ that the particle moves in time t, and its fluctuation $\sqrt{(Dt)}$ for $t = 1\,ms, 1\,s, 1\,hour, 1\,day$, and discuss when we need to consider the Brownian motion.

(6.2) Calculate the mean square displacement of the free Brownian particle using eqs. (6.29) and (6.30) and show that eq. (6.33) is correct for $t \gg \tau_v$.

(6.3) The Langevin equation for the Brownian particle in a harmonic potential (6.43) is solved for $x(t)$ as

$$x(t) = \int_{-\infty}^{t} dt_1 e^{-(t-t_1)/\tau} v_r(t_1) \qquad (6.83)$$

Answer the following questions.

(a) Calculate $\langle x(t)^2 \rangle$ using eqs. (6.40) and (6.83) and show that it is equal to $k_B T/k$.

(b) Calculate the time correlation function $\langle x(t)x(t') \rangle$ for $t > t'$, and $t < t'$.

(c) Calculate $\langle x(t)v_r(0) \rangle$ for $t > 0$, $t = 0$, and $t < 0$. (Notice that the time correlation function $\langle x(t)v_r(0) \rangle$ does not satisfy the time reversal symmetry relation (6.7) because $v_r(t) = \dot{x} + (kx/\zeta)$ does not satisfy the relation (6.6).)

(d) Calculate $\langle \dot{x}(t)\dot{x}(0) \rangle$ using eq. (6.43), and show that it is given by eq. (6.52).

(6.4) *The friction constant of a rod-like particle is not isotropic. If the rod moves parallel to the axis, the friction constant is $\zeta_\parallel$, while if the rod moves perpendicular to the axis, the friction constant is $\zeta_\perp$

(see Fig. 6.10). Let $\boldsymbol{u}$ be the unit vector parallel to the axis. Answer the following questions.

(a) The frictional force $F_{f\alpha}$ ($\alpha = x, y, z$) exerted by the fluid on the rod moving with velocity V_α is written as

$$F_{f\alpha} = -\zeta_{\alpha\beta} V_\beta \qquad (6.84)$$

Express $\zeta_{\alpha\beta}$ using $\zeta_\parallel$, $\zeta_\perp$, and $\boldsymbol{u}$, and show that the friction tensor $\zeta_{\alpha\beta}$ is symmetric. (Hint: Use the fact that the frictional forces are given by $-\zeta_\parallel \boldsymbol{V}_\parallel - \zeta_\perp \boldsymbol{V}_\perp$, where $\boldsymbol{V}_\parallel$ and $\boldsymbol{V}_\perp$ are the parallel and perpendicular components of $\boldsymbol{V}$ defined in Fig. 6.10.)

(b) Express the inverse matrix, $(\zeta^{-1})_{\alpha\beta}$, of the matrix $\zeta_{\alpha\beta}$ using $\zeta_\parallel$, $\zeta_\perp$, and u_α.

(c) Let $x_\alpha(t)$ be the α-component of the position vector of the centre of mass of the rod. Equation (6.58) is then written as

$$-\zeta_{\alpha\beta} \frac{dx_\beta}{dt} + F_{r\alpha} = 0 \qquad (6.85)$$

Show that this equation can be written as

$$\frac{dx_\alpha}{dt} = v_{r\alpha}(t) \qquad (6.86)$$

where $v_{r\alpha}(t) = (\zeta^{-1})_{\alpha\beta} F_{r\beta}$. Calculate the time correlation function $\langle v_{r\alpha}(t)v_{r\beta}(t') \rangle$.

(d) Show that if the angular distribution of the rod is isotropic, the mean square displacement of the centre of mass is given by

$$\langle (\boldsymbol{x}(t) - \boldsymbol{x}(0))^2 \rangle = 2k_B T\left(\frac{1}{\zeta_\parallel} + \frac{2}{\zeta_\perp}\right)t \quad (6.87)$$

(e) Calculate the mean square displacement $\langle (\boldsymbol{x}(t) - \boldsymbol{x}(0))^2 \rangle$ for a rod which is preferentially oriented along the z-direction, with the order parameter $S = (3/2)(\langle u_z^2 \rangle - 1/3)$

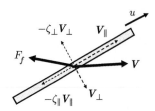

Fig. 6.10

(6.5) Consider the electrical circuit shown in Fig. 6.11(a). The circuit is made of a resistor (resistance R) and a capacitor (capacitance C). The charge Q stored in the capacitor is zero on average, but can be positive or negative due to the thermal fluctuations of charge carriers. Answer the following questions.

(a) The fluctuation of $Q(t)$ is described by the following equation

$$R\dot{Q} + \frac{Q}{C} - \psi_r(t) = 0 \qquad (6.88)$$

where $\psi_r(t)$ is the fluctuating voltage. Comparing eq. (6.88) with eq. (6.58), show that the fluctuating voltage satisfies the following relation (Nyquist theorem)

$$\langle \psi_r(t)\psi_r(0) \rangle = 2Rk_BT\delta(t) \qquad (6.89)$$

(b) Calculate the time correlation functions $\langle Q(t)Q(0) \rangle$ and $\langle I(t)I(0) \rangle$, where $I(t) = \dot{Q}(t)$ is the current flowing in the resistor.

(6.6) Consider an ohmic material connected to many electrodes attached on its surface as in Fig. 6.11(b). Let ψ_i be the voltage of the electrode i, and I_i be the electric current flowing in the resistor at the electrode i. The relation between ψ_i and I_i is obtained as follows. Let $\psi(\mathbf{r})$ be the electric potential at point $\mathbf{r}$ in the resistor. The electric current density $\mathbf{j}(\mathbf{r})$ is given by $\mathbf{j} = -\sigma\nabla\psi$, where $\sigma(\mathbf{r})$ is the conductivity at point $\mathbf{r}$. Since the current density has to satisfy the charge conservation equation $\nabla \cdot \mathbf{j} = 0$, the potential $\psi(\mathbf{r})$ satisfies the following equation:

$$\nabla(\sigma\nabla\psi) = 0 \qquad (6.90)$$

If this equation is solved under the boundary condition

$$\psi(\mathbf{r}) = \psi_i \qquad i = 1, 2, \ldots \qquad (6.91)$$

the current I_i is obtained as

$$I_i = \int_{\text{electrode } i} dS\mathbf{n} \cdot \sigma\nabla\psi \qquad (6.92)$$

where $\mathbf{n}$ is a unit vector normal to the surface. Answer the following questions. (Hint: Follow the argument given in Appendix D.1.)

(a) Show that ψ_i and J_i are related by a linear relation

$$\psi_i = \sum_j R_{ij} J_j \qquad (6.93)$$

The matrix R_{ij} is called the resistance matrix.

(b) Prove that the matrix R_{ij} is symmetric and positive definite.

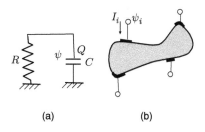

(a)　　　　　　　(b)

Fig. 6.11

(6.7) Consider the equation of motion of a Brownian particle with inertia term included:

$$m\frac{d^2x}{dt^2} = -\zeta\frac{dx}{dt} + F_r(t) + F^{(e)}(t) \qquad (6.94)$$

where $F_r(t)$ and $F^{(e)}(t)$ are the random force and the external force. Answer the following questions.

(a) Calculate the mean displacement of the particle $\langle x(t) - x(0) \rangle_{F^{(e)}}$ when the external force is applied in a step-wise fashion at time $t = 0$, i.e., $F^{(e)}(t) = F^{(e)}\Theta(t)$.

(b) Using the generalized Einstein relation, calculate the mean square displacement of the particle $\langle (x(t) - x(0))^2 \rangle_{eq,0}$.

(c) Calculate the velocity correlation function $\langle \dot{x}(t)\dot{x}(0) \rangle_{eq,0}$ at equilibrium, and confirm that it agrees with eq. (6.29).

(6.8) Consider two Brownian particles connected by a harmonic spring (see Fig. 6.12(a)). The Langevin equations for their positions x_1 and x_2 are written as

$$\zeta\frac{dx_1}{dt} = -k(x_1 - x_2) + F_{r1}(t) \qquad (6.95)$$

$$\zeta\frac{dx_2}{dt} = -k(x_2 - x_1) + F_{r2}(t) \qquad (6.96)$$

Answer the following questions.

(a) Suppose that a constant force $F_1^{(e)}$ is applied to the particle 1 for $t > 0$. Calculate the average displacements of the particles $\langle x_1(t) - x_1(0) \rangle_{F_1^{(e)}}$ and $\langle x_2(t) - x_2(0) \rangle_{F_1^{(e)}}$ in this situation.

(b) Calculate the time correlation function $\langle (x_1(t) - x_1(0))^2 \rangle$ and $\langle (x_1(t) - x_1(0))(x_2(t) - x_2(0)) \rangle$ using the fluctuation–dissipation theorem.

(c) Discuss the behaviour of $\langle (x_1(t) - x_1(0))^2 \rangle$ at short time $(t \ll \zeta/k)$ and at long time $(t \gg \zeta/k)$.

(6.9) *Figure 6.12(b) shows a simple model of a polymer chain. The polymer chain is represented by $N + 1$ beads connected by springs. To calculate the mean square displacement of each bead, we consider the situation that a constant external force $\boldsymbol{F}^{(e)}$ is applied on bead m for $t > 0$. The time evolution equation for the average position $\bar{\boldsymbol{r}}_n = \langle \boldsymbol{r}_n \rangle$ of the n-th bead is written as

$$\zeta \frac{d\bar{\boldsymbol{r}}_n}{dt} = k(\bar{\boldsymbol{r}}_{n+1} - 2\bar{\boldsymbol{r}}_n + \bar{\boldsymbol{r}}_{n-1}) + \boldsymbol{F}^{(e)}\delta_{nm}$$

$$n = 0, 1, \ldots, N \qquad (6.97)$$

where $\boldsymbol{r}_{-1}$ and $\boldsymbol{r}_{N+1}$ are defined by

$$\boldsymbol{r}_{-1} = \boldsymbol{r}_0, \qquad \boldsymbol{r}_{N+1} = \boldsymbol{r}_N \qquad (6.98)$$

If n is regarded as a continuous variable, eqs. (6.97) is written as a partial differential equation

$$\zeta \frac{\partial \bar{\boldsymbol{r}}_n}{\partial t} = k \frac{\partial^2 \bar{\boldsymbol{r}}_n}{\partial n^2} + \boldsymbol{F}^{(e)}\delta(n - m) \qquad (6.99)$$

and eq. (6.98) is written as the boundary conditions

$$\frac{\partial \boldsymbol{r}_n}{\partial n} = 0, \qquad \text{at} \quad n = 0, N \qquad (6.100)$$

Equation (6.99) is solved by using the Green function $G(n, m, t)$ which is a solution of the equation

$$\frac{\partial}{\partial t} G(n, m, t) = \frac{k}{\zeta} \frac{\partial^2}{\partial n^2} G(n, m, t) + \delta(t)\delta(n - m) \qquad (6.101)$$

under the boundary condition

$$\frac{\partial G(n, m, t)}{\partial n} = 0 \qquad \text{at} \quad n = 0, N \qquad (6.102)$$

The solution of eq. (6.99) is written as

$$\bar{\boldsymbol{r}}_n(t) = \frac{\boldsymbol{F}^{(e)}}{\zeta} \int_0^t dt' G(n, m; t - t') \qquad (6.103)$$

Answer the following questions.

(a) Show that the mean square displacement of the segment m is given by

$$\langle (\boldsymbol{r}_m(t) - \boldsymbol{r}_m(0))^2 \rangle_{eq} = \frac{6k_B T}{\zeta} \int_0^t dt' G(m, m, t') \qquad (6.104)$$

(b) Show that at short time, $G(n, m, t)$ is given by

$$G(n, m, t) = \frac{1}{\sqrt{2\pi\lambda t}} \exp\left[-\frac{(n - m)^2}{2\lambda t} \right] \qquad (6.105)$$

where $\lambda = k/\zeta$.

(c) Show that the mean square displacement of a bead at short time is proportional to $\sqrt{t}$.

(d) Show that the mean square displacement of bead n at long time is given by

$$\langle (\boldsymbol{r}_n(t) - \boldsymbol{r}_n(0))^2 \rangle_{eq} = 6\frac{k_B T}{N\zeta} t \qquad (6.106)$$

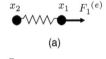

(a)

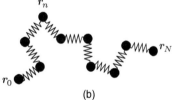

(b)

Fig. 6.12

7 Variational principle in soft matter dynamics

In this chapter, we shall discuss a variational principle which is useful for various problems in soft matter dynamics. The variational principle is most clearly stated for a system consisting of small particles and viscous fluids, e.g., particles moving in viscous fluids, or a fluid moving through a bed of particles. Here the principle essentially states that the time evolution of the system is determined by the balance of two forces, the potential force which drives the system to the state of a potential minimum, and the frictional force which resists this change. This principle is justified by hydrodynamics, especially, the reciprocal relation for the friction coefficient.

Onsager noticed that many phenomenological equations which describe the time evolution of non-equilibrium systems have the same structure as those of particle–fluid systems. He then proposed that since the reciprocal relation is generally guaranteed by the time reversal symmetry of fluctuations, the evolution law of such systems can be stated in the form of a variational principle. We shall call this variational principle the Onsager principle.

The Onsager principle is based on phenomenological equations, and is therefore valid only for a certain class of problems. However, most problems in soft matter (diffusion, flow, and rheology, etc.) belong to this class, and their basic equations are derived from the Onsager principle. In subsequent chapters we shall show this through the analysis of various examples.

In this chapter, we shall first explain the variational principle for particle–fluid systems. We then state the variational principle in a more general form, and show two applications which demonstrate the usefulness of the Onsager principle.

7.1 Variational principle for the dynamics of particle–fluid systems

7.1.1 Motion of particles in viscous fluids

Let us consider a system of particles moving in a viscous fluid driven by potential forces. For example, consider a collection of particles sedimenting under gravity (see Fig. 7.1(a)). The configuration of the particles is represented by a set of parameters $x = (x_1, x_2, \ldots, x_f)$ which may denote, for example, the positions and orientations of all particles. In the previous chapter, we have shown that the time evolution of such configurational parameters is described by the following equation

$$-\sum_j \zeta_{ij}(x)\dot{x}_j - \frac{\partial U(x)}{\partial x_i} = 0 \qquad (7.1)$$

where $\zeta_{ij}(x)$ are the associated friction coefficients, and $U(x)$ is the potential energy of the system. Here we have ignored the random forces assuming that the particles are macroscopic. The first term represents the frictional force (conjugate to x_i) when the particles move with velocity $\dot{x} = (\dot{x}_1, \dot{x}_2, \ldots, \dot{x}_f)$, and the second term represents the potential force. Equation (7.1) states that the time evolution of the system is determined by the condition that these two forces must balance. (The inertia forces are assumed to be negligibly small.)

Notice that eq. (7.1) is a set of nonlinear differential equations for $x(t) = (x_1(t), x_2(t), \ldots, x_f(t))$ since both ζ_{ij} and U are functions of x. Equation (7.1), therefore, can describe rather complex motion. For example, in the situation shown in Fig. 7.1(a), the particle cluster sediments, changing its shape and splitting into smaller clusters. Such complex structural evolution can be described by eq. (7.1).

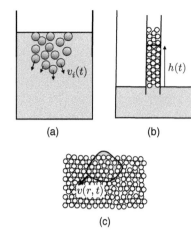

Fig. 7.1 Examples of particle–fluid systems. (a) A collection of particles sedimenting in a quiescent fluid. (b) A fluid rising in a capillary filled with small particles. (c) A fluid drop spreading and wetting dry sand.

7.1.2   Variational principle for particle motion

As shown in Section 6.4.2, the friction coefficients ζ_{ij} satisfy the reciprocal relation

$$\zeta_{ij}(x) = \zeta_{ji}(x) \qquad (7.2)$$

The reciprocal relation allows us to write the time evolution equation (7.1) in the form of a variational principle. We consider the following quadratic function of $\dot{x} = (\dot{x}_1, \ldots, \dot{x}_f)$:

$$R(\dot{x}; x) = \sum_i \frac{\partial U(x)}{\partial x_i}\dot{x}_i + \frac{1}{2}\sum_{i,j} \zeta_{ij}\dot{x}_i\dot{x}_j \qquad (7.3)$$

Since the matrix $\zeta_{ij}(x)$ is positive definite, $R(\dot{x}; x)$ has a unique minimum as a function of $\dot{x} = (\dot{x}_1, \ldots, \dot{x}_f)$ when the conditions $\partial R/\partial \dot{x}_i = 0$ $(i = 1, 2, \ldots, f)$ are satisfied. It is easy to see that this condition gives the time evolution equation (7.1).[1] Therefore the time evolution of x is determined by the condition that $R(\dot{x}; x)$ be minimized. This is the variational principle for the motion of particles in viscous fluids. The variational principle is a simple rewriting of the force balance equation (7.1), but writing the time evolution law in this form has various advantages as we shall see in later sections.

The function $R(\dot{x}; x)$ is called the Rayleighian. The Rayleighian consists of two terms. One term, which we shall denote $\dot{U}$, is given by the linear function of $\dot{x}_i$

[1] Notice that here the reciprocal relation is essentially important since the condition $\partial R/\partial \dot{x}_i = 0$ gives the relation $\partial U(x)/\partial x_i + \sum_j (1/2)(\zeta_{ij} + \zeta_{ji})\dot{x}_j = 0$. For this to be written in the form of (7.1), the reciprocal relation is needed.

$$\dot{U} = \sum_i \frac{\partial U(x)}{\partial x_i} \dot{x}_i \tag{7.4}$$

This term represents the rate of change of the potential energy when particles move at velocity $\dot{x}$. The other term is a quadratic function of $\dot{x}$

$$\Phi = \frac{1}{2} \sum_{ij} \zeta_{ij} \dot{x}_i \dot{x}_j \tag{7.5}$$

This term is related to the energy dissipation taking place in the system. When particles move with velocity $\dot{x}$ in a viscous fluid, the fluid exerts a frictional force $F_{fi} = -\sum_j \zeta_{ij} \dot{x}_j$. The work done to the fluid per unit time is given by $-\sum_i F_{fi} \dot{x}_i = \sum_{ij} \zeta_{ij} \dot{x}_i \dot{x}_j$. Since the fluid is a purely viscous medium, the work done to the fluid is changed into heat immediately. Therefore 2Φ represents the energy dissipated into heat in the fluid per unit time. The function Φ is called the energy dissipation function.[2]

[2] For historic reasons, the energy dissipation function is defined with the factor $1/2$.

7.1.3   Variational principle for fluid flow

The variational principle given above determines the motion of particles moving in a fluid. In such a treatment, the effects of the fluid are included in the friction coefficient ζ_{ij}. In fact, the motion of fluid is also determined by the variational principle. In Appendix D, it is shown that if the fluid flows at velocity $\boldsymbol{v}(\boldsymbol{r})$, it causes the following dissipation of energy in the fluid per unit time and per unit volume:

$$\frac{\eta}{2} \left(\frac{\partial v_\alpha}{\partial r_\beta} + \frac{\partial v_\beta}{\partial r_\alpha} \right)^2 \tag{7.6}$$

The energy dissipation taking place in the system is the integral of this over the entire fluid volume. Therefore the energy dissipation function Φ is given by[3]

[3] Since Φ is now a functional of the velocity field $\boldsymbol{v}(\boldsymbol{r})$, Φ should be called an energy dissipation functional. In this book, however, we shall keep calling Φ the energy dissipation function.

$$\Phi_h[\boldsymbol{v}(\boldsymbol{r})] = \frac{\eta}{4} \int d\boldsymbol{r} \left(\frac{\partial v_\alpha}{\partial r_\beta} + \frac{\partial v_\beta}{\partial r_\alpha} \right)^2 \tag{7.7}$$

If the fluid is subject to gravitational force, the potential term is given by

$$\dot{U}_h[\boldsymbol{v}(\boldsymbol{r})] = -\int d\boldsymbol{r} \rho \boldsymbol{g} \cdot \boldsymbol{v} \tag{7.8}$$

where ρ is the density of the fluid and $\boldsymbol{g}$ is the acceleration of gravity. The fluid velocity is obtained by minimizing the Rayleighian $R_h[\boldsymbol{v}(\boldsymbol{r})] = \Phi_h[\boldsymbol{v}(\boldsymbol{r})] + \dot{U}_h[\boldsymbol{v}(\boldsymbol{r})]$ with respect to $\boldsymbol{v}$, subject to the requirement that the fluid velocity $\boldsymbol{v}$ has to satisfy the incompressible condition

$$\nabla \cdot \boldsymbol{v} = 0 \tag{7.9}$$

This constraint can be accounted for by Lagrange's method of undetermined multipliers. Thus, we consider the Rayleighian

$$R_h[\boldsymbol{v}(\boldsymbol{r})] = \frac{\eta}{4}\int d\boldsymbol{r}\left(\frac{\partial v_\alpha}{\partial r_\beta}+\frac{\partial v_\beta}{\partial r_\alpha}\right)^2 - \int d\boldsymbol{r}\rho\boldsymbol{g}\cdot\boldsymbol{v} - \int d\boldsymbol{r}p(\boldsymbol{r})\nabla\cdot\boldsymbol{v} \quad (7.10)$$

Here $p(\boldsymbol{r})$ is introduced as a Lagrange multiplier, but it turns out to be the pressure in the fluid.

For the Rayleighian $R_h[\boldsymbol{v}]$ to be minimized, the functional derivative $\delta R/\delta \boldsymbol{v}$ must be zero. It is easy to show that this condition gives the following Stokes equation in hydrodynamics[4]

$$\eta\nabla^2\boldsymbol{v} = -\rho\boldsymbol{g} + \nabla p \quad (7.13)$$

The above variational principle is called the principle of minimum energy dissipation in hydrodynamics.

If there are particles in the fluid, the fluid velocity has to satisfy the non-slip boundary condition at the surface of the particles. This condition is written by using the geometrical function $\boldsymbol{G}_i(\boldsymbol{r};x)$ introduced in eq. (D.4) as

$$\boldsymbol{v}(\boldsymbol{r}) = \sum_i \boldsymbol{G}_i(\boldsymbol{r};x)\dot{x}_i, \qquad \text{at the surface of the particle} \quad (7.14)$$

Hence the Rayleighian is written in this case as

$$R[\dot{x},\boldsymbol{v};x] = R_h[\boldsymbol{v}] + \sum_i \frac{\partial U_p}{\partial x_i}\dot{x}_i + \int dS\,\boldsymbol{f}_h(\boldsymbol{r})\cdot\left[\boldsymbol{v}(\boldsymbol{r}) - \sum_i \boldsymbol{G}_i(\boldsymbol{r};x)\dot{x}_i\right] \quad (7.15)$$

where U_p represents the potential energy of the particles, and the last term represents the constraint of the non-slip boundary condition. Here $\boldsymbol{f}_h(\boldsymbol{r})$ is introduced as a Lagrange multiplier, but turns out to be equal to the frictional force acting on unit area of the surface at $\boldsymbol{r}$. Minimization of $R[\dot{x},\boldsymbol{v};x]$ with respect to $\boldsymbol{v}$ gives the Stokes equation (7.13), and the boundary condition (7.14), and minimization of $R[\dot{x},\boldsymbol{v};x]$ with respect to $\dot{x}_i$ gives the force balance equation (7.1). Therefore the Rayleighian (7.15) gives the set of equations we have been using for the particle dynamics in viscous fluids.

7.1.4 Example: fluid flow in porous media

Before proceeding further, we demonstrate the usefulness of the variational principle using the simple problem shown in Fig. 7.1(b). A capillary packed with dry particles having hydrophilic surfaces is inserted in a fluid. Fluid will rise in such a capillary. Equilibrium rise of the meniscus was discussed in Section 4.2.3. Here we consider how the meniscus rises in time.

We assume that the capillary is filled with spherical particles of radius a at number density n. The volume fraction of the particles is

$$\phi = \frac{4\pi}{3}a^3 n \quad (7.16)$$

[4]
$$\delta R_h = \int d\boldsymbol{r}\left[\frac{\eta}{2}\left(\frac{\partial v_\alpha}{\partial r_\beta}+\frac{\partial v_\beta}{\partial r_\alpha}\right)\right.$$
$$\left(\frac{\partial\delta v_\alpha}{\partial r_\beta}+\frac{\partial\delta v_\beta}{\partial r_\alpha}\right)$$
$$\left. -\rho g_\alpha\delta v_\alpha - p\frac{\partial\delta v_\alpha}{\partial r_\alpha}\right] \quad (7.11)$$

The right-hand side is rewritten using integration by parts as

$$\delta R_h = \int d\boldsymbol{r}\,\delta v_\alpha\left[-\eta\frac{\partial}{\partial r_\beta}\right.$$
$$\left(\frac{\partial v_\alpha}{\partial r_\beta}+\frac{\partial v_\beta}{\partial r_\alpha}\right)$$
$$\left. -\rho g_\alpha + \frac{\partial p}{\partial r_\alpha}\right] \quad (7.12)$$

With eq. (7.9), this equation gives eq. (7.13).

We assume that the remaining volume, of fraction $1 - \phi$, is occupied either by air (in the vicinity of dry particles) or by fluid (in the vicinity of wet particles).

Now the driving force for the fluid rise is the difference in the surface energy between the wet region and the dry region. If the fluid occupies height h in the capillary, nhS particles (S being the cross-section of the capillary) are wetted by the fluid. Let γ_S be the spreading coefficient (see eq. (4.21)). Then the surface energy is written as

$$U_\gamma = -4\pi a^2 \gamma_S nhS = -3\frac{\phi}{a}\gamma_S hS \qquad (7.17)$$

On the other hand, the gravitational potential energy of the fluid is given by

$$U_g = \frac{1}{2}(1 - \phi)\rho g h^2 S \qquad (7.18)$$

Thus $\dot{U}$ is given by

$$\dot{U} = \dot{U}_\gamma + \dot{U}_g = \left[-3\frac{\phi}{a}\gamma_S + (1 - \phi)\rho g h\right]\dot{h}S \qquad (7.19)$$

The energy dissipation function is obtained by the following consideration. When the fluid moves with velocity v relative to the particles, there will be a frictional drag. The drag acting on the fluid in the region of volume V is proportional to the fluid velocity v, and can be written as $-\xi vV$. Here ξ stands for the friction coefficient per unit volume. This quantity can be measured experimentally by the setup shown in Fig. 7.2(a). Also ξ can be calculated by Stokesian hydrodynamics. By dimensional analysis, ξ is written as

$$\xi = \frac{\eta}{a^2}f_\xi(\phi) \qquad (7.20)$$

where $f_\xi(\phi)$ is a certain function of ϕ. In the dilute limit of particles, ξ is given by $\xi = n\zeta$, where $\zeta = 6\pi\eta a$ is the friction constant of a particle:

$$\xi = 6\pi\eta an = \frac{9\eta}{2a^2}\phi \qquad \text{for} \quad \phi \ll 1 \qquad (7.21)$$

As the volume fraction ϕ increases, ξ increases steeply as shown in Fig. 7.2(b).

For the situation shown in Fig. 7.1(b), the fluid velocity v is equal to $\dot{h}$. Therefore the energy dissipation function is written as

$$\Phi = \frac{1}{2}\xi\dot{h}^2 hS \qquad (7.22)$$

The Rayleighian is then given by

$$R(\dot{h}; h) = \frac{1}{2}\xi\dot{h}^2 hS + \left[-3\frac{\phi}{a}\gamma_S + (1 - \phi)\rho g h\right]\dot{h}S \qquad (7.23)$$

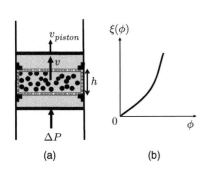

(a)   (b)

Fig. 7.2 (a) Experimental setup to measure the friction coefficient ξ. The particles are fixed in space bounded by porous walls (through which the fluid can flow freely), and the bottom piston is pushed by an extra pressure ΔP. If the fluid in the box moves with velocity v, the friction coefficient ξ is obtained by $\xi = \Delta P/(vh)$, where h is the thickness of the box. The fluid velocity v is obtained from the velocity of the piston v_{piston} by $v = v_{piston}/(1 - \phi)$. In this definition, particles are assumed to be fixed in space. Such a condition is satisfied for densely packed particles. In dilute solutions, care must be taken so that the applied pressure ΔP is low enough not to disturb the equilibrium distribution of the particles. (b) Typical behaviour of the friction coefficient ξ plotted against the volume fraction of the particle.

Hence $\partial R/\partial \dot{h} = 0$ gives

$$\xi h \dot{h} - 3\frac{\phi}{a}\gamma_S + (1 - \phi)\rho g h = 0 \qquad (7.24)$$

For short times, the gravitational force $(1 - \phi)\rho g h$ is small. Ignoring this term, eq. (7.24) is solved as

$$h(t) = \sqrt{\frac{6\phi\gamma_S}{\xi a}t} \qquad (7.25)$$

Therefore $h(t)$ increases in proportion to $t^{1/2}$ for short times. This is known as Washburn's law of wetting. At long times, $h(t)$ approaches the equilibrium value

$$h_{eq} = \frac{3\phi\gamma_S}{(1 - \phi)\rho g a} \qquad (7.26)$$

with the relaxation time

$$\tau = \frac{\xi h_{eq}}{(1 - \phi)\rho g} \qquad (7.27)$$

As the particle size a decreases, the equilibrium height h_{eq} increases. On the other hand, the rising speed decreases since ξ is proportional to a^{-2}.

The above analysis is for one-dimensional fluid motion. This can be easily generalized for three-dimensional motion. Consider the situation shown in Fig. 7.1(c) where a fluid drop starts to wet a dry bed of sand. In this case, we need to determine the time variation of the wet front.

Let $\boldsymbol{v}(\boldsymbol{r}, t)$ be the velocity of the fluid at point $\boldsymbol{r}$ at time t. The energy dissipation function is written as

$$\Phi = \frac{\xi}{2}\int d\boldsymbol{r}\,\boldsymbol{v}^2 \qquad (7.28)$$

and the $\dot{U}$ term is written as

$$\dot{U} = \int dS\,\boldsymbol{v} \cdot \boldsymbol{n}\left[-\frac{3\phi}{a}\gamma_S - (1 - \phi)\rho\boldsymbol{g} \cdot \boldsymbol{r}\right] \qquad (7.29)$$

where $\boldsymbol{n}$ is a unit vector normal to the wet front, and the integral is done over the surface of the wet front.

Since the particles and the fluid are both incompressible, the velocity $\boldsymbol{v}$ has to satisfy the constraint

$$\nabla \cdot \boldsymbol{v} = 0 \qquad (7.30)$$

The Rayleighian is therefore given by

$$R[\boldsymbol{v}(\boldsymbol{r})] = \frac{\xi}{2}\int d\boldsymbol{r}\,\boldsymbol{v}^2 + \int dS\left[-\frac{3\phi}{a}\gamma_S - (1 - \phi)\rho\boldsymbol{g} \cdot \boldsymbol{r}\right]\boldsymbol{v} \cdot \boldsymbol{n}$$
$$- \int d\boldsymbol{r}\,p(\boldsymbol{r})\nabla \cdot \boldsymbol{v} \qquad (7.31)$$

The last term on the right-hand side accounts for the constraint (7.30). The condition of $\delta R/\delta \boldsymbol{v} = 0$ therefore gives the following equation

$$\boldsymbol{v} = -\frac{1}{\xi}\nabla p, \quad \text{in the bulk} \tag{7.32}$$

and the following boundary condition

$$p = -\gamma_S\frac{3\phi}{a} - (1-\phi)\rho\boldsymbol{g}\cdot\boldsymbol{r}, \quad \text{at the moving front} \tag{7.33}$$

Equations (7.30), (7.32), and (7.33) determine the dynamics of the wet front. Equation (7.32) is known as Darcy's law for fluid flow in porous materials.

7.2   Onsager principle

7.2.1   Kinetic equations for state variables

We have seen that in the particle–fluid system, the time evolution equation (7.1) can be written in the form of a variational principle due to the reciprocal relation (7.2). Here x represents a set of parameters specifying the configuration of the particles.

As we have seen in Section 6.5, the reciprocal relation can be proven without invoking hydrodynamics. Let us therefore consider that $x = (x_1, x_2, \ldots, x_f)$ represents a set of parameters which specify the non-equilibrium state of a system, and assume that the time evolution of the system is described by the following kinetic equation

$$\frac{dx_i}{dt} = -\sum_j \mu_{ij}(x)\frac{\partial A}{\partial x_j} \tag{7.34}$$

where $A(x)$ is the free energy of the system,[5] and $\mu_{ij}(x)$ are certain functions of x. The coefficients $\mu_{ij}(x)$ are called kinetic coefficients, and correspond to the inverse of the friction coefficients in eq. (7.1). If we define ζ_{ij} by

$$\sum_k \zeta_{ik}\mu_{kj} = \delta_{ij} \tag{7.35}$$

eq. (7.34) is written in the same form as eq. (7.1):

$$-\sum_j \zeta_{ij}(x)\dot{x}_j - \frac{\partial A}{\partial x_i} = 0 \tag{7.36}$$

As shown in Section 6.4.2, as long as the time evolution equation for x is written in the form of eq. (7.34) or eq. (7.36), Onsager's reciprocal relation must hold:[6]

$$\zeta_{ij}(x) = \zeta_{ji}(x), \quad \mu_{ij}(x) = \mu_{ji}(x) \tag{7.37}$$

This is the result of the time reversal symmetry of the fluctuations at equilibrium. Therefore the time evolution equations of type (7.34)

[5] Strictly speaking, $A(x)$ is the restricted free energy, the free energy of the system which is subject to the constraints that the parameters are fixed at x. Further discussion of the restricted free energy is given in Appendix B.

[6] Here it is important to note that for the reciprocal relation (7.37) to hold, the time evolution equation must be written in the form of eq. (7.34) with kinetic coefficients that are independent of the potential $A(x)$. This condition is needed in the proof of the reciprocal relation described in Appendix D.3.

or (7.36) are always translated into the variational principle, i.e., the time evolution of the system is determined by the condition that the Rayleighian

$$R(\dot{x}; x) = \frac{1}{2} \sum_{i,j} \zeta_{ij} \dot{x}_i \dot{x}_j + \sum_i \frac{\partial A(x)}{\partial x_i} \dot{x}_i \qquad (7.38)$$

be minimized with respect to $\dot{x}_i$. We shall call this Onsager's variational principle, or simply the Onsager principle.[7]

In this argument, x can be any set of variables which specify the non-equilibrium state of the system. For example, it can represent the concentration profile in the problem of diffusion, or the order parameters in the dynamics of liquid crystals. Many equations known to describe the time evolution of the non-equilibrium state are written in the form of eq. (7.34).

Why are so many time evolution equations written in the form of (7.34)? A reason for this may be seen in the argument given in Section 6.5.2. We have seen that the force exerted by the system when external parameters x_i are changed at rate $\dot{x}_i$ is given by eq. (6.69). If the correlation time of the random force $F_{ri}(t)$ is very short and if $x_i(t)$ changes very slowly (i.e., if the change of $\dot{x}_i(t)$ during the correlation time of $F_{ri}(t)$ is negligible), eq. (6.69) is written as

$$\langle F_i(t) \rangle = -\frac{\partial A(x)}{\partial x_i} - \sum_j \zeta_{ij} \dot{x}_j \qquad (7.39)$$

where

$$\zeta_{ij} = \int_0^\infty dt\, \alpha_{ij}(t) \qquad (7.40)$$

If the variable x_i is free to change, the time evolution for x_i is determined by the condition $\langle F_i(t) \rangle = 0$, which is eq. (7.36).

The essential point of the above argument is that if a certain set of variables $x = (x_1, \ldots, x_f)$ changes much more slowly than the other variables, their time evolution equations are written in the form of eq. (7.36). We shall call $x = (x_1, \ldots, x_f)$ slow variables, and the others fast variables. The argument developed in Section 6.5.2 shows that as long as the fast variables are not far from their equilibrium state, and as long as the relaxation times of the fast variables are much shorter than the characteristic times of the slow variables, the time evolution equations of the slow variables are written in the form of eq. (7.36).

7.2.2   Forces needed to control external parameters

Equation (7.39) represents the force exerted by the system when the slow variables are changed at the rate $\dot{x}_i$. Therefore, the force $F_i^{(e)}$ needed to change the slow variable x_i at rate $\dot{x}_i$ is given by $-\langle F_i(t) \rangle$, i.e.,

$$F_i^{(e)} = \frac{\partial A}{\partial x_i} + \sum_j \zeta_{ij} \dot{x}_j \qquad (7.41)$$

[7] Onsager actually formulated his variational principle using the entropy $S(x)$ and the entropy production rate. Here we are considering the isothermal system. For such system, his variational principle is written in the form presented here.

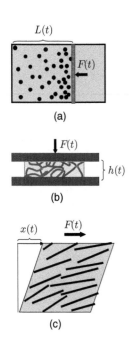

$L(t)$

$F(t)$

(a)

$F(t)$

$h(t)$

(b)

$x(t)$ $F(t)$

(c)

Fig. 7.3 Examples of external para-meters and their conjugate forces. See the text for explanation.

This force is written as

$$F_i^{(e)} = \frac{\partial R}{\partial \dot{x}_i} \qquad (7.42)$$

For example, to change the particle position x of a Brownian particle with velocity $\dot{x}$ (see Fig. 6.6), we need to apply a force

$$F^{(e)} = \zeta \dot{x} + \frac{\partial U}{\partial x} \qquad (7.43)$$

which is equal to $\partial R/\partial \dot{x}$.

Examples which demonstrate the usefulness of eq. (7.42) are shown in Fig. 7.3. In Fig. 7.3(a), a semi-permeable membrane separating the particle solution from the pure solvent is moved with velocity $\dot{L}$. The force $F(t)$ needed to cause such motion is given by $\partial R/\partial \dot{L}$. In Fig. 7.3(b), a porous wall sandwiching a gel is pressed to squeeze out the solvent in the gel. The force $F(t)$ needed to move the wall with velocity $\dot{h}$ is given by $\partial R/\partial \dot{h}$. In Fig. 7.3(c), the top wall sandwiching a solution of rod-like particles is moved horizontally with velocity $\dot{x}$. The shearing force needed to cause such motion is given by $\partial R/\partial \dot{x}$. Actual calculation of these quantities will be given later.

7.3   Diffusion of particles in dilute solutions

7.3.1   Particle diffusion and Brownian motion

As a first application of the Onsager principle, we shall discuss the diffusion of Brownian particles. In the previous chapter, we used the Langevin equation to discuss the random motion of Brownian particles. This is not the only way to describe Brownian motion; the Brownian motion can also be described by a deterministic equation which does not involve any stochastic variables. The deterministic equation is called the Smoluchowskii equation and describes the time evolution of the probability of finding the Brownian particles in a certain configuration.

In the case of one-dimensional Brownian motion in a potential field $U(x)$, the Smoluchowskii equation for the probability density $\psi(x,t)$ of finding the Brownian particle at position x is written as

$$\frac{\partial \psi}{\partial t} = D \frac{\partial}{\partial x} \left[\frac{\partial \psi}{\partial x} + \frac{\psi}{k_B T} \frac{\partial U}{\partial x} \right] \qquad (7.44)$$

This equation describes the same motion as that described by the Langevin equation (6.39). Indeed as shown in Appendix F, the Smoluch-owskii equation (7.44) can be derived from the Langevin equation (6.43).

The Smoluchowskii equation (7.44) can also be derived directly by physical argument. In the following, we shall first show such a derivation, and then show another derivation based on the Onsager principle.

The function $\psi(x,t)$ represents the probability of finding the Brownian particle at position x at time t. This probability can be interpreted as follows. Let us imagine a dilute solution of Brownian particles, where many Brownian particles are moving independently of each other. Let $n(x,t)$ be the number density of Brownian particles at position x. Then the distribution function $\psi(x,t)$ is given by $\psi(x,t) = n(x,t)/N_{tot}$ where N_{tot} is the total number of particles in the system. Therefore, we may focus on the number density $n(x,t)$ of many Brownian particles instead of the probability density $\psi(x,t)$ of a Brownian particle.

7.3.2 Derivation of the diffusion equation from macroscopic force balance

If $n(x,t)$ is not homogeneous, the particles will diffuse from a high concentration region to a low concentration region. The driving force for this phenomenon is the osmotic pressure.

To see the point clearly, let us imagine that the Brownian particles are confined by a box made of two hypothetical semi-permeable membranes, one located at x and the other located at $x+dx$ (see Fig. 7.4). The outer solution exerts osmotic pressure $S\Pi(x)$ from the left (S being the cross-section of the box), and $S\Pi(x+dx)$ from the right to the hypothetical box. If $n(x)$ is larger than $n(x+dx)$, $\Pi(x)$ is larger than $\Pi(x+dx)$, and therefore the box will be pushed to the right with a certain velocity v.

The velocity v is obtained by considering the force balance acting on the hypothetical box. If the box moves with velocity v in the fluid, the fluid exerts a frictional force $-Sdx\xi v$, where ξ is the friction coefficient per unit volume. The external potential (for example, the gravitational potential) $U(x)$ exerts a potential force $-\partial U/\partial x$ for each particle, and the force $nSdx(-\partial U/\partial x)$ on the collection of particles in the box. Therefore the force balance equation is written as

$$\Pi(x) - \Pi(x+dx) - ndx\frac{\partial U}{\partial x} - \xi vdx = 0 \tag{7.45}$$

which gives the velocity

$$v = -\frac{1}{\xi}\left(\frac{\partial \Pi}{\partial x} + n\frac{\partial U}{\partial x}\right) \tag{7.46}$$

In a dilute solution, Π is given by nk_BT, and ξ is given by $n\zeta$. Therefore eq. (7.46) is written as

$$v = -\frac{1}{n\zeta}\left[k_BT\frac{\partial n}{\partial x} + n\frac{\partial U}{\partial x}\right]$$

$$= -\frac{D}{n}\left[\frac{\partial n}{\partial x} + \frac{n}{k_BT}\frac{\partial U}{\partial x}\right] \tag{7.47}$$

where the Einstein relation $D = k_BT/\zeta$ has been used. Given the velocity v, the time evolution of n is obtained by the conservation equation

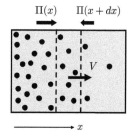

$\Pi(x)$ $\Pi(x+dx)$

Fig. 7.4 Derivation of the Smoluchowskii equation by consideration of the forces acting on Brownian particles.

$$\frac{\partial n}{\partial t} = -\frac{\partial (nv)}{\partial x} \qquad (7.48)$$

Equations (7.47) and (7.48) give

$$\frac{\partial n}{\partial t} = D\frac{\partial}{\partial x}\left[\frac{\partial n}{\partial x} + \frac{n}{k_B T}\frac{\partial U}{\partial x}\right] \qquad (7.49)$$

which is the same as eq. (7.44). Notice that in this derivation, the diffusion term $D\partial^2 n/\partial x^2$ arises from the osmotic pressure $\Pi = nk_B T$.

7.3.3   Derivation of the diffusion equation by the Onsager principle

We now derive eq. (7.49) using the Onsager principle. To do this, we regard the particle density $n(x)$ as the set of slow variables which describes the non-equilibrium state of the diffusing particles.[8] In principle, the Rayleighian can be written as a functional of $\dot{n} = \partial n/\partial t$. However it is more convenient to express the Rayleighian as a functional of $v(x)$, the average velocity of particles at x. Since $\dot{n}$ and v are related by

$$\dot{n} = -\frac{\partial nv}{\partial x} \qquad (7.50)$$

we can use $v(x)$ as a substitute of $\dot{n}$.[9]

The energy dissipation function is given by

$$\Phi = \frac{1}{2}\int_{-\infty}^{\infty} dx\ \xi v^2 \qquad (7.51)$$

The free energy is expressed as a functional of $n(x)$

$$A[n(x)] = \int_{-\infty}^{\infty} dx\ [k_B T n \ln n + nU(x)] \qquad (7.52)$$

Hence $\dot{A}$ is calculated as

$$\dot{A} = \int dx\ [k_B T(\ln n + 1) + U(x)]\ \dot{n} \qquad (7.53)$$

Using eq. (7.50), and integration by parts, we have

$$\begin{aligned} \dot{A} &= -\int dx\ [k_B T(\ln n + 1) + U(x)]\frac{\partial nv}{\partial x} \\ &= \int dx\ v\left(k_B T\frac{\partial n}{\partial x} + n\frac{\partial U}{\partial x}\right) \end{aligned} \qquad (7.54)$$

Hence

$$R = \frac{1}{2}\int dx\ \xi v^2 + \int dx\ v\left(k_B T\frac{\partial n}{\partial x} + n\frac{\partial U}{\partial x}\right) \qquad (7.55)$$

The condition $\delta R/\delta v = 0$ gives

$$v = -\frac{1}{\xi}\left(k_B T\frac{\partial n}{\partial x} + n\frac{\partial U}{\partial x}\right) \qquad (7.56)$$

Equations (7.56), (7.50) and the relation $\xi = n\zeta$ give eq. (7.49).

[8] This statement may become clear if we represent the function $n(x)$ by a set of variables n_i $(i = 1, 2, \ldots)$ which represent the value of $n(x)$ at the discretized points $x = i\Delta x$. The slow variables in this case are n_i. In the limit of $\Delta x \to 0$, the set of variables n_i $(i = 1, 2, \ldots)$ defines the function $n(x)$.

[9] There is a more fundamental reason to choose v instead of $\dot{n}$. Equation (7.50) represents the constraint that the particle density n can change only by the actual transport of particles, not by chemical reaction or other teleportational processes.

7.3.4 Diffusion potential

The above argument shows that the Smoluchowskii equation (7.44) is derived from two equations, i.e., the conservation equation

$$\frac{\partial \psi}{\partial t} = -\frac{\partial}{\partial x}(v\psi) \tag{7.57}$$

and the force balance equation for the particles

$$-\zeta v - \frac{\partial}{\partial x}(k_B T \ln \psi + U) = 0 \tag{7.58}$$

Equation (7.58) indicates that if the particle is undergoing Brownian motion, the effective potential acting on the particle is $k_B T \ln \psi + U$: the Brownian motion gives an extra potential $k_B T \ln \psi$. We shall call this potential the diffusion potential. As seen in the discussion in the previous section, the diffusion potential arises from the entropic term $k_B T \psi \ln \psi$ in the free energy.

If the diffusion potential is taken into account, the time evolution of the distribution function ψ is obtained by macroscopic equations, i.e., the force balance equation (7.58), and the conservation euqation (7.57).

7.4 Diffusion of particles in concentrated solutions

7.4.1 Collective diffusion

Let us now consider particle diffusion in a concentrated solution. In concentrated solutions, particles are interacting with each other. If particles repel each other, they spread faster, i.e., the diffusion takes place faster. This effect can be naturally taken into account by generalizing the argument developed in Section 7.3.3.

Let $f(n)$ be the free energy density of the solution of particles of number density n. The total free energy of the solution which has concentration profile $n(x)$ is given by

$$A = \int dx[f(n) + nU(x)] \tag{7.59}$$

With the use of eq. (7.50), $\dot{A}$ is expressed as

$$\dot{A} = \int dx \, [f'(n) + U(x)]\dot{n} = -\int dx \, [f'(n) + U(x)]\frac{\partial(vn)}{\partial x}$$

$$= \int dx \, vn\frac{\partial}{\partial x}[f'(n) + U(x)] \tag{7.60}$$

where $f'(n) = \partial f/\partial n$. Now the osmotic pressure Π is expressed by the free energy density $f(n)$ (see eq. (2.23)),

$$\Pi(n) = -f(n) + nf'(n) + f(0) \tag{7.61}$$

This gives the identity

$$n\frac{\partial f'(n)}{\partial x} = \frac{\partial \Pi(n)}{\partial x} = \Pi'\frac{\partial n}{\partial x} \tag{7.62}$$

where $\Pi' = \partial \Pi/\partial n$. Hence $\dot{A}$ is written as

$$\dot{A} = \int dx \; v \left[\Pi'\frac{\partial n}{\partial x} + n\frac{\partial U}{\partial x} \right] \tag{7.63}$$

On the other hand, the energy dissipation function $\Phi[v]$ is given by the same form as eq. (7.51), (where ξ is now a function of n). Minimizing the Rayleighian $\Phi + \dot{A}$, we have

$$v = -\frac{1}{\xi} \left[\Pi'\frac{\partial n}{\partial x} + n\frac{\partial U}{\partial x} \right] \tag{7.64}$$

Equations (7.50) and (7.64) give the following diffusion equation

$$\frac{\partial n}{\partial t} = \frac{\partial}{\partial x} \left[D_c \left(\frac{\partial n}{\partial x} + \frac{n}{\Pi'}\frac{\partial U}{\partial x} \right) \right] \tag{7.65}$$

where

$$D_c(n) = \frac{n\Pi'}{\xi} = \frac{n}{\xi}\frac{\partial \Pi}{\partial n} \tag{7.66}$$

is the diffusion constant in concentrated solution.

The diffusion constant D_c represents the speed of spreading of a collection of particles placed in a solvent, and is called the collective diffusion constant. The collective diffusion constant is different from the diffusion constant appearing in eq. (6.32) or eq. (6.33). The diffusion constant in eq. (6.32) represents the spreading speed of a single particle in a medium and is called the self-diffusion constant. In concentrated solutions, the self-diffusion constant D_s is obtained by selecting a particle in the solution, and measuring its mean square displacement in a given time.

The two diffusion constants D_c and D_s agree with each other in the limit of dilute solutions, but they deviate from each other with the increase of concentration. For example, the self-diffusion constant D_s of a polymer in a solution always decreases with the increase of polymer concentration (since polymers entangle each other and become less mobile), while the collective diffusion constant D_c usually increases with concentration in a good solvent since polymers repel each other in a good solvent.

7.4.2   Sedimentation

Under gravity, a colloidal particle has the potential energy

$$U(x) = mgx \tag{7.67}$$

where mg represents the gravitational force (with the correction of the buoyancy force) acting on the particle. The time evolution equation (7.65) now becomes

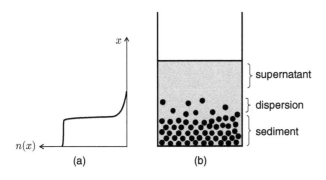

$$\frac{\partial n}{\partial t} = \frac{\partial}{\partial x}\left[D_c\left(\frac{\partial n}{\partial x} + \frac{n}{\Pi'}mg\right)\right] \tag{7.68}$$

The equilibrium solution of eq. (7.68) is given by the solution of the
following equation

$$\frac{\partial n}{\partial x} + \frac{nmg}{\Pi'(n)} = 0 \tag{7.69}$$

In a dilute solution, the osmotic pressure is given by $\Pi = nk_BT$. In
this case, eq. (7.69) becomes

$$\frac{\partial n}{\partial x} + \frac{nmg}{k_BT} = 0 \tag{7.70}$$

This gives the exponential distribution

$$n_{eq}(x) \propto \exp\left(-\frac{mgx}{k_BT}\right) \tag{7.71}$$

On the other hand, in a concentrated solution, the solution of eq. (7.69)
can take various forms depending on the functional form of $\Pi'(n)$. An
example is shown in Fig. 7.5. At the bottom, there is a sediment layer
where the particle density is almost constant, nearly equal to the closest
packing density. In charge stabilized colloids, crystalline phases are often
seen in this layer. On top of the sediment, there is a dispersion layer,
where the particle distribution is approximately given by the exponential
law (7.71). Above this layer, there is a clear solvent. Measurement of such
a concentration profile gives the osmotic pressure curve $\Pi(n)$.

7.4.3   Forces acting on a semi-permeable membrane

Let us now discuss the force needed to control the position of the semi-
permeable membrane shown in Fig. 7.3(a). Since the integral in eq. (7.59)
is done between 0 and $L(t)$, $\dot{A}$ is calculated as[10]

$$\dot{A} = \dot{L}[f(n) + nU(x)]_{x=L} + \int_0^{L(t)} dx\ \dot{n}[f'(n) + U(x)] \tag{7.72}$$

[10] Here we have taken the free energy
of the pure solvent $f(0)$ equal to zero. If
$f(0)$ is not equal to zero, the free energy
of the system is given by

$$A = \int_0^{L(t)} dx\ [f(n) + nU(x)]$$

$$+ \int_{L(t)}^{L_{max}} dx\ f(0)$$

The second integral gives the contribution of pure solvent. For such an
expression, eq. (7.72) has an extra term
$-\dot{L}f(0)$, and the final result (7.73) is
not changed.

Using eq. (7.50) and integrating by parts, we have

$$\dot{A} = \dot{L}[f(n) + nU(x)]_{x=L} - [vn(f'(n) + U(x))]_{x=L}$$

$$+ \int_0^{L(t)} dx \, vn \frac{\partial}{\partial x}[f'(n) + U(x)]$$

$$= -\dot{L}\Pi(n)_{x=L} + \int_0^{L(t)} dx \, vn \frac{\partial}{\partial x}[f'(n) + U(x)] \qquad (7.73)$$

where we have used the expression for the osmotic pressure $\Pi(n) = nf'(n) - f(n)$ and the condition that the velocity v satisfies $v(L) = \dot{L}$.

On the other hand, the energy dissipation function is given by

$$\Phi = \frac{1}{2}\int_0^{L(t)} dx \, \xi v^2 + \frac{1}{2}\zeta_m \dot{L}^2 \qquad (7.74)$$

The second term represents the extra energy dissipation caused by the solvent flow through the membrane (ζ_m is the friction constant of the membrane). The force $F(t)$ needed to move the semi-permeable membrane is therefore given by

$$F(t) = \frac{\partial}{\partial \dot{L}}(\Phi + \dot{A}) = \zeta_m \dot{L} - \Pi(n(L)) \qquad (7.75)$$

If the membrane is not moving, i.e., if $\dot{L} = 0$, the force is given by $-\Pi(n(L))$, the osmotic pressure at the surface of the membrane. If the membrane is displaced in a stepwise fashion by ΔL at time $t = 0$, the particle concentration at the surface of the membrane changes (either becomes larger or smaller than that in the bulk depending on the direction of the displacement). The particles then start to diffuse. Eventually the particle concentration becomes homogeneous in the solution, but this will take a very long time. For the particle concentration at the membrane to become equal to that in the bulk, it will take a time of the order of L^2/D_c. Notice that the potential $U(x)$ does not enter in the expression of $F(t)$: the potential exerts a force on the particles and affects the concentration at the membrane $n(L)$, but does not exert direct force on the membrane.

7.5   Rotational Brownian motion of rod-like particles

7.5.1   Description of rotational Brownian motion

A rod-like particle in solution changes its orientation randomly by Brownian motion. This is called rotational Brownian motion. The orientation of a rod can be specified by two angular coordinates θ and ϕ defined in Fig. 7.6. Alternatively, the orientation can be specified by a unit vector $\boldsymbol{u}$ parallel to the rod. The rotational Brownian motion can be described by the time variation of the probability distribution functions $\psi(\theta, \phi; t)$ or $\psi(\boldsymbol{u}; t)$. The Smoluchowskii equations for these functions can

Fig. 7.6 The orientation of a rod-like particle can be specified by two angular coordinates θ and ϕ, or by a unit vector $\boldsymbol{u}$ parallel to the rod axis.

be derived straightforwardly by the Onsager principle. Here we shall first derive the Smoluchowskii equation for $\psi(\boldsymbol{u};t)$. The equation for $\psi(\theta,\phi;t)$ is given in Section 7.5.6.

7.5.2   Conservation equation in the configurational space

The vector $\boldsymbol{u}$ is confined to the two-dimensional space on the surface of the unit sphere $|\boldsymbol{u}|=1$. The distribution function $\psi(\boldsymbol{u};t)$ is normalized as

$$\int d\boldsymbol{u}\ \psi(\boldsymbol{u},t)=1 \tag{7.76}$$

where the integral over $\boldsymbol{u}$ involves the integral on the surface of the sphere $|\boldsymbol{u}|=1$.

The conservation equation for such a system is written as follows. Suppose that the rod rotates with angular velocity $\boldsymbol{\omega}$. Then $\boldsymbol{u}$ changes as

$$\dot{\boldsymbol{u}}=\boldsymbol{\omega}\times\boldsymbol{u}\qquad\text{or}\qquad \dot{u}_\alpha=e_{\alpha\beta\gamma}\omega_\beta u_\gamma \tag{7.77}$$

where $e_{\alpha\beta\gamma}$ is the Levi-Civita symbol defined by

$$e_{\alpha\beta\gamma}=\begin{cases}1 & (\alpha,\beta,\gamma)=(x,y,z)\ \text{or}\quad (y,z,x)\quad\text{or}\quad (z,x,y)\\-1 & (\alpha,\beta,\gamma)=(y,x,z)\quad\text{or}\quad (x,z,y)\quad\text{or}\quad (z,y,x)\\0 & \text{otherwise}\end{cases} \tag{7.78}$$

The conservation equation for $\psi(\boldsymbol{u};t)$ is written as

$$\dot{\psi}=-\frac{\partial}{\partial\boldsymbol{u}}\cdot(\dot{\boldsymbol{u}}\psi)=-\frac{\partial}{\partial u_\alpha}(e_{\alpha\beta\gamma}\omega_\beta u_\gamma\psi)$$
$$=-u_\gamma e_{\gamma\alpha\beta}\frac{\partial}{\partial u_\alpha}(\omega_\beta\psi)=-\left(\boldsymbol{u}\times\frac{\partial}{\partial\boldsymbol{u}}\right)\cdot(\boldsymbol{\omega}\psi) \tag{7.79}$$

This can be written using the differential operator

$$\mathcal{R}=\boldsymbol{u}\times\frac{\partial}{\partial\boldsymbol{u}} \tag{7.80}$$

as

$$\dot{\psi}=-\mathcal{R}\cdot(\boldsymbol{\omega}\psi) \tag{7.81}$$

The operator $\mathcal{R}$ corresponds to the nabla ($\boldsymbol{\nabla}$) operator in three-dimensional space. Indeed $\mathcal{R}$ satisfies the following identity which corresponds to the formula of integration by parts

$$\int d\boldsymbol{u}\psi(\boldsymbol{u})\mathcal{R}\phi(\boldsymbol{u})=-\int d\boldsymbol{u}(\mathcal{R}\psi(\boldsymbol{u}))\phi(\boldsymbol{u}) \tag{7.82}$$

7.5.3   Smoluchowskii equation

We use the Onsager principle to determine $\boldsymbol{\omega}$ in eq. (7.81). If the rod rotates with angular velocity ω around the axis normal to $\boldsymbol{u}$, the fluid exerts a frictional torque T_f proportional to ω

$$T_f = -\zeta_r \omega \tag{7.83}$$

The constant ζ_r is called the rotational friction constant. For a rod of length L and diameter b, a hydrodynamic calculation gives (see e.g., *The Theory of Polymer Dynamics* by M. Doi and S. Edwards listed in the Further reading)

$$\zeta_r = \frac{\pi \eta L^3}{3(\ln(L/b) - 0.8)} \tag{7.84}$$

The energy dissipation caused by the rotation is given by $-T_f \omega = \zeta_r \omega^2$. Hence the energy dissipation function (per particle) is given by

$$\Phi = \frac{1}{2} \int d\boldsymbol{u} \ \zeta_r \omega^2 \psi \tag{7.85}$$

On the other hand, the free energy of the system (per particle) is given by

$$A = \int d\boldsymbol{u} \ [k_B T \psi \ln \psi + \psi U(\boldsymbol{u})] \tag{7.86}$$

where $U(\boldsymbol{u})$ is the external potential acting on the rod. From eqs. (7.81) and (7.86), $\dot{A}$ is calculated as

$$\dot{A} = -\int d\boldsymbol{u} \ \mathcal{R} \cdot (\boldsymbol{\omega}\psi) \left[k_B T (\ln \psi + 1) + U(\boldsymbol{u}) \right] \tag{7.87}$$

Using integration by parts (eq. (7.82)), this can be rewritten as

$$\dot{A} = \int d\boldsymbol{u} \boldsymbol{\omega}\psi \cdot \mathcal{R}\tilde{U} \tag{7.88}$$

where

$$\tilde{U} = k_B T \ln \psi + U(\boldsymbol{u}) \tag{7.89}$$

represents the effective potential which is the sum of the external potential and the diffusion potential (see eq. (7.58)).

Therefore the Rayleighian becomes

$$R = \frac{1}{2} \int d\boldsymbol{u} \ \zeta_r \omega^2 \psi + \int d\boldsymbol{u} \ \psi \boldsymbol{\omega} \cdot \mathcal{R}\tilde{U} \tag{7.90}$$

The extremum condition $\delta R / \delta \boldsymbol{\omega} = 0$ gives

$$\boldsymbol{\omega} = -\frac{1}{\zeta_r} \mathcal{R}\tilde{U} = -\frac{1}{\zeta_r} \mathcal{R}(k_B T \ln \psi + U) = -D_r \left(\mathcal{R} \ln \psi + \frac{\mathcal{R}U}{k_B T} \right) \tag{7.91}$$

where

$$D_r = \frac{k_B T}{\zeta_r} \tag{7.92}$$

is the rotational diffusion constant. From eqs. (7.81) and (7.91), the time evolution equation becomes

$$\frac{\partial \psi}{\partial t} = D_r \mathcal{R} \cdot \left(\mathcal{R}\psi + \frac{\psi}{k_B T} \mathcal{R}U \right) \tag{7.93}$$

This is the Smoluchowskii equation for rotational Brownian motion.

7.5.4   Time correlation functions

As a special case of the above equation, let us consider the rotational Brownian motion in free space ($U = 0$). Let us assume that $\boldsymbol{u}(t)$ was $\boldsymbol{u}_0$ at time $t = 0$. The average of $\boldsymbol{u}$ at time t is calculated by

$$\langle \boldsymbol{u}(t) \rangle = \int d\boldsymbol{u} \psi(\boldsymbol{u}, t) \boldsymbol{u} \tag{7.94}$$

To obtain the time evolution of $\langle \boldsymbol{u}(t) \rangle$, we multiply both sides of eq. (7.93) by $\boldsymbol{u}$, and integrate for $\boldsymbol{u}$. This gives

$$\frac{\partial}{\partial t} \langle \boldsymbol{u}(t) \rangle = D_r \int d\boldsymbol{u} \; (\mathcal{R}^2 \psi) \boldsymbol{u} \tag{7.95}$$

The right-hand side can be rewritten by use of eq. (7.82):

$$\int d\boldsymbol{u} \; (\mathcal{R}^2 \psi) \boldsymbol{u} = - \int d\boldsymbol{u} \; (\mathcal{R}\psi) \cdot \mathcal{R}\boldsymbol{u} = \int d\boldsymbol{u} \; \psi \mathcal{R}^2 \boldsymbol{u} \tag{7.96}$$

Using the definition of eq. (7.80), one can show (after some calculation)

$$\mathcal{R}^2 \boldsymbol{u} = -2\boldsymbol{u} \tag{7.97}$$

Hence eq. (7.95) gives

$$\frac{\partial}{\partial t} \langle \boldsymbol{u}(t) \rangle = D_r \int d\boldsymbol{u} \; (-2\boldsymbol{u}) \psi = -2D_r \langle \boldsymbol{u}(t) \rangle \tag{7.98}$$

Solving this equation under the initial condition $\langle \boldsymbol{u}(0) \rangle = \boldsymbol{u}_0$ we finally have

$$\langle \boldsymbol{u}(t) \rangle = \boldsymbol{u}_0 \exp(-2D_r t) \tag{7.99}$$

Therefore the time correlation function of $\boldsymbol{u}(t)$ can be calculated as

$$\langle \boldsymbol{u}(t) \boldsymbol{u}(0) \rangle = \langle \boldsymbol{u}_0^2 \rangle \exp(-2D_r t) = \exp(-2D_r t) \tag{7.100}$$

The mean square displacement of $\boldsymbol{u}(t)$ in time t is calculated from eq. (7.100) as

$$\langle (\boldsymbol{u}(t) - \boldsymbol{u}(0))^2 \rangle = \langle \boldsymbol{u}(t)^2 + \boldsymbol{u}(0)^2 - 2\boldsymbol{u}(t) \cdot \boldsymbol{u}(0) \rangle = 2[1 - \exp(-2D_r t)] \tag{7.101}$$

In particular, for very short times, $tD_r \ll 1$

$$\langle (\boldsymbol{u}(t) - \boldsymbol{u}_0)^2 \rangle = 4D_r t \tag{7.102}$$

This is equal to the mean square displacement of a point which is moving in a two-dimensional plane with diffusion constant D_r.

7.5.5   Magnetic relaxation

Certain rod-like particles have a permanent magnetic moment μ_m along the long axis of the particle as shown in Fig. 6.9(a). Solutions of such particles will have a magnetic moment

$$\boldsymbol{m} = \mu_m \boldsymbol{u} \tag{7.103}$$

per particle. If there is no magnetic field, the orientation of the particle is isotropic, and the average magnetic moment $\langle \boldsymbol{m} \rangle$ is zero. If a magnetic field $\boldsymbol{H}$ is applied in the z-direction, the particle orients to the z-direction, and there appears a macroscopic magnetic moment $\langle m_z \rangle$. We shall now calculate the time dependence of $\langle m_z \rangle$ when the field is switched on as shown in Fig. 6.9(b).

Under the magnetic field $\boldsymbol{H}$ applied in the z-direction, the particle has a potential

$$U(\boldsymbol{u}) = -\mu_m H u_z \tag{7.104}$$

The Smoluchowskii equation (7.93) then becomes

$$\frac{\partial \psi}{\partial t} = D_r \mathcal{R} \cdot \left(\mathcal{R}\psi + \frac{\mu_m \boldsymbol{u} \times \boldsymbol{H}}{k_B T} \psi \right) \tag{7.105}$$

To obtain $\langle u_z \rangle$, we multiply both sides of eq. (7.105) by u_z, and integrate with respect to $\boldsymbol{u}$. The resulting equation can be written, again through integration by parts, as

$$\frac{\partial \langle u_z \rangle}{\partial t} = -2D_r \left[\langle u_z \rangle - \frac{H\mu_m}{2k_B T} \underline{\langle 1 - u_z^2 \rangle} \right] \tag{7.106}$$

Let us assume that the magnetic field is weak, and consider the lowest order perturbation in H. Since the underlined term in the above equation involves H, we may replace $\langle 1 - u_z^2 \rangle$ by $\langle 1 - u_z^2 \rangle_0$, the average in the absence of H. Since the distribution of $\boldsymbol{u}$ is isotropic in the absence of H, $\langle u_z^2 \rangle_0 = 1/3$. Hence eq. (7.106) can be approximated as

$$\frac{\partial \langle u_z \rangle}{\partial t} = -2D_r \left[\langle u_z \rangle - \frac{\mu_m H}{3k_B T} \right] \tag{7.107}$$

Solving the equation under the initial condition $\langle u_z \rangle = 0$ at $t = 0$, we have

$$\langle u_z \rangle = \frac{\mu_m H}{3k_B T} [1 - \exp(-2D_r t)] \tag{7.108}$$

Therefore

$$\langle m_z(t) \rangle = \frac{\mu_m^2 H}{3k_B T} [1 - \exp(-2D_r t)] \tag{7.109}$$

Comparing this with eq. (6.79), we have the response function

$$\chi(t) = \frac{\mu_m^2}{3k_B T} [1 - \exp(-2D_r t)] \tag{7.110}$$

This is shown in Fig. 6.9(c).

The result of eq. (7.110) can also be obtained by using the fluctuation–dissipation theorem. According to Eq. (6.80),

$$\chi(t) = \frac{1}{2k_B T} \langle [m_z(t) - m_z(0)]^2 \rangle_0 \tag{7.111}$$

The right-hand side of eq. (7.111) can be calculated by using eq. (7.101)

$$\langle[m_z(t) - m_z(0)]^2\rangle_0 = \frac{\mu_m^2}{3}\langle[\boldsymbol{u}(t) - \boldsymbol{u}(0)]^2\rangle_0 = \frac{2\mu_m^2}{3}[1 - \exp(-2D_r t)]$$
(7.112)

Equations (7.111) and (7.112) give (7.110).

7.5.6   Diffusion equation in angular space

Here we derive the diffusion equation in angular space. We define the distribution function $\tilde{\psi}(\theta, \phi; t)$ in such a way that the probability of finding the angular coordinates θ', ϕ' of a rod in the region $\theta < \theta' < \theta + d\theta$ and $\phi < \phi' < \phi + d\phi$ is given by $\tilde{\psi}(\theta, \phi; t)d\theta d\phi$. This probability is normalized as

$$\int_0^\pi d\theta \int_0^{2\pi} d\phi \tilde{\psi}(\theta, \phi; t) = 1$$
(7.113)

We shall derive the diffusion equation for the case that there is no external potential. In this case, the equilibrium distribution ψ_{eq} is given by[11]

$$\tilde{\psi}_{eq}(\theta, \phi) = \frac{\sin\theta}{4\pi}$$
(7.114)

To derive the diffusion equation for $\tilde{\psi}(\theta, \phi; t)$, we follow the same procedure as in Section 7.3.3. We start from the conservation equation

$$\dot{\tilde{\psi}} = -\frac{\partial(\dot{\theta}\tilde{\psi})}{\partial\theta} - \frac{\partial(\dot{\phi}\tilde{\psi})}{\partial\phi}$$
(7.115)

and determine $\dot{\theta}$ and $\dot{\phi}$ using the Onsager principle.

The norm of the angular velocity ω is expressed by $\dot{\theta}$ and $\dot{\phi}$ as

$$\omega^2 = \dot{\theta}^2 + (\sin\theta\dot{\phi})^2$$
(7.116)

Therefore the energy dissipation function is given by

$$\Phi = \frac{1}{2}\int_0^\pi d\theta \int_0^{2\pi} d\phi\, \zeta_r[\dot{\theta}^2 + (\sin\theta\dot{\phi})^2]\tilde{\psi}$$
(7.117)

On the other hand, the free energy of the present system is given by

$$A[\tilde{\psi}] = k_B T \int_0^\pi d\theta \int_0^{2\pi} d\phi\, \tilde{\psi}\ln\left(\frac{4\pi\tilde{\psi}}{\sin\theta}\right)$$
(7.118)

Notice that the integrand is not equal to $\tilde{\psi}\ln\tilde{\psi}$. The extra term $\tilde{\psi}\ln(4\pi/\sin\theta)$ arises from the fact that at equilibrium (i.e., at the state where $A[\tilde{\psi}]$ becomes a minimum) $\tilde{\psi}$ is given by eq. (7.114).

[11] The factor $\sin\theta$ comes from the fact that the area of the region of $\theta < \theta' < \theta + d\theta$, $\phi < \phi' < \phi + d\phi$ is $\sin\theta d\theta d\phi$.

From eqs. (7.115) and (7.118), $\dot{A}$ is calculated as

$$
\dot{A} = -k_B T \int d\theta d\phi \left(\frac{\partial \dot{\theta} \tilde{\psi}}{\partial \theta} + \frac{\partial \dot{\phi} \tilde{\psi}}{\partial \phi} \right) \left[\ln \left(\frac{4\pi \tilde{\psi}}{\sin \theta} \right) + 1 \right]
$$

$$
= k_B T \int d\theta d\phi \; \tilde{\psi} \left(\dot{\theta} \frac{\partial}{\partial \theta} + \dot{\phi} \frac{\partial}{\partial \phi} \right) \ln \left(\frac{\tilde{\psi}}{\sin \theta} \right) \qquad (7.119)
$$

Minimization of $\Phi + \dot{A}$ with respect to $\dot{\theta}$ and $\dot{\phi}$ gives

$$
\dot{\theta} = -\frac{k_B T}{\zeta_r \tilde{\psi}} \left(\frac{\partial \tilde{\psi}}{\partial \theta} - \cot \theta \tilde{\psi} \right) \qquad (7.120)
$$

$$
\dot{\phi} = -\frac{k_B T}{\zeta_r \tilde{\psi} \sin^2 \theta} \frac{\partial \tilde{\psi}}{\partial \phi} \qquad (7.121)
$$

The time evolution equation for $\tilde{\psi}$ is determined by eqs. (7.115), (7.120), and (7.121):

$$
\frac{\partial \tilde{\psi}}{\partial t} = D_r \left[\frac{\partial}{\partial \theta} \left(\frac{\partial \tilde{\psi}}{\partial \theta} - \cot \theta \tilde{\psi} \right) + \frac{1}{\sin^2 \theta} \frac{\partial^2 \tilde{\psi}}{\partial \phi^2} \right] \qquad (7.122)
$$

This is the rotational diffusion equation in angular space.

In many publications, the rotational diffusion equation is written for the probability distribution function defined by

$$
\psi(\theta, \phi) = \frac{1}{\sin \theta} \tilde{\psi}(\theta, \phi) \qquad (7.123)
$$

which is normalized as

$$
\int_0^\pi d\theta \int_0^{2\pi} d\phi \sin \theta \psi(\theta, \phi; t) = 1 \qquad (7.124)
$$

The equation for $\psi(\theta, \phi; t)$ becomes

$$
\frac{\partial \psi}{\partial t} = D_r \left[\frac{1}{\sin \theta} \frac{\partial}{\partial \theta} \left(\sin \theta \frac{\partial \psi}{\partial \theta} \right) + \frac{1}{\sin^2 \theta} \frac{\partial^2 \psi}{\partial \phi^2} \right] \qquad (7.125)
$$

The rotational diffusion can be discussed using either eq. (7.122) or eq. (7.125).

7.6   Summary of this chapter

The time evolution of the configuration of particles moving in viscous fluids can be formulated by the variational principle that the Rayleighian of the system is minimized with respect to the particle velocity. The variational principle represents the condition that the viscous force is balanced with the potential force, and is generally guaranteed by the reciprocal relation for the friction matrix.

Onsager showed that the same structure exists in many time evolution equations for the general non-equilibrium state, and therfore they can be formulated by the variational principle. This variational principle, called the Onsager principle, gives us a powerful tool in deriving the time evolution for the system, and also in calculating the force needed to change the external parameters

The usefulness of the Onsager principle has been demonstrated for problems such as (i) the dynamics of wetting in a porous medium, (ii) diffusion of particles in dilute and concentrated solutions, and (iii) rotational diffusion of rod-like particles.

Further reading

(1) *The Theory of Polymer Dynamics*, Masao Doi and Sam F. Edwards, Oxford University Press (1986).

(2) *Colloidal Dispersions*, William B. Russel, D. A. Saville, W. R. Schowalter, Cambridge University Press (1992).

Exercises

(7.1) Calculate the time evolution of the meniscus height $h(t)$ when a hollow cylindrical capillary is inserted in a pool of fluid (see Fig. 4.7(a)) by answering the following questions.

(a) Assume that the fluid velocity in the capillary has a vertical component only, and is written as $v(r)$, where r is the distance from the axis of the capillary. Show that the energy dissipation function of the system is written as

$$\Phi = \frac{\eta}{2} h \int_0^a dr\, 2\pi r \left(\frac{dv}{dr}\right)^2 \quad (7.126)$$

where η is the viscosity of the fluid, and a is the radius of the capillary.

(b) If the meniscus is rising at rate $\dot{h}$, $v(r)$ has to satisfy the constraint

$$\dot{h} = \frac{1}{\pi a^2} \int_0^a dr\, 2\pi r v(r) \quad (7.127)$$

Minimize Φ under this constraint, and express Φ as a function of $\dot{h}$. (Answer $\Phi = 4\pi\eta h \dot{h}^2$.)

(c) Show that $h(t)$ satisfies the following equation,

$$\dot{h} = \frac{2a\gamma_s}{\eta h}\left(1 - \frac{h}{h_e}\right) \quad (7.128)$$

where $h_e = 2\gamma_s/\rho g a$. Discuss the behaviour of $h(t)$ in the short and the long time limits.

(7.2) Consider a colloidal suspension contained in a box made of a porous wall as shown in Fig. 7.2(a). Suppose that the fluid is flowing with velocity v from bottom to top. Answer the following questions.

(a) Show that the number density of particles in a steady state is given by

$$\frac{\partial n}{\partial x} = \frac{vn}{D_c(n)} \quad (7.129)$$

(b) Estimate the difference in the number density Δn at the top and that at the bottom when $v = 1\,\mu m$ for particles of sizes $a = 0.01, 0.1, 1, 10\,\mu m$. (Assume that the suspension is dilute.)

(7.3) Calculate the time evolution of the movement of the piston in the situation illustrated in Fig. 2.11 by answering the following questions.

(a) First consider the case that the energy dissipation in the system is mainly due to the fluid flow across the membrane. In this case, we may assume that the concentration in each chamber is uniform and that the energy dissipation function Φ is written as

$$\Phi = \frac{1}{2}\xi_m A\dot{x}^2 \quad (7.130)$$

Derive the equation for $x(t)$ and discuss the motion of the piston.

(b) Next consider the case that the energy dissipation is mainly caused by the diffusion of solute in the top chamber. Discuss the time needed for the piston to achieve the equilibrium position.

(7.4) The diffusion equation (7.49) can be written in the form of Onsager's kinetic equation (7.34). In fact, eq. (7.49) can be written as

$$\dot{n}(x) = -\int dx' \mu(x, x') \frac{\delta A}{\delta n(x')} \qquad (7.131)$$

Here the discrete indices i, j are replaced by continuous variables x, x', the summation is replaced by the integral, and the matrix μ_{ij} is replaced by a function $\mu(x, x')$. Answer the following questions.

(a) Show that if $\mu(x, x')$ is given by

$$\mu(x, x') = -\frac{\partial}{\partial x} \left[\frac{n(x)}{\zeta} \frac{\partial}{\partial x} \delta(x - x') \right] \qquad (7.132)$$

eq. (7.131) gives the diffusion equation (7.49).

(b) Show that the function $\mu(x, x')$ is symmetric, i.e., for any functions $n_1(x)$ and $n_2(x)$, the following identity holds.

$$\int dx \int dx' \, \mu(x, x') n_1(x) n_2(x')$$
$$= \int dx \int dx' \, \mu(x', x) n_1(x) n_2(x')$$
$$\qquad (7.133)$$

(7.5) The rotational diffusion equation of a rod-like particle can be derived by considering the distribution function $\psi(\mathbf{r}; t)$ for the vector $\mathbf{r}$ which joins the two ends of the rod. Answer the following questions.

(a) Express the energy dissipation function Φ using the rotational friction constant ζ_r and $\dot{\mathbf{r}}$

(b) The free energy of the system is written as

$$A = \int d\mathbf{r} k_B T \psi \ln \psi \qquad (7.134)$$

Show that the Rayleighian of the system is written as

$$R = \frac{\zeta_r}{2} \int d\mathbf{r} \frac{\dot{\mathbf{r}}^2}{r^2} \psi + \int d\mathbf{r} k_B T \dot{\psi} (\ln \psi + 1)$$
$$+ \int d\mathbf{r} \lambda \dot{\mathbf{r}} \cdot \mathbf{r} \qquad (7.135)$$

where λ is the Lagrangean constant representing the constraint $\dot{\mathbf{r}} \cdot \mathbf{r} = 0$, and $\dot{\psi}$ is given by

$$\dot{\psi} = -\frac{\partial}{\partial \mathbf{r}} \cdot (\dot{\mathbf{r}} \psi) \qquad (7.136)$$

(c) Using the Onsager principle, derive the following time evolution equation for ψ

$$\frac{\partial \psi}{\partial t} = D_r \frac{\partial}{\partial \mathbf{r}} \cdot \left[(r^2 - \mathbf{r}\mathbf{r}) \cdot \frac{\partial \psi}{\partial \mathbf{r}} \right] \qquad (7.137)$$

(d) Show that eq. (7.137) can be written as

$$\frac{\partial \psi}{\partial t} = D_r \mathcal{R}^2 \psi \qquad (7.138)$$

where $\mathcal{R}$ is defined by

$$\mathcal{R} = \mathbf{r} \times \frac{\partial}{\partial \mathbf{r}} \qquad (7.139)$$

(7.6) Let $\mathbf{u}(t)$ be a unit vector parallel to the rod in rotational Brownian motion. Show that the time correlation function $\langle (\mathbf{u}(t) \cdot \mathbf{u}(0))^2 \rangle$ is given by

$$\langle (\mathbf{u}(t) \cdot \mathbf{u}(0))^2 \rangle = \frac{2}{3} e^{-6D_r t} + \frac{1}{3} \qquad (7.140)$$

Explain why the correlation time of $\langle (\mathbf{u}(t) \cdot \mathbf{u}(0))^2 \rangle$ is shorter than that of $\langle \mathbf{u}(t) \cdot \mathbf{u}(0) \rangle$.

Diffusion and permeation in soft matter

8

In this chapter, we shall discuss the phenomena of diffusion and permeation in soft matter. Diffusion and permeation are the processes taking place in multi-component systems, where certain components (diffusing components) move relative to the other components (matrix components). This phenomenon is related to many situations seen in our everyday life such as dissolving, soaking, and drying, etc. If the phenomena take place under the action of gravitational or other external forces, it is equivalent to sedimentation, squeezing, or consolidation. Although these phenomena have different names, the physics is the same, and they can be treated in the same framework.

Diffusion was already discussed in the previous chapter for simple cases. In this chapter, we shall discuss two new topics which are important in soft matter.

First we shall discuss the effect of spatial correlation in the diffusing components. A common feature of soft matter systems is that they consist of large structural units. We shall discuss how to characterize such structures, and how it affects the speed of diffusion and permeation.

Next we shall discuss the motion of the matrix component in diffusion and permeation. Usually, diffusion and permeation are treated as if one component is moving in a background matrix which is at rest. In reality, when one component is moving, other components are also moving. Therefore the full description of the phenomena requires the motion of all components. We will discuss several cases where this effect is important, namely (i) sedimentation of colloidal particles, (ii) phase separation of simple molecules, and (iii) kinetics of solvent permeation in gels. It will be shown that the basic equations for such phenomena are conveniently derived from the Onsager principle.

8.1 Spatial correlation in soft matter solutions

8.1.1 Solutions with long-range correlation

As we have discussed in Chapter 1, a common feature of soft matter is that their structural units are large. For example, in polymer solutions and colloidal solutions, the size of solute (polymer molecules or

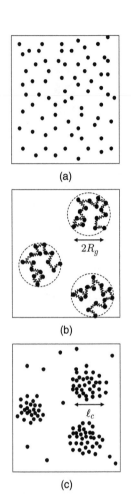

(a)

(b)

(c)

Fig. 8.1 Schematic view of solute concentration in solutions. (a) Solutions of simple molecules: the solute segments are distributed nearly uniformly. (b) Polymer solutions: there is a strong correlation in the distribution of polymer segments, i.e., the segments are seen within a polymer coil, but not outside of the polymer coil. (c) Solutions of surfactants or solutions near the critical point: strong correlation is seen even in systems of small molecules.

colloidal particles) is much larger than solvent molecules. In surfactant solutions, surfactant molecules assemble to form potentially large micelles.

Such characteristics are graphically illustrated in Fig. 8.1. Figure 8.1(a) shows a solution of simple molecules. Figure 8.1(b) shows the polymer solution. Here the dots represent solute molecules in (a), and polymer segments in (b). Figure 8.1(c) shows the surfactant solution where the surfactant molecules (again represented by dots) assemble to form micelles. The structure shown in Fig. 8.1(c) can also be found in solutions of simple molecules near the critical point, where the solute molecules tend to form large clusters as a precursory phenomena of phase separation.

As indicated in Fig. 8.1(b) and (c), a characteristic feature of soft matter is that solute segments are unevenly distributed: they form clusters, and their concentration is high in the cluster, but low outside of the cluster. To characterize this, we consider the fluctuations of the local density of solute segments.

Let us assume that the solution includes N solute segments in volume V, and let $\boldsymbol{r}_n$ ($n = 1, 2, \ldots, N$) be their position vectors. The average number density $\bar{n}$ of the solute segment is given by

$$\bar{n} = \frac{N}{V} \tag{8.1}$$

but the local number density fluctuates around this value.

Caution is required in discussing the fluctuation in the 'local number density'. The local number density is defined as the number of solute segments in a certain local region divided by the volume of the region. If we consider a spherical region of radius a, the local number density at point $\boldsymbol{r}$ can be defined by

$$n_a(\boldsymbol{r}) = \frac{1}{(4\pi/3)a^3} \int_{|\boldsymbol{r}-\boldsymbol{r}'|<a} d\boldsymbol{r}' \sum_n \delta(\boldsymbol{r}' - \boldsymbol{r}_n) \tag{8.2}$$

Here the integral on the right-hand side represents the number of solute segments in the spherical region $|\boldsymbol{r} - \boldsymbol{r}'| < a$. Clearly, the value of the local number density depends on a, the radius of the sphere.

In the following we consider the limit $a \to 0$, and denote the local density in this limit by $\hat{n}(\boldsymbol{r})$. Therefore

$$\hat{n}(\boldsymbol{r}) = \sum_n \delta(\boldsymbol{r} - \boldsymbol{r}_n) \tag{8.3}$$

The average of $\hat{n}(\boldsymbol{r})$ is equal to $\bar{n}$.

$$\langle \hat{n}(\boldsymbol{r}) \rangle = \bar{n} \tag{8.4}$$

8.1.2 Density correlation functions

The uneven distribution of solute segments is characterized by the spatial correlation function for the local number density.

$$\langle \hat{n}(\boldsymbol{r})\hat{n}(\boldsymbol{r}')\rangle = \sum_{n,m} \langle \delta(\boldsymbol{r}-\boldsymbol{r}_n)\delta(\boldsymbol{r}'-\boldsymbol{r}_m)\rangle \tag{8.5}$$

For a macroscopic sample at equilibrium, the correlation $\langle \hat{n}(\boldsymbol{r})\hat{n}(\boldsymbol{r}')\rangle$ has translational invariance, i.e., for an arbitrary vector $\boldsymbol{r}''$, the following identity holds:

$$\langle \hat{n}(\boldsymbol{r})\hat{n}(\boldsymbol{r}')\rangle = \langle \hat{n}(\boldsymbol{r}+\boldsymbol{r}'')\hat{n}(\boldsymbol{r}'+\boldsymbol{r}'')\rangle \tag{8.6}$$

We define the pair correlation function by

$$g(\boldsymbol{r}) = \frac{1}{\bar{n}}\langle \hat{n}(\boldsymbol{r})\hat{n}(0)\rangle = \frac{1}{\bar{n}}\sum_{n,m}\langle \delta(\boldsymbol{r}-\boldsymbol{r}_n)\delta(\boldsymbol{r}_m)\rangle \tag{8.7}$$

By eq. (8.6), $g(\boldsymbol{r})$ is written as

$$g(\boldsymbol{r}) = \frac{1}{\bar{n}}\sum_{n,m}\langle \delta(\boldsymbol{r}-\boldsymbol{r}_n+\boldsymbol{r}')\delta(\boldsymbol{r}_m+\boldsymbol{r}')\rangle \tag{8.8}$$

Since the right-hand side of eq. (8.8) is independent of $\boldsymbol{r}'$, it can be rewritten as

$$\langle \delta(\boldsymbol{r}-\boldsymbol{r}_n+\boldsymbol{r}')\delta(\boldsymbol{r}_m+\boldsymbol{r}')\rangle = \frac{1}{V}\int d\boldsymbol{r}' \langle \delta(\boldsymbol{r}-\boldsymbol{r}_n+\boldsymbol{r}')\delta(\boldsymbol{r}_m+\boldsymbol{r}')\rangle$$

$$= \frac{1}{V}\langle \delta(\boldsymbol{r}-\boldsymbol{r}_n-\boldsymbol{r}_m)\rangle \tag{8.9}$$

Hence

$$g(\boldsymbol{r}) = \frac{1}{\bar{n}V}\sum_{n,m}\langle \delta(\boldsymbol{r}-\boldsymbol{r}_n+\boldsymbol{r}_m)\rangle = \frac{1}{N}\sum_{n,m}\langle \delta(\boldsymbol{r}-\boldsymbol{r}_n+\boldsymbol{r}_m)\rangle \tag{8.10}$$

The meaning of $g(\boldsymbol{r})$ may be seen more clearly if $g(\boldsymbol{r})$ is written as

$$g(\boldsymbol{r}) = \frac{1}{N}\sum_{m} g_m(\boldsymbol{r}), \quad \text{with} \quad g_m(\boldsymbol{r}) = \sum_{n}\langle \delta(\boldsymbol{r}-\boldsymbol{r}_n+\boldsymbol{r}_m)\rangle \tag{8.11}$$

$g_m(\boldsymbol{r})$ represents the average density of segments at the point which is separated by vector $\boldsymbol{r}$ from a particular segment m. Therefore $g(\boldsymbol{r})$ represents the average density of segments at the point which is separated by $\boldsymbol{r}$ from an arbitrary chosen segment.[1]

At large distance $|\boldsymbol{r}|$, $g(\boldsymbol{r})$ becomes equal to the average density $\bar{n}$. At small distance $g(\boldsymbol{r})$ differs from $\bar{n}$ since segments are attracted or repelled by the segment at the centre. In a solution of simple molecules, $g(\boldsymbol{r})$ differs from $\bar{n}$ only over the range of atomic sizes (less than $1\,\text{nm}$). On the other hand, in the solutions of soft matter illustrated in Fig. 8.1(b) and (c), the correlation persists over much larger distances.

[1] The correlation function $g(\boldsymbol{r})$ is equivalent to the radial distribution $\tilde{g}(r)$ used in liquid theory except for the normalization constant. The radial distribution $\tilde{g}(r)$ is normalized in such a way that $\tilde{g}(r)$ approaches unity for large $|\boldsymbol{r}|$. Therefore $g(\boldsymbol{r})$ and $\tilde{g}(r)$ are related by $g(\boldsymbol{r}) = \bar{n}\tilde{g}(r)$.

If the segments form clusters, $g(\boldsymbol{r})$ is larger than $\bar{n}$ over the distance of the cluster size. The number of excess segments in a cluster can therefore be estimated by

$$N_c = \int d\boldsymbol{r}[g(\boldsymbol{r}) - \bar{n}] \tag{8.12}$$

N_c represents the number of correlated segments, and is called the correlation mass. In a dilute solution of polymers, N_c corresponds to the number of segments in a polymer chain. In micellar solutions, N_c corresponds to the average number of surfactant molecules in a micelle. The cluster size can be estimated by the length ℓ_c defined by

$$\ell_c^2 = \frac{\int d\boldsymbol{r}\ r^2[g(\boldsymbol{r}) - \bar{n}]}{6\int d\boldsymbol{r}\ [g(\boldsymbol{r}) - \bar{n}]} = \frac{1}{6N_c}\int d\boldsymbol{r}\ r^2[g(\boldsymbol{r}) - \bar{n}] \tag{8.13}$$

The length ℓ_c is called the correlation length. (The reason for the factor $1/6$ is explained later.)

Let $\delta n(\boldsymbol{r}) = \hat{n}(\boldsymbol{r}) - \bar{n}$ be the fluctuation of the number density. Then

$$\langle \hat{n}(\boldsymbol{r})\hat{n}(\boldsymbol{0}) \rangle = \bar{n}^2 + \langle \delta n(\boldsymbol{r})\delta n(\boldsymbol{0}) \rangle \tag{8.14}$$

Hence the quantity $g(\boldsymbol{r}) - \bar{n}$ appearing in eqs. (8.12) and (8.13) can be written as the spatial correlation of the fluctuation in the number density:

$$g(\boldsymbol{r}) - \bar{n} = \frac{1}{\bar{n}}\langle \delta n(\boldsymbol{r})\delta n(\boldsymbol{0}) \rangle \tag{8.15}$$

8.1.3   Scattering functions

The correlation function $g(\boldsymbol{r})$ can be measured by various scattering experiments. The principle of scattering experiments is explained in Fig. 8.2.

In general, a beam of radiation (light, X-rays, or neutrons) propagates straight in a homogeneous medium. If the medium has spatial inhomogeneity, part of the beam is scattered by the medium. If the scattering

Fig. 8.2 Principle of scattering experiments. (a) The incident beam is scattered by solute segments. The intensity of the scattered beam is measured by the detector as a function of the scattering angle θ. (b) The scattering intensity can be represented as a function of the scattering vector defined by $\boldsymbol{k} = \boldsymbol{k}_{out} - \boldsymbol{k}_{in}$, where $\boldsymbol{k}_{in}$ and $\boldsymbol{k}_{out}$ are the wavevectors of the incident beam and that of the scattered beam. Since the magnitudes of $\boldsymbol{k}_{in}$ and $\boldsymbol{k}_{out}$ are expressed in terms of the wavelength λ of the beam as $|\boldsymbol{k}_{in}| = |\boldsymbol{k}_{out}| = 2\pi/\lambda$, the magnitude of the scattering vector $\boldsymbol{k}$ is expressed as $|\boldsymbol{k}| = (4\pi/\lambda)\sin(\theta/2)$.

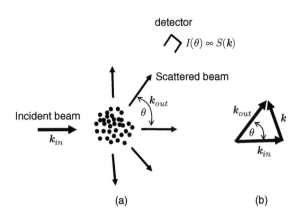

is caused by point-like objects located at r_n $(n = 1, 2, \ldots, N)$, the scattering intensity is proportional to the following quantity

$$S(k) = \frac{1}{N} \sum_{n,m} \left\langle e^{ik \cdot (r_n - r_m)} \right\rangle \qquad (8.16)$$

where k is the scattering vector defined in Fig. 8.2(b).

$S(k)$ is called the scattering function. By eq. (8.11), $S(k)$ is related to $g(r)$ by

$$S(k) = \int dr \, g(r) e^{ik \cdot r} \qquad (8.17)$$

For $k \neq 0$, this can be written as

$$S(k) = \int dr [g(r) - \bar{n}] e^{ik \cdot r} \qquad (8.18)$$

In isotropic materials, $g(r)$ depends on $r = |r|$ only. Therefore the integral on the right-hand side is calculated as

$$S(k) = \int_0^\infty dr r^2 \int_0^\pi d\theta \, 2\pi \sin\theta [g(r) - \bar{n}] e^{ikr \cos\theta}$$
$$= \int_0^\infty dr \, 4\pi r^2 [g(r) - \bar{n}] \frac{\sin(kr)}{kr} \qquad (8.19)$$

For small k, $\sin(kr)/(kr)$ is approximated as $1 - (kr)^2/6 + \cdots$. Hence, eq. (8.18) is written as

$$S(k) = \int dr \, [g(r) - \bar{n}] \left[1 - \frac{(kr)^2}{6} + \cdots \right] = N_c \left[1 - (k\ell_c)^2 + \cdots \right] \qquad (8.20)$$

where eqs. (8.12) and (8.13) have been used. Therefore N_c and ℓ_c are obtained from $S(k)$ in the small k region.

The schematic behaviour of the correlation function $g(r)$ and the scattering function $S(k)$ are shown in Fig. 8.3. The scattering function $S(k)$ starts with N_c at $k \to 0$, and decreases with k for $k\ell_c > 1$. This behaviour is often fitted by the following function (often called the Ornstein–Zernike form)

$$S(k) = \frac{N_c}{1 + (k\ell_c)^2} \qquad (8.21)$$

The correlation function $g(r)$ is obtained from the scattering function $S(k)$ by the inverse Fourier transform of eq. (8.18). For the scattering function of eq. (8.21), the inverse Fourier transform gives the following correlation function:

$$g(r) - \bar{n} = \frac{1}{(2\pi)^3} \int dk \, S(k) e^{-ik \cdot r} = \frac{N_c}{4\pi \ell_c^2} \frac{e^{-r/\ell_c}}{r} \qquad (8.22)$$

Therefore $g(r)$ decreases exponentially with r.

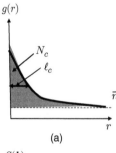

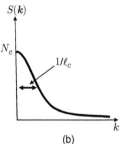

Fig. 8.3 Typical behaviour of the density correlation function $g(r)$, and the scattering function $S(k)$. Panels (a) and (b) schematically show how the number of correlated segments N_c and the correlation length ℓ_c are obtained from these functions.

8.1.4   Space–time correlation functions

The correlation function $g(\boldsymbol{r})$ defined above represents the spatial correlation of segments at a certain time. A time correlation function can also be defined. Let $\boldsymbol{r}_n(t)$ be the position of the segment n at time t. The local segment density at time t is defined by

$$\hat{n}(\boldsymbol{r}, t) = \sum_n \delta(\boldsymbol{r} - \boldsymbol{r}_n(t)) \tag{8.23}$$

The space–time correlation function of the segment density is defined by

$$g_d(\boldsymbol{r}, t) = \frac{1}{\bar{n}} \langle \hat{n}(\boldsymbol{r}, t)\hat{n}(0, 0)\rangle$$

$$= \frac{1}{N} \sum_{n,m} \langle \delta(\boldsymbol{r} - \boldsymbol{r}_n(t) + \boldsymbol{r}_m(0))\rangle \tag{8.24}$$

These are the generalization of eqs. (8.7) and (8.11). The Fourier transform of $g_d(\boldsymbol{r}, t)$ is called the dynamic scattering function

$$S_d(\boldsymbol{k}, t) = \int d\boldsymbol{r} \; g_d(\boldsymbol{r}, t)e^{i\boldsymbol{k}\cdot\boldsymbol{r}} = \frac{1}{N} \sum_{n,m} \left\langle e^{i\boldsymbol{k}\cdot[\boldsymbol{r}_n(t) - \boldsymbol{r}_m(0)]}\right\rangle \tag{8.25}$$

The dynamic scattering function can also be measured by scattering experiments (for example by analysing the time correlations of the intensity of the scattered beam). The dynamic scattering function $S_d(\boldsymbol{k}, t)$ agrees with the static scattering function $S(\boldsymbol{k})$ at $t = 0$:

$$S_d(\boldsymbol{k}, t = 0) = S(\boldsymbol{k}) \tag{8.26}$$

If we define the Fourier transform of the segment density at time t by

$$n_{\boldsymbol{k}}(t) = \frac{1}{V} \int d\boldsymbol{r} \; \hat{n}(\boldsymbol{r}, t)e^{i\boldsymbol{k}\cdot\boldsymbol{r}} = \frac{1}{V} \sum_n e^{i\boldsymbol{k}\cdot\boldsymbol{r}_n(t)} \tag{8.27}$$

the dynamic scattering function (8.25) is written as a time correlation of $n_{\boldsymbol{k}}(t)$:

$$S_d(\boldsymbol{k}, t) = \frac{V^2}{N} \langle n_{\boldsymbol{k}}(t)n_{\boldsymbol{k}}^*(0)\rangle = \frac{V}{\bar{n}} \langle n_{\boldsymbol{k}}(t)n_{\boldsymbol{k}}^*(0)\rangle \tag{8.28}$$

8.1.5   The long wavelength limit

The behaviour of the dynamic scattering function in the small $\boldsymbol{k}$ limit can be expressed in terms of macroscopic material parameters. To show this, we use the fluctuation–dissipation theorem discussed in Section 6.5.3. Suppose that a constant external field $h(\boldsymbol{r})$ is applied to the solute segments at time $t = 0$. Under the field $h(\boldsymbol{r})$, the free energy of the system is written as[2]

$$A[n(\boldsymbol{r})] = \int d\boldsymbol{r} \; [f(n(\boldsymbol{r})) - h(\boldsymbol{r})n(\boldsymbol{r})] \tag{8.30}$$

where $f(n)$ is the free energy density of the solution. By the same argument as given in Section 7.3, it can be shown that under the external

[2] Notice that an expression such as eq. (8.30) is valid only when $n(\boldsymbol{r})$ represents the macroscopic segment density; i.e., when $n(\boldsymbol{r})$ represents the density smeared over a length-scale larger than the correlation length ℓ_c, i.e., the density with a taken to be larger than ℓ_c. Alternatively, the smeared density $n(\boldsymbol{r})$ may be defined by

$$n(\boldsymbol{r}) = \sum_{|\boldsymbol{k}|<1/\ell_c} n_{\boldsymbol{k}}e^{-i\boldsymbol{k}\cdot\boldsymbol{r}} \tag{8.29}$$

where the summation is taken for $\boldsymbol{k}$ which satisfies $|\boldsymbol{k}| < 1/\ell_c$.

potential $h(\boldsymbol{r})$, the number density $n(\boldsymbol{r}, t)$ obeys the following diffusion equation (see eq. (7.65))

$$\frac{\partial n}{\partial t} = \nabla \cdot D_c(n) \left[\nabla n - \frac{n}{\Pi'(n)} \nabla h \right] \tag{8.31}$$

where D_c is the collective diffusion constant. Equation (8.31) can be solved for small h. The result can be written by use of the response function $\chi(\boldsymbol{r}, \boldsymbol{r}', t)$ as

$$\delta n(\boldsymbol{r}, t) = \int d\boldsymbol{r}' \, \chi(\boldsymbol{r}, \boldsymbol{r}', t) h(\boldsymbol{r}') \tag{8.32}$$

where $\delta n(\boldsymbol{r}, t) = n(\boldsymbol{r}, t) - \bar{n}$.

According to eq. (6.80), the response function is related to the time correlation function at equilibrium

$$\chi(\boldsymbol{r}, \boldsymbol{r}', t) = \frac{1}{2k_B T} \big\langle \, [\hat{n}(\boldsymbol{r}, t) - \hat{n}(\boldsymbol{r}, 0)] \, [\hat{n}(\boldsymbol{r}', t) - \hat{n}(\boldsymbol{r}', 0)] \, \big\rangle \tag{8.33}$$

The right-hand side can be expressed by $g_d(\boldsymbol{r}, t)$. Such a calculation gives the following expression for the dynamic scattering function (see problem (8.4))

$$S_d(\boldsymbol{k}, t) = \frac{k_B T}{\Pi'} e^{-D_c k^2 t} \tag{8.34}$$

Equation (8.34) indicates that the dynamic scattering function in the long wavelength limit ($\|\boldsymbol{k}\| \to 0$) is represented by macroscopic material parameters such as Π' and D_c.

Comparing eqs. (8.20) and (8.34), we see that the number of correlated segments is given by

$$N_c = \frac{k_B T}{\Pi'} \tag{8.35}$$

In the case of dilute polymer solutions, Π is given by $(n/N)k_B T$, where N is the number of segments in a polymer chain. Therefore N_c is equal to N, and the correlation effect is very large. With the increase of polymer concentration, however, the osmotic pressure increases quickly, and N_c decreases with the increase of polymer concentration. At high concentration, N_c becomes about the same order as that of simple liquids.

The correlation effect can be important in solutions of simple molecules, if Π' is small. This happens near the critical point. At the critical point, Π' becomes zero (see Section 2.3.4). Near the critical point, Π' is not zero, but becomes very small. Therefore there is a strong correlation effect. This creates various anomalies in the static and dynamic properties of the solutions, which are generally called critical phenomena.

8.1.6   Correlation effects in friction constants and diffusion constants

As discussed in Section 7.4.1, the diffusion speed is determined by the balance of two forces: the thermodynamic force which drives the spreading of solute molecules, and the frictional force which resists the spreading. The former is represented by $\Pi' = \partial\Pi/\partial n$, and the latter is represented by the friction constant ξ. Both parameters are affected by the spatial correlation of solute segments. The effect of correlation on Π' is represented by eq. (8.35). We now discuss the effect of correlation on the friction constant ξ.

The correlation in the particle configuration strongly affects the friction constant ξ. To see this, let us compare the two situations shown in Fig. 8.4. Here the dots represent the obstacles fixed in space. In (a), the obstacles are distributed homogeneously, while in (b) the same number of obstacles are distributed unevenly. If the segments are uniformly distributed as in the case of (a), each segment acts as a hydrodynamic friction centre, and ξ can be estimated by $\xi_0 = \bar{n}6\pi\eta a$. On the other hand, if segments form clusters as in (b), it is the cluster which acts as a hydrodynamic friction centre. Since the fluid velocity is strongly damped within a cluster, the fluid cannot pass through the clusters. Accordingly, an individual cluster behaves as a solid friction centre. Let N_c be the mean number of obstacles in a cluster. The number density of clusters is $\bar{n}/N_c$ and the radius of the cluster is ℓ_c. Therefore ξ is estimated by

$$\xi \simeq \frac{\bar{n}}{N_c}6\pi\eta\ell_c \qquad (8.36)$$

which is less than ξ_0 by the factor $\ell_c/(aN_c)$. Usually N_c and ℓ_c are related by

$$N_c \approx \left(\frac{\ell_c}{a}\right)^d \qquad (8.37)$$

The exponent d is called the fractal dimension. The fractal dimension d is equal to 3 if segments are densely packed in the cluster, but is less than 3 if they are loosely packed. According to eq. (8.36), the friction constant ξ is written as

$$\xi \approx \xi_0 N_c^{-(1-1/d)} \qquad (8.38)$$

Therefore the friction constant decreases with the increase of the cluster size. Many phenomena are explained by this result. (i) Water moves faster through pebbles than sand since the cluster size is larger in pebbles than sand, (ii) colloidal particles sediment faster when they aggregate, (iii) tiny water droplets in a cloud start to fall when their size becomes large, and (iv) the sedimentation speed of a polymer in a solution increases with the increase of the molecular weight.

An approximate theory has been developed for the friction constant for a system made of point-like obstacles. According to one such theory, ξ is given by

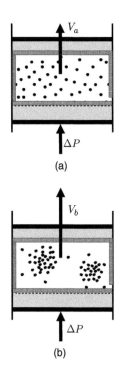

Fig. 8.4 Comparison of the friction constant ξ of a bed of obstacles (represented by dots) for two situations. (a) The obstacles are placed uniformly in space, and (b) the obstacles are placed non-uniformly forming clusters. The friction constant ξ in (a) is larger than that in (b).

$$\frac{1}{\xi} = \frac{2}{3\bar{n}\eta} \int_0^\infty dr \; r[g(r) - \bar{n}] \qquad (8.39)$$

If eq. (8.22) is used for $g(r)$, eq. (8.39) gives

$$\xi = \frac{\bar{n}}{N_c} 6\pi\eta\ell_c \qquad (8.40)$$

Therefore the collective diffusion constant (eq. (7.66)) is given by

$$D_c = \frac{\bar{n}\Pi'}{(\bar{n}/N_c)6\pi\eta\ell_c} = \frac{k_B T}{6\pi\eta\ell_c} \qquad (8.41)$$

This expression is known to hold in many solutions which have long-range correlation such as polymer solutions, micellar solutions, and solutions near the critical point.

8.1.7 Density correlation at small length-scales

The above discussion does not hold for small length-scales, those smaller than the correlation length ℓ_c. At such length-scales, the free energy is not written as in eq. (8.30). The free energy now involves gradient terms ∇n (see Section 8.3.1), and the friction constant ξ now depends on the wavevector $\boldsymbol{k}$.

In the large-wavelength limit, a concentration fluctuation of wavevector $\boldsymbol{k}$ always decays with rate $D_c\boldsymbol{k}^2$ (see eq. (8.34)). When the wavelength $1/k$ becomes less than ℓ_c, this is not true any more. It has been shown that the decay rate becomes proportional to k^3 for wavelengths between the correlation length ℓ_c and atomic size a. If the wavelength $1/k$ becomes comparable to the atomic size, the details of the molecular structure matter. At such small length-scales, the density correlation function $g(\boldsymbol{r})$ shows an oscillation reflecting the microscopic structure of molecules.

8.2 Diffusio-mechanical coupling in particle sedimentation

8.2.1 Motion of a medium in diffusion

So far we have been considering the motion of solute only, assuming that the background medium is at rest. However, in multi-component systems, when one component moves, the other components move also. For example, when particles sediment like those shown in Fig. 8.5, the solvent must flow from bottom to top in order to fill the space evacuated by the falling particles. The volume conservation condition indicates that if the particles move down with average velocity $\bar{v}_p$, the solvent has to move up with average velocity $\bar{v}_s = -\phi/(1-\phi)\bar{v}_p$. This velocity is non-negligible unless ϕ is very small. Therefore to describe such phenomena completely, it is necessary to determine the velocity of all components.

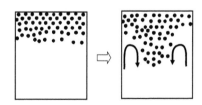

Fig. 8.5 When a container of particulate slurry is placed upside down, the particles start to sediment. If a portion of the slurry starts to sediment faster than the other part, it causes a convective flow, and destabilizes homogeneous sedimentation.

$$v_m = \frac{\rho_p \phi v_p + (1-\phi)\rho_s v_s}{\rho_p \phi + (1-\phi)\rho_s} \quad (8.42)$$

where ρ_p and ρ_s are the density of particles and solvent. Such a velocity has a non-zero value even if the sedimentation takes place uniformly. It is therefore reasonable to use the volume average velocity rather than the mass average velocity.

Let us consider the case of two-component systems. Let $v_p(r)$ and $v_s(r)$ be the average velocities of particles (or solute) and solvent at point r. A natural definition of the velocity of the medium (or the background velocity) is the volume average velocity[3]

$$v = \phi v_p + (1-\phi)v_s \quad (8.43)$$

We shall call this the medium velocity. When sedimentation takes place uniformly in a container, the medium velocity v is zero. However, if a certain portion of particles start to sediment faster than the others, a macroscopic flow is induced as shown in Fig. 8.5. In this case, we have to consider how the medium velocity v is coupled with the particle velocity.

The coupling between the diffusive flow (i.e., the relative motion between components) and the convective flow (i.e., the flow of the medium) is called diffusio-mechanical coupling. The diffusio-mechanical coupling is important in both fluids and gels. In the following, we shall first discuss the diffusio-mechanical coupling in fluids. The diffusio-mechanical coupling in gels will be discussed in Section 8.4.

8.2.2   Continuum description of particle sedimentation

Let us now derive the equations that describe the diffusio-mechanical coupling in particle sedimentation using the Onsager principle. In the flow of two-component fluids, there are essentially two mechanisms for energy dissipation. One is associated with the relative velocity between components, and the other is associated with the spatial gradient of the velocity.

The energy dissipation function for the first mechanism is written as

$$\Phi_1 = \frac{1}{2} \int dr \tilde{\xi}(v_p - v_s)^2 \quad (8.44)$$

where $\tilde{\xi}$ represents the friction coefficient for the relative motion between solute and solvent. The velocity difference $v_p - v_s$ can be expressed by $v_p - v$, the particle velocity relative to the background by use of eq. (8.43)

$$v_p - v = (1-\phi)(v_p - v_s) \quad (8.45)$$

Therefore the energy dissipation function is written as

$$\Phi_1 = \frac{1}{2} \int dr \xi(v_p - v)^2 \quad (8.46)$$

where

$$\xi = \frac{\tilde{\xi}}{(1-\phi)^2} \quad (8.47)$$

The second mechanism of energy dissipation is due to the spatial gradient of the fluid velocity. The existence of such a term is clear for a pure solvent. If the solution can be regarded as a viscous fluid with viscosity $\eta(\phi)$, the energy dissipation function is written as[4]

[4] The energy dissipation function can include other terms such as $\nabla \boldsymbol{v}_p : \nabla \boldsymbol{v}_p$, and $\nabla \boldsymbol{v}_p : \nabla \boldsymbol{v}$, etc. To simplify the equations, such terms are ignored here.

$$\Phi_2 = \frac{1}{4} \int d\boldsymbol{r} \eta \left(\frac{\partial v_\beta}{\partial r_\alpha} + \frac{\partial v_\alpha}{\partial r_\beta} \right)^2$$

$$= \frac{1}{4} \int d\boldsymbol{r} \eta \left(\nabla \boldsymbol{v} + (\nabla \boldsymbol{v})^t \right) : \left(\nabla \boldsymbol{v} + (\nabla \boldsymbol{v})^t \right) \qquad (8.48)$$

Here $\boldsymbol{a}^t$ represents the transpose of a tensor $\boldsymbol{a}$, and $\boldsymbol{a} : \boldsymbol{a}$ represents the trace of the tensor product $\boldsymbol{a} \cdot \boldsymbol{a}$, i.e., $(\boldsymbol{a}^t)_{\alpha\beta} = (\boldsymbol{a})_{\beta\alpha}$ and $\boldsymbol{a} : \boldsymbol{a} = \mathrm{Tr}(\boldsymbol{a} \cdot \boldsymbol{a}) = a_{\alpha\beta}a_{\beta\alpha}$).

On the other hand, the free energy of the system is written in the same form as eq. (7.59)

$$A = \int d\boldsymbol{r} \left[f(\phi) - \rho_1 \phi \boldsymbol{g} \cdot \boldsymbol{r} \right] \qquad (8.49)$$

where $f(\phi)$ is the free energy density of the solution (expressed as a function of volume fraction), and ρ_1 is the difference in the density of solute and solution. Since ϕ satisfies the conservation law

$$\dot{\phi} = -\nabla \cdot (\boldsymbol{v}_p \phi) \qquad (8.50)$$

$\dot{A}$ is calculated in the same way as in eq. (7.56):

$$\dot{A} = \int d\boldsymbol{r} \, \dot{\phi} \left[f'(\phi) - \rho_1 \boldsymbol{g} \cdot \boldsymbol{r} \right]$$

$$= -\int d\boldsymbol{r} \, \nabla \cdot (\phi \boldsymbol{v}_p) \left[f'(\phi) - \rho_1 \boldsymbol{g} \cdot \boldsymbol{r} \right]$$

$$= \int d\boldsymbol{r} \, \boldsymbol{v}_p \cdot \left[\nabla \Pi - \phi \rho_1 \boldsymbol{g} \right] \qquad (8.51)$$

Note that the medium velocity $\boldsymbol{v}$ has to satisfy the incompressible condition.

$$\nabla \cdot \boldsymbol{v} = 0 \qquad (8.52)$$

Summing up these equations, we have the Rayleighian

$$R = \frac{1}{2} \int d\boldsymbol{r} \xi (\boldsymbol{v}_p - \boldsymbol{v})^2 + \frac{1}{4} \int d\boldsymbol{r} \eta (\nabla \boldsymbol{v} + (\nabla \boldsymbol{v})^t) : (\nabla \boldsymbol{v} + (\nabla \boldsymbol{v})^t)$$

$$+ \int d\boldsymbol{r} \, \boldsymbol{v}_p \cdot \left[\nabla \Pi - \phi \rho_1 \boldsymbol{g} \right] - \int d\boldsymbol{r} \, p \nabla \cdot \boldsymbol{v} \qquad (8.53)$$

The last term represents the constraint (8.52).

Setting the variation of R with respect to $\boldsymbol{v}_p$ and $\boldsymbol{v}$ equal to zero, we have the following equations

$$\xi (\boldsymbol{v}_p - \boldsymbol{v}) = -\nabla \Pi + \phi \rho_1 \boldsymbol{g} \qquad (8.54)$$

$$\nabla \cdot \eta \left[\nabla \boldsymbol{v} + (\nabla \boldsymbol{v})^t \right] + \xi (\boldsymbol{v}_p - \boldsymbol{v}) = \nabla p \qquad (8.55)$$

Equation (8.54) represents the force balance for the solute: the frictional force exerted on the solute by the medium is balanced by the thermodynamic driving force. Equation (8.55) represents the force balance for the medium.

The two equations, (8.54) and (8.55) can be written as

$$\boldsymbol{v}_p = \boldsymbol{v} - \frac{1}{\xi}(\nabla\Pi - \rho_1\phi\boldsymbol{g}) \qquad (8.56)$$

$$\nabla \cdot \eta \left[\nabla\boldsymbol{v} + (\nabla\boldsymbol{v})^t\right] = \nabla(p + \Pi) - \phi\rho_1\boldsymbol{g} \qquad (8.57)$$

For given $\phi(\boldsymbol{r})$, $(\boldsymbol{v}_p, \boldsymbol{v}, p)$ are obtained by solving eqs. (8.52), (8.56), and (8.57). Once $\boldsymbol{v}_p$ is obtained, $\dot{\phi}$ is determined by eq. (8.50). Therefore the above set of equations determine the time evolution of the concentration field $\phi(\boldsymbol{r})$. It should be noted that this complex set of equations can still be written in the form of Onsager's kinetic equation (7.34) since the equations are derived from the Onsager principle (see problem (7.4)).

The unstable sedimentation shown in Fig. 8.5 can be described by these equations. In fact, eq. (8.57) is equivalent to the Stokes equation for a fluid which has inhomogeneous density $\phi\rho_1$. The situation shown in Fig. 8.5(b) is equivalent to the situation that a heavy fluid is placed on top of a light fluid. Such a configuration is unstable, and the instability, known as Rayleigh–Taylor instability, takes place.

Notice that the osmotic pressure Π does not cause such instability. If there is no effect of gravity (i.e., if $\rho_1\boldsymbol{g} = 0$), and if the fluid is in a closed container, the medium velocity remains at zero even if the osmotic pressure gradient is very large. This can be seen as follows. In the velocity-determining equations (eqs. (8.52) and (8.57)), the osmotic pressure is absorbed in the pressure term: it can be eliminated by defining the pressure $p' = p + \Pi$. If the fluid is in a closed container, $\boldsymbol{v}$ must be zero at the wall. Under such boundary conditions, the solution of eqs. (8.52) and (8.57) is that the medium velocity $\boldsymbol{v}$ is zero everywhere in the container.[5]

[5] This does not mean that the osmotic pressure does not induce the macroscopic motion of fluid. In an open container, osmotic pressure can induce convective flow. For example, in the standard U-tube experiments to measure the osmotic pressure, convective flow is induced by the osmotic pressure.

8.3   Kinetics of phase separation

8.3.1   Phase separation from a thermodynamically unstable state

As a topic related to diffusive transport, we now discuss the kinetics of phase separation. Phase separation can be thought of as an inverse diffusion process. When a homogeneous solution made of components A and B is brought into a thermodynamically unstable state (by, for example, changing the temperature), the solution starts to separate into two phases, the A-rich solution and the B-rich solution. This process is called phase separation. Since this process is caused by the relative motion between components driven by thermodynamic force, it can be treated in the same framework as that of diffusion.

Let $f(\phi)$ be the free energy density of the solution in which A and B are mixed uniformly with volume ratio $\phi : (1 - \phi)$. As discussed in Chapter 2, the solution is unstable in the concentration region where $f''(\phi) = \partial^2 f/\partial \phi^2$ is negative. A homogeneous solution having such an initial concentration ϕ_0 eventually turns into two solutions with different volume fractions ϕ_a and ϕ_b which are determined by the common tangent construction shown in Fig. 8.6(a) (see also Fig. 2.4).

In the unstable region, the collective diffusion constant $D_c(\phi)$ is negative because $D_c(\phi)$ is given by

$$D_c = \frac{1}{\xi} \frac{\partial \Pi}{\partial \phi} = \frac{\phi}{\xi} \frac{\partial^2 f}{\partial \phi^2} \tag{8.58}$$

Therefore solute moves from low concentration region to high concentration. As a result, if there is a small concentration inhomogeneity, it grows in time as shown in Fig. 8.6(b) (I to II). With the passage of time, droplets of concentrated solution (with volume fraction ϕ_a) appear in the sea of dilute solution (with volume fraction ϕ_b) as shown in Fig. 8.6(b) III.

Such a process can be discussed using the same theoretical framework as in the previous section. A modification, however, is needed for phase separation.

In the final stage of phase separation, the A-rich phase coexists with the B-rich phase, and there is an interface between the two phases. As we have seen in Chapter 4, the interface has an extra free energy, the interfacial free energy, that is not accounted for in the free energy expression used in the previous section (eq. (8.49)). According to eq. (8.49), the free energy A of the phase separated state depends on the volume of each phase only (i.e., $A = V_a f(\phi_a) + V_b f(\phi_b)$), and does not depend on the area of the interface.

The interfacial free energy is accounted for by modifying eq. (8.49) as follows.

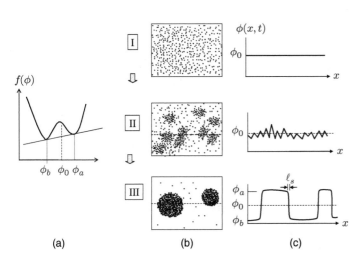

(a) (b) (c)

Fig. 8.6 (a) The free energy density $f(\phi)$ for solutions undergoing phase separation. If the curve $f(\phi)$ is upper convex (i.e., if $f''(\phi) < 0$) at the initial concentration ϕ_0, the solution is thermodynamically unstable, and eventually separates into two solutions having concentration ϕ_a and ϕ_b. (b) An example of the structural change in phase separation. When phase separation starts, the solute molecules assemble to form locally concentrated regions (I to II). These regions grow and form macroscopic droplets of phase separated solutions (II to III). (c) Time evolution of the density profile in the three stages shown in (b).

$$A[\phi(\boldsymbol{r})] = \int d\boldsymbol{r} \left[f(\phi) - \rho_1 \phi \boldsymbol{g} \cdot \boldsymbol{r} + \frac{1}{2} \kappa_s (\boldsymbol{\nabla}\phi)^2 \right] \tag{8.59}$$

where κ_s is a positive constant. At the interface, $\phi(\boldsymbol{r})$ changes steeply from ϕ_a to ϕ_b along the direction normal to the interface. Therefore the gradient term $\kappa_s(\boldsymbol{\nabla}\phi)^2$ becomes very large at the interface. Let ℓ_s be the thickness of the interface (see Fig. 8.6(c) III). Then $\boldsymbol{\nabla}\phi$ is of the order of $(\phi_a - \phi_b)/\ell_s$ in the interfacial region. Therefore the integral of the gradient term in eq. (8.59) is estimated as

$$\int d\boldsymbol{r}\kappa_s(\boldsymbol{\nabla}\phi)^2 \approx \kappa_s \left(\frac{\phi_a - \phi_b}{\ell_s} \right)^2 \ell_s \times \text{the interfacial area} \tag{8.60}$$

Therefore, the interfacial tension γ is estimated by

$$\gamma \approx \kappa_s \frac{(\phi_a - \phi_b)^2}{\ell_s} \tag{8.61}$$

For the free energy expression of eq. (8.59), $\dot{A}$ is calculated as

$$\dot{A} = \int d\boldsymbol{r} \,\, \dot{\phi}\frac{\delta A}{\delta \phi} = -\int d\boldsymbol{r} \,\, \boldsymbol{\nabla} \cdot (\boldsymbol{v}_p \phi)\frac{\delta A}{\delta \phi} = \int d\boldsymbol{r} \,\, \phi \boldsymbol{v}_p \cdot \boldsymbol{\nabla}\left(\frac{\delta A}{\delta \phi} \right) \tag{8.62}$$

If this expression is used, eq. (8.54) is now replaced by

$$\xi(\boldsymbol{v}_p - \boldsymbol{v}) = -\phi \boldsymbol{\nabla} \left(\frac{\delta A}{\delta \phi} \right) \tag{8.63}$$

The variational derivative $\delta A/\delta \phi$ is given by

$$\frac{\delta A}{\delta \phi} = f'(\phi) - \rho_1 \boldsymbol{g} \cdot \boldsymbol{r} - \kappa_s \boldsymbol{\nabla}^2 \phi \tag{8.64}$$

Using this equation, we finally have the following equations

$$\frac{\partial \phi}{\partial t} = -\boldsymbol{\nabla} \cdot (\phi \boldsymbol{v}) + \boldsymbol{\nabla} \cdot \left[\frac{\phi}{\xi} \left(\boldsymbol{\nabla}\Pi - \rho_1 \phi \boldsymbol{g} - \kappa_s \phi \boldsymbol{\nabla}\boldsymbol{\nabla}^2 \phi \right) \right] \tag{8.65}$$

$$\eta \boldsymbol{\nabla}^2 \boldsymbol{v} = \boldsymbol{\nabla}p + \boldsymbol{\nabla}\Pi - \rho_1 \phi \boldsymbol{g} + \kappa_s \phi \boldsymbol{\nabla}\boldsymbol{\nabla}^2 \phi \tag{8.66}$$

Here we have assumed that the viscosity η is constant independent of concentration. Notice that the term involving κ_s cannot be included in the pressure term. Therefore, with the presence of the κ_s term, convective flow may be induced in a closed container even if there is no effect of gravity.

Equations (8.65) and (8.66) have been solved numerically in many phase separation problems, and have clarified the characteristic features of phase separation kinetics. Here we summarize the main results of such studies

8.3.2   Early stage of phase separation

In the early stage of phase separation, the deviation from the homogeneous state is small. Therefore we can solve the equations expressed in terms of this small deviation from the initial state. Let $\delta\phi(\boldsymbol{r}, t)$ be defined by

$$\delta\phi(\boldsymbol{r}, t) = \phi(\boldsymbol{r}, t) - \phi_0 \qquad (8.67)$$

Equation (8.65) is then linearized with respect to $\delta\phi$. The result is[6]

$$\frac{\partial \delta\phi}{\partial t} = -\phi_0 \boldsymbol{\nabla} \cdot \boldsymbol{v} + \frac{\phi_0}{\xi} \boldsymbol{\nabla} \cdot \left[\left(K - \kappa_s \phi_0 \boldsymbol{\nabla}^2 \right) \boldsymbol{\nabla} \delta\phi \right] \qquad (8.68)$$

where

$$K = \left. \frac{\partial \Pi}{\partial \phi} \right|_{\phi_0} \qquad (8.69)$$

which is negative in the unstable region.

The first term on the right-hand side of eq. (8.68) is zero due to the incompressible condition (8.52). Equation (8.68) can be easily solved by use of the Fourier transform

$$\delta\phi(\boldsymbol{r}, t) = \sum_{\boldsymbol{k}} \delta\phi_{\boldsymbol{k}}(t) e^{i\boldsymbol{k} \cdot \boldsymbol{r}} \qquad (8.70)$$

The time evolution equation for $\delta\phi_{\boldsymbol{k}}(t)$ becomes

$$\frac{\partial \delta\phi_{\boldsymbol{k}}}{\partial t} = \alpha_{\boldsymbol{k}} \delta\phi_{\boldsymbol{k}} \qquad (8.71)$$

with

$$\alpha_{\boldsymbol{k}} = -\frac{\phi_0}{\xi} \boldsymbol{k}^2 \left(K + \kappa_s \phi_0 \boldsymbol{k}^2 \right) \qquad (8.72)$$

$\alpha_{\boldsymbol{k}}$ is plotted as a function of $k = |\boldsymbol{k}|$ in Fig. 8.7. Since K is negative, $\alpha_{\boldsymbol{k}}$ first increases as k increases, taking a maximum at $k^* = \sqrt{|K|/(2\kappa_s\phi_0)}$ and then decreases. As long as $\alpha_{\boldsymbol{k}}$ is positive, $\phi_{\boldsymbol{k}}$ increases in time, i.e., the concentration inhomogeneity grows in time. This process is called spinodal decomposition (which means phase separation from a thermodynamically unstable state). The growth rate $\alpha_{\boldsymbol{k}}$ is largest at $k = k^*$. Therefore the structure resulting from the spinodal decomposition has characteristic length $1/k^*$.

[6] In eq. (8.68). we have assumed that the gravitational term $\rho_1 \boldsymbol{g}$ is negligible. Notice that $\boldsymbol{v}$ is zero in the homogeneous state and is therefore of the same order as $\delta\phi$.

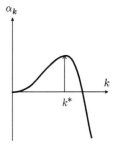

Fig. 8.7 The growth rate $\alpha_{\boldsymbol{k}}$ plotted against the wavevector $k = |\boldsymbol{k}|$.

8.3.3 Late stage of phase separation

As phase separation proceeds, the growth of the concentration inhomogeneity stops when $\phi(\boldsymbol{r}, t)$ becomes equal to ϕ_a or ϕ_b. When this state is attained, the solution has separated into domains in which $\phi(\boldsymbol{r}, t)$ takes the value of either ϕ_a or ϕ_b. Within each domain, $\phi(\boldsymbol{r}, t)$ does not change any more, but the size and the shape of the domains change in time. The driving force for this change is the interfacial tension γ. If the characteristic size of the domains is ℓ, the interfacial energy per unit volume is about γ/ℓ. The structural change of the domains takes place to decrease this free energy, i.e., to increase ℓ. This process is called coarsening. Both diffusion and convection are important in the coarsening process.

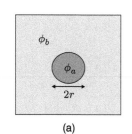

(a)

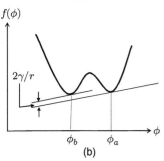

(b)

Fig. 8.8 (a) A droplet of concentrated phase of radius r is in equilibrium in the sea of dilute phase. (b) The graphical solution for the equilibrium volume fraction ϕ_a and ϕ_b for the droplet of finite radius r.

Diffusion

The diffusion is caused by the fact that due to the interfacial tension γ, the concentrations ϕ_a and ϕ_b around a droplet are slightly different from that given by the common tangent construction. Consider the situation shown in Fig. 8.8(a). A droplet of A-rich solution with concentration ϕ_a and radius r exists in a sea of the B-rich solution with concentration ϕ_b. The total free energy of the system is given by

$$A = \frac{4\pi}{3}r^3 f(\phi_a) + \left(V - \frac{4\pi}{3}r^3\right) f(\phi_b) + 4\pi r^2 \gamma \qquad (8.73)$$

Equilibrium is obtained by minimizing eq. (8.73) under the constraint

$$\frac{4\pi}{3}r^3 \phi_a + \left(V - \frac{4\pi}{3}r^3\right)\phi_b = V_A \qquad (8.74)$$

where V_A is the total volume of the A component. Minimizing eq. (8.73) with respect to ϕ_b, ϕ_a, and r, we have

$$f'(\phi_b) = f'(\phi_a) = \frac{1}{\phi_a - \phi_b}\left[f(\phi_a) - f(\phi_b) + \frac{2\gamma}{r}\right] \qquad (8.75)$$

The last term $2\gamma/r$ in the bracket $[\cdots]$ represents the effect of interfacial tension. If this term is negligible, eq. (8.75) gives the common tangent construction. If this term is non-negligible, the equilibrium concentration outside of the droplet changes by

$$\delta\phi_b = \frac{1}{(\phi_a - \phi_b)f''(\phi_b)}\frac{2\gamma}{r} \qquad (8.76)$$

Since $f''(\phi_b)$ is positive, $\delta\phi_b$ is positive, i.e., the concentration is higher than that obtained by the common tangent construction.

Now consider the situation shown in Fig. 8.9(a) where two droplets D_1 and D_2 of different sizes coexist in the solution. According to eq. (8.76), the concentration outside of the smaller droplet (D_2) is higher than that outside of the larger droplet (D_1). Therefore, there will be a flux of solute diffusing from D_2 to D_1. Near the droplet D_2, as the solute diffuses away, more solute molecules are taken out from the droplet. Accordingly the smaller droplet D_2 becomes even smaller and the larger droplet D_1 becomes even larger.

If there are many droplets, this mechanism works for all droplets. Solute molecules are taken from smaller droplets, transported to larger droplets by diffusion, which increases the volume of the larger droplets. As a result, larger droplets become larger at the expense of smaller droplets. This phenomenon is seen in the coarsening of water droplets formed in a closed chamber filled with over–saturated air (e.g., in a bathroom). The process is called the evaporation–condensation process or Lifshitz–Slyozov process.

The growth law for the characteristic droplet size ℓ can be obtained as follows. If the average size ℓ changes with rate $\dot{\ell}$, the energy dissipation function per unit volume is $\Phi \approx \xi \dot{\ell}^2$ On the other hand, the free energy per unit volume is $A \approx \gamma/\ell$. Therefore the Onsager principle $\partial(\Phi + \dot{A})/\partial \dot{\ell} = 0$ gives

$$\xi \dot{\ell} \approx \frac{\gamma}{\ell^2} \tag{8.77}$$

This gives

$$\ell \simeq \left(\frac{\gamma}{\xi} t\right)^{1/3} \tag{8.78}$$

Therefore, the droplet size increases in proportion to $t^{1/3}$.

Convection

Convection is a much faster process than diffusion. A non-spherical droplet can become spherical by the diffusion process described above. However such a process is very slow. If the material can flow, the transformation from non-spherical to spherical droplet usually takes place by convection (i.e., by the motion in which the solute and solvent move together) as shown in Fig. 8.9(b).

Once droplets become spherical, convective flow stops and plays no role in the coarsening. However, there are situations for which convective flow keeps working. The followings are two typical cases.

(i) In some cases, spherical droplets are not formed. For example, in a 50:50 mixture, it can happen that both the A-rich phase and the B-rich phase are continuous, forming a labyrinth-like pattern. In this case, the convective flow remains as the dominant mechanism in the coarsening process. The growth law for this mechanism can be obtained by a similar argument as above. Since the dissipation is dominated by the viscosity η of the solution, the energy dissipation function is given by $\Phi \approx \eta(\dot{\ell}/\ell)^2$. On the other hand, the energy term $\dot{A}$ is given by the same expression as above. Accordingly, the time evolution equation for ℓ becomes

$$\eta \frac{\dot{\ell}}{\ell^2} \approx \frac{\gamma}{\ell^2} \tag{8.79}$$

This gives

$$\ell \approx \frac{\gamma}{\eta} t \tag{8.80}$$

i.e., ℓ increases linearly with t.

(ii) More commonly, convective flow is important due to the effect of gravity. If the density of the A-rich phase is less than that of the B-rich phase, the droplet will move up. The density difference between the two phases is $\rho_{eff} = \rho_1(\phi_b - \phi_a)$. Even if ρ_{eff} is small, gravity becomes important as the droplet size becomes larger. The rising speed of a droplet of size r is estimated to be $v = (4\pi/3)r^3\rho_{eff}g/(6\pi\eta r) \approx \rho_{eff}gr^2/\eta$ which increases with the

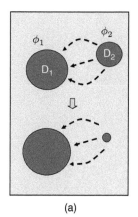

(a)

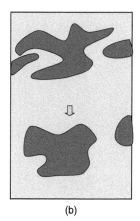

(b)

Fig. 8.9 Two mechanisms for structural coarsening in the late stage of phase separation. (a) The evaporation condensation process. (b) Convective flow driven by interfacial tension.

increase of the droplet size. Therefore, larger droplets move up faster and assemble at the top surface of the solution. Once the droplets touch each other, they coagulate and form larger droplets.

8.4   Diffusio-mechanical coupling in gels

8.4.1   Coupling of permeation and deformation

As the last topic of this chapter, we shall discuss the dynamics of gels. In Chapter 3, we have seen that gels change their volume by taking in (or expelling) solvent from (or to) the surroundings. The volume change is caused by the change of environment (e.g., the change of temperature or outer solvent), or by mechanical forces. Such phenomena are commonly seen in daily life: dried beans swell in water, salt is added to take out water from plums, oranges are squeezed to take out juices, etc. The final state, i.e., the equilibrium state of such processes, was discussed in Chapter 3. In this section, we shall discuss the dynamics, i.e., how such processes take place in time.

To clarify the argument, we consider polymeric gels, gels made of a network polymer and a solvent of simple molecules. The dynamics of gels can be discussed in the same theoretical framework as that of solutions except that one component, the polymer component, is not a fluid, but an elastic material. This gives a modification to the theory.

In solutions, the non-equilibrium state is specified by the concentration field $\phi(\boldsymbol{r})$. The free energy is written as a functional of $\phi(\boldsymbol{r})$, and the time evolution of $\phi(\boldsymbol{r})$ is determined by the Onsager principle. On the other hand, as gels are elastic materials, their free energy depends on the deformation from the reference state. Therefore the non-equilibrium state of a gel must be specified by the displacement of the gel network with respect to the reference state. Let $\boldsymbol{u}(\boldsymbol{r})$ be the displacement of the gel network which was located at $\boldsymbol{r}$ in the reference state. The free energy is written as a functional of $\boldsymbol{u}(\boldsymbol{r})$, and the time evolution of $\boldsymbol{u}(\boldsymbol{r})$ is determined by the Onsager principle as we shall show in the following.

8.4.2   Basic equations in gel dynamics

The general expression for the free energy of a gel was discussed in Section 3.4. Here we shall restrict ourselves to the case of small deformation. We assume that the gel was at equilibrium in the reference state, and that the deformation from this state is small. Let $\boldsymbol{u}(\boldsymbol{r})$ be the displacement vector: the point on the gel network which was located at $\boldsymbol{r}$ is displaced to $\boldsymbol{r} + \boldsymbol{u}(\boldsymbol{r})$. We define the displacement gradient tensor $\epsilon_{\alpha\beta}$ by

$$\epsilon_{\alpha\beta} = \frac{\partial u_\alpha}{\partial r_\beta} \tag{8.81}$$

The free energy of the gel is generally written as a functional of $\epsilon_{\alpha\beta}$. For small deformation, the free energy is written as a quadratic function of $\epsilon_{\alpha\beta}$. For an isotropic material, the free energy is written as (see Appendix A)

$$A[\boldsymbol{u}(\boldsymbol{r})] = \int d\boldsymbol{r} \left[\frac{1}{2} K \left(w - \alpha(T) \right)^2 + \frac{G}{4} \left(\epsilon_{\alpha\beta} + \epsilon_{\beta\alpha} - \frac{2}{3} w \delta_{\alpha\beta} \right)^2 . \right]$$
(8.82)

where

$$w = \epsilon_{\alpha\alpha} = \boldsymbol{\nabla} \cdot \boldsymbol{u} \qquad (8.83)$$

is the volume strain, i.e., $w = \Delta V / V$ (ΔV is the volume change, and V is the original volume), $\alpha(T)$ represents the equilibrium volume change caused by temperature change, and K and G are elastic constants.

Equation (8.82) has the same form as the expression for the elastic free energy in the theory of elasticity. In the theory of elasticity, K and G represent the bulk modulus and shear modulus, respectively. In gels, G has the same meaning (the shear modulus), but K has a slightly different meaning.

As discussed in Chapter 3, gels can be regarded as incompressible: gels cannot be compressed if solvent is not allowed to move out. The compression of a gel can take place only when solvent is allowed to move out of the gel. In this sense, K stands for the osmotic stress when the polymer network is compressed relative to the solvent, and corresponds to $\partial \Pi / \partial \phi$ in solution. For this reason, K is called the osmotic bulk modulus.

The time derivative of A is calculated as

$$
\begin{aligned}
\dot{A} &= \int d\boldsymbol{r} \left[K(w - \alpha) \frac{\partial \dot{u}_\alpha}{\partial r_\alpha} \right. \\
&\quad \left. + \frac{G}{2} \left(\epsilon_{\alpha\beta} + \epsilon_{\beta\alpha} - \frac{2}{3} w \delta_{\alpha\beta} \right) \left(\frac{\partial \dot{u}_\alpha}{\partial r_\beta} + \frac{\partial \dot{u}_\beta}{\partial r_\alpha} - \frac{1}{3} \delta_{\alpha\beta} \frac{\partial \dot{u}_\gamma}{\partial r_\gamma} \right) \right] \\
&= \int d\boldsymbol{r} \, \sigma_{\alpha\beta} \frac{\partial \dot{u}_\alpha}{\partial r_\beta}
\end{aligned}
\qquad (8.84)
$$

where $\sigma_{\alpha\beta}$ is defined by

$$\sigma_{\alpha\beta} = K(w - \alpha(T))\delta_{\alpha\beta} + G \left(\epsilon_{\alpha\beta} + \epsilon_{\beta\alpha} - \frac{2}{3} w \delta_{\alpha\beta} \right) \qquad (8.85)$$

Equation (8.85) corresponds to the stress tensor in the theory of elasticity. In the case of gels, $\sigma_{\alpha\beta}$ represents the stress acting on the polymer network, and is called the osmotic stress. As we shall see later, the total mechanical stress includes the contribution from the solvent pressure p, and is given by $\sigma_{\alpha\beta} - p\delta_{\alpha\beta}$.

Next we consider the dissipation function. The major contribution to the energy dissipation is the relative motion between polymer and solvent. Let $v_s(r)$ be the solvent velocity at point r. Then the energy dissipation is given by

$$\Phi[\dot{u}, v_s] = \frac{1}{2} \int dr \tilde{\xi}(\dot{u} - v_s)^2 \tag{8.86}$$

where $\tilde{\xi}$ is the friction constant per unit volume.

It is important to note that the solvent velocity v_s has to be accounted for in the energy dissipation function. If v_s is ignored, it leads to the incorrect conclusion that the force needed to deform a gel is proportional to $\tilde{\xi}\dot{u}$ and very large. In reality, we can deform a gel easily because the solvent and gel network are moving together.

Volume conservation for polymer component is written as

$$\frac{\partial \phi}{\partial t} = -\boldsymbol{\nabla} \cdot (\dot{u}\phi) \tag{8.87}$$

On the other hand, $\dot{u}$ and v_s must satisfy the incompressible condition

$$\boldsymbol{\nabla} \cdot [\dot{u}\phi + v_s(1 - \phi)] = 0 \tag{8.88}$$

Hence the Rayleighian is given by

$$R = \int dr \left[\frac{1}{2}\tilde{\xi}(\dot{u} - v_s)^2 + \boldsymbol{\sigma} : \boldsymbol{\nabla}\dot{u} - p\boldsymbol{\nabla} \cdot [\phi\dot{u} + (1 - \phi)v_s] \right] \tag{8.89}$$

The conditions $\delta R/\delta \dot{u} = 0$ and $\delta R/\delta v_s = 0$ give

$$\tilde{\xi}(\dot{u} - v_s) = \boldsymbol{\nabla} \cdot \boldsymbol{\sigma} - \phi\boldsymbol{\nabla}p \tag{8.90}$$

$$\tilde{\xi}(v_s - \dot{u}) = -(1 - \phi)\boldsymbol{\nabla}p \tag{8.91}$$

Equations (8.90) and (8.91) represent the balance of the forces acting on the polymer and solvent, respectively.

Summation of both sides of eqs. (8.90) and (8.91) gives

$$\boldsymbol{\nabla} \cdot (\boldsymbol{\sigma} - p\boldsymbol{I}) = 0 \tag{8.92}$$

This represents the balance of the total mechanical force in the gel. Thus $\boldsymbol{\sigma} - p\boldsymbol{I}$ is the total mechanical stress tensor in the gel. The stress consists of two parts, the osmotic stress $\boldsymbol{\sigma}$ which arises from the free energy of the gel, and the pressure p which arises from the incompressibility of the gel. Notice that in the usual theory of elasticity, there is no term which corresponds to p. The pressure p represents the effect of the solvent included in the gel. By eq. (8.85), the force balance equation (8.92) is written as

$$\left(K + \frac{1}{3}G\right)\frac{\partial w}{\partial r_\alpha} + G\frac{\partial^2 u_\alpha}{\partial r_\beta^2} = \frac{\partial p}{\partial r_\alpha} \tag{8.93}$$

On the other hand, eq. (8.91) represents Darcy's law known for fluid permeation in a porous material (see eq. (7.32)). From eq. (8.91), v_s is given by

$$v_s = \dot{u} - \frac{(1-\phi)}{\tilde{\xi}} \nabla p \qquad (8.94)$$

Using this for eq. (8.88), we have

$$\nabla \cdot \left[\dot{u} - \frac{(1-\phi)^2}{\xi} \nabla p \right] = 0 \qquad (8.95)$$

or

$$\frac{\partial w}{\partial t} = \kappa \nabla^2 p \qquad (8.96)$$

where

$$\kappa = \frac{(1-\phi)^2}{\tilde{\xi}} = \frac{1}{\xi} \qquad (8.97)$$

The set of equations (8.83), (8.93), and (8.96) determine the time evolution of u, w, and p.

8.4.3   Concentration fluctuations

Taking the divergence of (8.93), we have

$$\nabla^2 p = \left(K + \frac{4}{3} G \right) \nabla^2 w \qquad (8.98)$$

Equations (8.96) and (8.98) give the following diffusion equation for w:

$$\frac{\partial w}{\partial t} = D_c \nabla^2 w \qquad (8.99)$$

where D_c is given by

$$D_c = \kappa \left(K + \frac{4}{3} G \right) \qquad (8.100)$$

For small deformation, w is related to the change of the volume fraction $\delta\phi = \phi - \phi_0$ by $\delta\phi = -\phi_0 w$. Therefore $\delta\phi$ also satisfies the diffusion equation:

$$\frac{\partial \delta\phi}{\partial t} = D_c \nabla^2 \delta\phi \qquad (8.101)$$

Equation (8.101) indicates that if there is a concentration fluctuation $\phi_k e^{i k \cdot r}$ in the gel, it will relax with relaxation time $1/D_c k^2$, the same formula as in solution (see eq. (8.34)). The expression for the collective diffusion constant D_c is slightly different. In the case of solutions, D_c is given by eq. (7.66) which is similar to eq. (8.100). (Notice that $n(\partial\Pi/\partial n)$ is equal to the osmotic modulus K in solutions.) In the case of gels, the osmotic modulus K is replaced by $K + (4/3)G$, which indicates that, in gels, the shear modulus contributes to the relaxation of concentration fluctuations.

Equation (8.101) is useful in discussing problems which do not involve boundary conditions. If boundary conditions are important (for example, in the problems of how fast the solvent permeates through a gel membrane), eq. (8.101) is not very useful. This is because the boundary conditions for $\delta\phi$ are not known. Usually, the boundary conditions are given for the displacement or the mechanical forces acting at the boundary. Therefore to solve the problem, we need to go back to the basic equations (eqs. (8.83), (8.93), and (8.96)). In the following we shall show an example.

8.4.4   Mechanical relaxation by solvent diffusion in a stretched gel sheet

As discussed in Section 3.4.4, when a weight is placed on top of a gel (see Fig. 3.11), the gel first deforms keeping its volume unchanged, then shrinks while expelling the solvent, and finally relaxes to the equilibrium state. We shall now discuss the dynamics of such solvent transport induced by mechanical forces (i.e., the process of squeezing). Such phenomena can be discussed by solving the set of equations (8.83), (8.93), and (8.96). Here we demonstrate an analysis for the situation shown in Fig. 8.10. (Analysis for squeezing is done in problem (8.6).)

A weight W is hung from the bottom edge of a gel sheet of thickness $2a$, width b, and length L at time $t = 0$. We will investigate how the length of the gel changes in time. We choose the coordinate system shown in Fig. 8.10(a). The origin is taken at the centre of the sheet, and the x- and y-axes are taken to be normal and tangent to the sheet, respectively.

For a thin sheet, the strain must be uniform in the y- and z-directions, but the strain in the x-direction may vary with position. Therefore the displacement vector $\boldsymbol{u}$ can be written as

$$u_x = u(x,t), \qquad u_y = \epsilon_y(t)y, \qquad u_z = \epsilon_z(t)z \tag{8.102}$$

Using the volume strain

$$w(x,t) = \frac{\partial u}{\partial x} + \epsilon_y(t) + \epsilon_z(t) \tag{8.103}$$

the osmotic stress tensor (8.85) is written as

$$\sigma_{xx} = \left(K + \frac{4}{3}G\right)w - 2G(\epsilon_y + \epsilon_z) \tag{8.104}$$

$$\sigma_{yy} = \left(K - \frac{2}{3}G\right)w + 2G\epsilon_y \tag{8.105}$$

$$\sigma_{zz} = \left(K - \frac{2}{3}G\right)w + 2G\epsilon_z \tag{8.106}$$

Since there is no external force acting normal to the sheet, $\sigma_{xx} - p$ is identically zero everywhere in the gel. Therefore

$$p(x,t) = \sigma_{xx} = \left(K + \frac{4}{3}G\right)w - 2G(\epsilon_y + \epsilon_z) \tag{8.107}$$

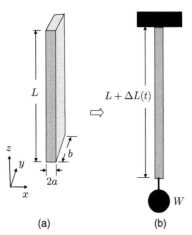

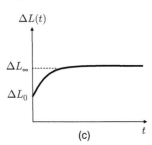

Fig. 8.10 (a) A thin gel sheet. (b) When a weight W is hung from the bottom of the gel sheet, the sheet first elongates by ΔL_0, then elongates further gradually. (c) Time dependence of $\Delta L(t)$.

Equations (8.96) and (8.107) give

$$\frac{\partial w}{\partial t} = D_c \frac{\partial^2 w}{\partial x^2} \tag{8.108}$$

To determine $\epsilon_y(t)$ we use the condition that the total force acting on the cross-section normal to the y-axis is zero:

$$\int_{-a}^{a} dx(\sigma_{yy} - p) = 0 \tag{8.109}$$

Using eqs. (8.105) and (8.107), $\sigma_{yy} - p$ is written as $-2Gw + 2G(2\epsilon_y + \epsilon_z)$. Hence, eq. (8.109) gives

$$\bar{w} - (2\epsilon_y + \epsilon_z) = 0 \tag{8.110}$$

where $\bar{w}$ stands for the average of w for x:

$$\bar{w}(t) = \frac{1}{2a} \int_{-a}^{a} dx w(x, t) \tag{8.111}$$

Similarly the total forces acting on the cross-section normal to the z-axis must be equal to W:

$$\frac{1}{2a} \int_{-a}^{a} dx(\sigma_{zz} - p) = \sigma_w \tag{8.112}$$

where $\sigma_w = W/(2ab)$ is the external force acting on unit area of the cross-section of the gel sheet. Again this is rewritten by using eqs. (8.106) and (8.107) as

$$\bar{w} - (\epsilon_y + 2\epsilon_z) = -\frac{\sigma_w}{2G} \tag{8.113}$$

At $x = \pm a$, p is equal to zero. Hence eq. (8.107) gives

$$\left(K + \frac{4}{3}G\right) w = 2G(\epsilon_y + \epsilon_z) \qquad \text{at } x = \pm a \tag{8.114}$$

By eqs. (8.110) and (8.113), this can be written as

$$\left(K + \frac{4}{3}G\right) w = \frac{4}{3}G\bar{w} + \frac{\sigma_w}{3} \qquad \text{at } x = \pm a \tag{8.115}$$

Equations (8.108) and (8.115) determine the time evolution of $w(x, t)$. Notice that the right-hand side of eq. (8.115) is not constant but changes with time as $\bar{w}(t)$ varies with time.

To determine the long-time behaviour, we assume the following form for the solution

$$w(x, t) = w_{eq} + g(x)e^{-t/\tau} \tag{8.116}$$

where w_{eq} is the equilibrium value given by[7]

$$w_{eq} = \frac{\sigma_w}{3K} \tag{8.117}$$

[7] At equilibrium, $w(x, t) = \bar{w} = w_{eq}$. Equation (8.115) then gives eq. (8.117).

and $g(x)$ and τ are the solutions of the following equations

$$-\frac{g}{\tau} = D_c \frac{\partial^2 g}{\partial x^2} \tag{8.118}$$

and

$$\left(K + \frac{4}{3}G\right)g = \frac{4}{3}G\bar{g}, \qquad \text{at } x = \pm a \tag{8.119}$$

where $\bar{g}$ stands for the average of g for x:

$$\bar{g} = \frac{1}{2a} \int_{-a}^{a} dx g(x) \tag{8.120}$$

Equations (8.118)–(8.120) set an eigenvalue problem: the task is to find the value of τ which gives the non-zero function $g(x)$ that satisfies both eqs. (8.118) and (8.119). The solution of this eigenvalue problem is given by

$$\tau = \frac{a^2}{D_c \chi^2} \tag{8.121}$$

where χ is the solution of the equation

$$\frac{1}{\chi} \tan(\chi) = 1 + \frac{3}{4}\frac{K}{G} \tag{8.122}$$

The solution of eq. (8.122) is $\chi = \pi/2$ for $K/G \gg 1$ and $\chi = \sqrt{K/G}$ for $K/G \ll 1$. Hence the relaxation time is given by

$$\tau = \begin{cases} \dfrac{4a^2}{\pi^2 \kappa K} & \text{if } K/G \gg 1 \\[2mm] \dfrac{3a^2}{4\kappa K} & \text{if } K/G \ll 1 \end{cases} \tag{8.123}$$

If $K/G \gg 1$, the relaxation time can be estimated by a^2/D_c. On the other hand, if $K/G \ll 1$, this estimation becomes quite wrong. In this case, although a^2/D_c remains finite, τ diverges as K goes to zero. This demonstrates that the gel dynamics cannot be handled by the diffusion equation alone.

8.4.5   Mechanical instabilities caused by solvent diffusion

The linear theory presented in the previous sections is limited to small deformation. When the deformation becomes large, nonlinear effects become important. Examples of such effects are shown in Fig. 8.11.

When a gel tends to swell strongly, taking in the solvent from the surroundings, wrinkles often appear on the surface of the gel (see Fig. 8.11(a)). The wrinkles start as tiny irregularities on the smooth surface, but then the amplitude and the wavelength of the wrinkles grow as swelling proceeds. The amplitude of the wrinkles starts to decrease when the wavelength becomes comparable to the radius of

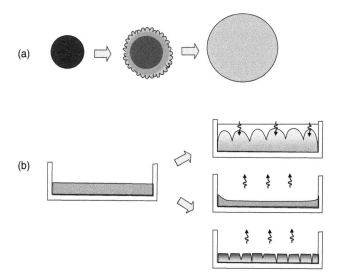

Fig. 8.11 (a) Surface roughening when a spherical gel undergoes a large volume change. (b) Surface instability of a gel in a dish. The gel is stuck to the wall of the dish. When the gel swells, surface roughening can take place. On the other hand, when the gel shrinks, the surface remains flat. However, if the volume change is too large, cracks appear on the surface of the gel.

the gel. Eventually, the wrinkles disappear and a smooth surface is recovered when the swelling is completed.

This phenomenon is caused by the fact that the gel is an elastic material. When the swelling starts, the outer part of the gel absorbs solvent and swells, while the inner part is unswollen. To accommodate the solvent taken from the surroundings, wrinkles appear near the surface. Notice that wrinkles are seen only in the intermediate state where the swelling of the outer part is different from that of the inner part. When the whole gel is swollen uniformly, the wrinkles disappear.

The wrinkles can be seen in the equilibrium situation shown in Fig. 8.11(b), where the gel is placed in a dish with the bottom and the side surfaces fixed to the dish. When solvent is poured on top of such a gel, solvent permeates through the gel and creates the wrinkles at the surface. On the other hand, if such a gel is dried, the gel shrinks, and if the drying speed is fast, cracks often appear at the surface. Such phenomena can be described by the diffusio-mechanical coupling model for large deformation.

8.5   Summary of this chapter

Diffusion and permeation are processes in which a certain component (say, the solute) moves relative to the others. The relative speed is determined by the balance of two forces, the thermodynamic force which drives the motion, and the frictional force which resists the motion.

Spatial correlation of the diffusing component affects the speed of diffusion and permeation. The spatial correlation is represented by the density correlation function $g(\boldsymbol{r})$, or the scattering function $S(\boldsymbol{k})$ and is typically characterized by two parameters, N_c, the correlation mass, and ℓ_c, the correlation length. The osmotic modulus K and the friction constant ξ are the relevant parameters representing the thermodynamic force and the frictional force, which are in turn related to N_c and ℓ_c, respectively.

The relative motion between the components generally induces a macroscopic velocity in the solution. This effect, called diffusio-mechanical coupling, is important in sedimentation of colloidal particles, phase separation kinetics, and in the dynamics of gels. Basic equations for the diffusio-mechanical coupling are obtained from the Onsager principle for solutions and gels.

Further reading

(1) *Gel dynamics*, Masao Doi, J. Phys. Soc. Jpn **78** 052001 (2009).

(2) *Phase Transition Dynamics*, Akira Onuki, Cambridge University Press (2002).

(3) *Colloidal Dispersions*, William B. Russel, D. A. Saville, William Raymond Schowalter, Cambridge University Press (1989).

Exercises

(8.1) Consider a single polymer chain in a dilute solution. The scattering function of the polymer chain is given by

$$S(\mathbf{k}) = \frac{1}{N} \sum_{n.m} \left\langle e^{i\mathbf{k} \cdot (\mathbf{r}_n - \mathbf{r}_m)} \right\rangle \qquad (8.124)$$

where n and m denote the segments constituting the polymer chain and N is the number of segments in the chain. Answer the following questions.

(a) For small $\mathbf{k}$, the right-hand side of eq. (8.124) can be expanded as a power series of $\mathbf{k}$. Show that in the isotropic state, $S(\mathbf{k})$ is written as

$$S(\mathbf{k}) = N \left[1 - \frac{1}{3} k^2 R_g^2 + \cdots \right] \qquad (8.125)$$

where R_g^2 is defined by

$$R_g^2 = \frac{1}{2N^2} \sum_{n,m} \left\langle (\mathbf{r}_n - \mathbf{r}_m)^2 \right\rangle \qquad (8.126)$$

R_g is called the radius of gyration of the polymer. (R_g is the same as the correlation length ℓ_c defined by eq. (8.20). In polymer science, it is customary to use R_g rather than ℓ_c.)

(b) Calculate the radius of gyration R_g of a linear Gaussian polymer. (Hint: use the fact that $\langle (\mathbf{r}_n - \mathbf{r}_m)^2 \rangle = |n - m| b^2$.)

(c) Calculate R_g for a rigid rod-like polymer in which the segments are uniformly distributed along a line of length L. Also calculate R_g for a globular polymer in which the segments are uniformly distributed in a sphere of radius R.

(d) Show that the fractal dimension of a flexible polymer is 2, while that of rod-like polymer is 1.

(8.2) For a Gaussian polymer, the vector $\mathbf{r}_{nm} = \mathbf{r}_n - \mathbf{r}_m$ obeys the following Gaussian distribution (see eq. (3.29))

$$\psi(\mathbf{r}_{nm}) = \left(\frac{3}{2\pi |n - m| b^2} \right)^{3/2} \exp \left(-\frac{3 r_{nm}^2}{2 |n - m| b^2} \right) \qquad (8.127)$$

Answer the following questions

(a) Show that $S(\mathbf{k})$ is given by

$$S(\mathbf{k}) = \frac{1}{N} \int_0^N dn \int_0^N dm \exp \left(-\frac{|n - m| k^2 b^2}{6} \right) \qquad (8.128)$$

(b) Show that for large $|\boldsymbol{k}|$, $S(\boldsymbol{k})$ has the following asymptotic form

$$S(\boldsymbol{k}) = \frac{12}{k^2 b^2} \qquad (8.129)$$

(c) Show that the pair correlation function $g(r)$ for the polymer segments behaves as $g(r) \propto 1/r$ for small r.

(8.3) Consider a rigid molecule made of N segments. Let ζ be the friction constant of the molecule, i.e., the frictional force exerted on the molecule when it moves with velocity v in a solvent is given by $-\zeta v$. Consider a dilute solution of such molecules. The friction constant ξ per unit volume of such solution is given by $\xi = \frac{\bar{n}}{N}\zeta$. Assuming that the approximation of eq. (8.39) is valid, answer the following questions.

(a) Show that the friction constant ζ per molecule is given by

$$\frac{1}{\zeta} = \frac{1}{6\pi\eta N} \int dr \frac{g(r)}{|r|} \qquad (8.130)$$

(b) Let $\boldsymbol{r}_n$, $(n = 1, 2, \ldots, N)$ be the position vector of the segments. Show that ζ can be written as

$$\frac{1}{\zeta} = \frac{1}{6\pi\eta N^2} \sum_n \sum_{n \neq m} \frac{1}{|\boldsymbol{r}_n - \boldsymbol{r}_m|} \qquad (8.131)$$

(c) Calculate the friction constant ζ of a sphere of radius a using eq. (8.131) and show that it is given by $\zeta = 6\pi\eta a$. (Hint: assume that the sphere is made of friction points which are uniformly distributed on the surface, then

$$\frac{1}{\zeta} = \frac{1}{6\pi\eta(4\pi a^2)^2} \int dS \int dS' \frac{1}{|\boldsymbol{r} - \boldsymbol{r}'|}) \qquad (8.132)$$

(d) Calculate the friction constant ζ of a rod-like particle of length L and diameter b assuming that the rod is made of $N = L/b$ friction points. (Answer: the result is

$$\zeta = \frac{3\pi\eta L}{\ln(L/b)} \qquad \text{for} \quad L \gg b) \qquad (8.133)$$

(8.4) Derive eq. (8.34) answering the following questions.

(a) Let $\delta n(\boldsymbol{r}, t) = n(\boldsymbol{r}, t) - \bar{n}$ be the change of the number density caused by the external field $h(\boldsymbol{r})$. Show that for small $h(\boldsymbol{r})$, $\delta n(\boldsymbol{r}, t)$ is the solution of the following equation.

$$\frac{\partial \delta n}{\partial t} = D_c(\bar{n})\nabla \cdot \left[\nabla \delta n - \frac{\bar{n}}{\Pi'(\bar{n})}\nabla h\right] \qquad (8.134)$$

(b) Solve eq. (8.134) using the Fourier transform

$$n_k(t) = \int dr \delta n(\boldsymbol{r}, t)e^{i\boldsymbol{k}\cdot\boldsymbol{r}},$$

$$h_k = \int dr h(\boldsymbol{r})e^{i\boldsymbol{k}\cdot\boldsymbol{r}} \qquad (8.135)$$

and show that the response function is given by

$$\chi(\boldsymbol{r}, \boldsymbol{r}', t) = \frac{1}{(2\pi)^3}\int dk \frac{\bar{n}}{\Pi'}\left(1 - e^{-D_c k^2 t}\right)$$
$$\times e^{i\boldsymbol{k}\cdot(\boldsymbol{r}-\boldsymbol{r}')} \qquad (8.136)$$

(c) Prove that the response function $\chi(\boldsymbol{r}, \boldsymbol{r}', t)$ is related to the dynamic correlation function $g_d(\boldsymbol{r}, t)$ by

$$\chi(\boldsymbol{r}, \boldsymbol{r}', t) = \frac{\bar{n}}{k_B T}\left[g_d(\boldsymbol{r} - \boldsymbol{r}', 0)\right.$$
$$\left. - g_d(\boldsymbol{r} - \boldsymbol{r}', t)\right] \qquad (8.137)$$

(d) Derive eq. (8.34).

(8.5) Consider a two-component solution near the critical point. Let ϕ_c and T_c be the concentration and the temperature at the critical point. The free energy of the system is given by eq. (8.49) (with $\rho_1 = 0$). If the concentration ϕ is not far from ϕ_c, the free energy density $f(\phi)$ can be written as

$$f(\phi) = c_0(T) + c_1(T)(\phi - \phi_c) + \frac{1}{2}c_2(T)(\phi - \phi_c)^2 + \cdots \qquad (8.138)$$

Answer the following questions.

(a) Show that $c_2(T)$ becomes zero at $T = T_c$ and therefore $c_2(T)$ is written as

$$c_2(T) = a(T - T_c) \qquad (8.139)$$

where a is a constant independent of T.

(b) If $\phi - \phi_c$ is small, the free energy of the system is written as

$$A = \int dr \left[\frac{1}{2}a(T - T_c)(\phi - \phi_c)^2 + \frac{1}{2}\kappa_s(\nabla\phi)^2\right] \qquad (8.140)$$

where the gravity term is ignored. Explain why the term linear in $\phi - \phi_c$ can be ignored in eq. (8.140).

(c) Consider that a hypothetical external field $h(\boldsymbol{r})$ is applied to the solute segment. Show that $\delta\phi(\boldsymbol{r}) = \phi(\boldsymbol{r}) - \phi_c$ satisfies the following equation

$$\frac{\partial \delta\phi}{\partial t} = \frac{\phi_c}{\xi}\nabla^2 \left[a(T - T_c)\phi_c\delta\phi - \phi_c\kappa_s\nabla^2\delta\phi \right. \\ \left. - \phi_c h(\boldsymbol{r}) \right] \qquad (8.141)$$

(d) Solve eq. (8.141) using the Fourier transform, and show that the dynamic scattering function $S_d(\boldsymbol{k}, t)$ is given by

$$S_d(\boldsymbol{k}, t) = \frac{k_B T}{a(T - T_c) + \kappa_s \boldsymbol{k}^2} \exp[-\alpha_k t] \qquad (8.142)$$

where α_k is given by

$$\alpha_k = \frac{\phi_c^2 \boldsymbol{k}^2}{\xi} \left[a(T - T_c) + \kappa_s \boldsymbol{k}^2 \right] \qquad (8.143)$$

(e) Confirm that the static scattering function $S(\boldsymbol{k}) = S_d(\boldsymbol{k}, 0)$ is written in the form of eq. (8.21), and show that the parameters N_c and ℓ_c diverge as

$$N_c \propto \frac{1}{T - T_c}, \qquad \ell_c \propto \frac{1}{(T - T_c)^{1/2}} \qquad (8.144)$$

(8.6) A gel of thickness h is squeezed by two porous walls with normal stress w (see Fig. 8.12). Assume that the gel is thin and their surfaces are mechanically fixed to the wall. In this situation, the displacement of the gel network can take place only in the direction normal to the wall. Let $u(x, t)$ be the displacement of the gel network, where x $(0 < x < h)$ is the coordinate taken normal to the wall. Answer the following questions.

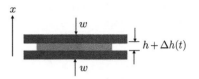

Fig. 8.12

(a) Show that the elastic energy of deformation of the gel is written as

$$A = \int_0^h dx \frac{K_e}{2}\left(\frac{\partial u}{\partial x}\right)^2 + [u(h) - u(0)]w \qquad (8.145)$$

where $K_e = K + (4/3)G$. The last term represents the energy due to the force by the wall.

(b) Assume that the volume average velocity is zero: $\phi\dot{u} + (1 - \phi)v_s = 0$. Show that the energy dissipation function is written as

$$\Phi = \frac{1}{2}\int_0^h dx \frac{\dot{u}^2}{\kappa} \qquad (8.146)$$

(c) Using the Onsager principle, show that $u(x, t)$ satisfies the following diffusion equation

$$\frac{\partial u}{\partial t} = \kappa K_e \frac{\partial^2 u}{\partial x^2} \qquad (8.147)$$

with the boundary conditions

$$K_e \frac{\partial u}{\partial x} = -w \qquad \text{at} \quad x = 0, h \qquad (8.148)$$

(d) Solve the above equation under the initial condition $u(x, 0) = 0$, and obtain the time variation of the gap distance $\Delta h(t) = u(h, t) - u(0, t)$.

Flow and deformation of soft matter

The most apparent characteristic of soft matter is that it is soft, i.e., easily deformed. This characteristic, however, is rather complex.

Consider for example, bubble gum. Bubble gum is a material which does not have its own shape: it can be deformed into any shape (a ball of bubble gum can be made into a thin flat sheet). Also a piece of bubble gum can be separated into many pieces, but these can be made into one piece again. Such an operation is not possible for elastic materials like rubbers and gels. Bubble gum is a kind of fluid, but it has elastic features. Indeed if bubble gum is stretched and released quickly, it tends to recover the original shape. The degree of recovery depends on time: if the gum is kept stretched for a long time, it stays in the stretched state when released.

The science that studies the flow and deformation of such complex materials is called rheology. Rheology is an important subject in soft matter. Many applications of soft matter rely on their unique mechanical properties. For example, one can make balloons with bubble gum, but not with usual chewing gum. This fact is important in industry. Plastic bottles are made by blowing molten polymers, but this is possible only when the molecular weights of the polymers are suitably controlled.

In this chapter we shall discuss the rheological properties of soft matter. Soft matter involves a large class of materials, and their rheological properties are quite diverse. We shall therefore focus the discussion on polymeric materials (polymer solutions, melts, and cross-linked polymers). Discussions of other materials can be found in the literature cited at the end of this chapter.

We shall first discuss how to characterize the rheological properties of materials. The resistance to deformation of solid materials is characterized by the elastic constants (the ratio between the force and the deformation) and the resistance to deformation of fluids is characterized by the viscosity (the ratio between the force and the rate of deformation). Soft matter usually has both viscosity and elasticity, a general attribute called viscoelasticy. We shall discuss how to characterize the viscoelasticity of materials.

We shall then discuss how to understand viscoelasticity from the molecular viewpoint. It will be shown that the viscoelasticity of polymeric materials is directly connected to the conformational relaxation of polymers. The last part of this chapter is devoted to solutions of rod-like polymers, which can form liquid crystalline phases.

9.1   Mechanical properties of soft matter

9.1.1   Viscosity, elasticity, and viscoelasticity

As discussed in Section 3.1, the mechanical behaviour of a simple elastic material (Hookean elastic material) is represented by the linear relation between strain and stress. In the shear deformation shown in Fig. 3.2, this is represented by the linear relation between the shear strain γ and the shear stress σ (see eq. (3.3)):

$$\sigma = G\gamma \tag{9.1}$$

The coefficient G is the shear modulus of the material.

On the other hand, the mechanical behaviour of a simple fluid (Newtonian fluid) is represented by the linear relation between the shear rate $\dot{\gamma} = d\gamma(t)/dt$ and the shear stress σ (see eq. (3.4)) :

$$\sigma = \eta\dot{\gamma} \tag{9.2}$$

The coefficient η is the viscosity of the material.

Soft matter generally has both viscosity and elasticity, an attribute that is referred to as viscoelasticity. The viscoelasticity of soft matter can be studied by the instrument shown in Fig. 3.2, which creates a homogeneous shear deformation in the sample. This instrument, however, is not suited to study fluid soft matter (such as polymer solutions) since a large strain cannot be applied. More standard instruments are shown in Fig. 9.1. They are designed to create a homogeneous shear deformation in the sample. With such instruments, one can apply a controlled, time-dependent strain $\gamma(t)$ to the material, and can measure the

Fig. 9.1 Examples of the instruments used to study the viscoelasticity of fluid soft matter. (a) Concentric type. The sample is set between two concentric cylinders, and is sheared by the rotation of the outer cylinder with the inner cylinder fixed. If the gap between the cylinders is taken to be small, uniform shear deformation is realized. The shear rate $\dot{\gamma}$ is obtained from the angular velocity Ω of the outer cylinder, and the shear stress σ is obtained from the torque T acting on the inner cylinder. (b) Cone–plate type. The sample is set between a plate and a cone, and is sheared by the rotation of the bottom plate with the top cone fixed. Uniform shear deformation is created in the material. The shear rate $\dot{\gamma}$ and the shear stress σ are obtained from the angular velocity Ω of the plate and the torque T acting on the cone, respectively.

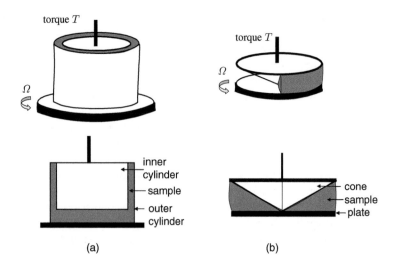

response of the stress $\sigma(t)$ induced in the material. Alternatively, one can measure the strain response $\gamma(t)$ for a given time-dependent stress $\sigma(t)$.

Figure 9.2 shows the schematic behaviour of the stress response when a step strain, described by

$$\gamma(t) = \gamma_0 \Theta(t) = \begin{cases} 0 & t < 0 \\ \gamma_0 & t > 0 \end{cases} \tag{9.3}$$

is applied to a sample. The stress induced in the material is a function of γ_0 and time t, and can be written as $\sigma(t; \gamma_0)$. For small γ_0, $\sigma(t; \gamma_0)$ is proportional to γ_0, and can be written as

$$\sigma(t) = \gamma_0 G(t) \tag{9.4}$$

The function $G(t)$ is called the relaxation modulus.

Examples of $G(t)$ are illustrated in Fig. 9.2(b). If the material is ideally elastic, $G(t)$ is constant, while in many polymeric materials, $G(t)$ decreases in time. In polymer melts, which do not have their own shape, $G(t)$ eventually goes to zero. Such a material is called a viscoelastic fluid. Polymer melts and polymer solutions show such behaviour. On the other hand, in rubbers and gels which have their own shape, $G(t)$ does not go to zero: the material maintains a finite stress while it is in the deformed state. Such a material is called a viscoelastic solid.

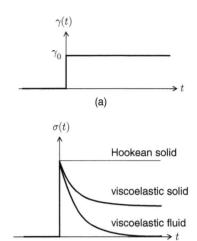

Fig. 9.2 Stress relaxation experiment: a step shear shown in (a) is applied to the sample, and the stress induced in the material is measured. (b) shows examples of the stress response.

9.1.2   Linear viscoelasticity

The stress of viscoelastic materials depends on the strain (or the strain rate) applied to the material in the past. Therefore characterization of mechanical properties of viscoelastic materials generally becomes rather complex. However, when the stress is small, the relation between the strain and the stress becomes simple. This is due to the superposition principle which generally holds for systems not strongly perturbed from the equilibrium state.[1] In the present context, this principle can be stated as follows.

Suppose that when a time-dependent strain $\gamma_1(t)$ is applied to a material, the stress response is $\sigma_1(t)$, and that when $\gamma_2(t)$ is applied, the stress is $\sigma_2(t)$. The superposition principle states that when the superposed strain $\gamma(t) = \gamma_1(t) + \gamma_2(t)$ is applied, the stress response is $\sigma(t) = \sigma_1(t) + \sigma_2(t)$.

If the superposition principle holds, the stress response to any time-dependent strain $\gamma(t)$ can be expressed by the relaxation modulus $G(t)$. This is because any time-dependent strain $\gamma(t)$ can be regarded as a superposition of step strains $\Delta\gamma_i = \dot{\gamma}(t_i)\Delta t$ applied at time t_i (see Fig. 9.3), i.e.,

$$\gamma(t) = \sum_i \Delta\gamma_i \Theta(t - t_i) \tag{9.5}$$

The step strain $\Delta\gamma_i$ applied at time t_i creates the stress $G(t - t_i)\Delta\gamma_i$ at a later time t. Summing up all the stresses created by the step strains applied in the past, we have the stress at time t:

[1] The superposition principle generally holds when small external stimuli are applied to a system at equilibrium. In the context of viscoelasticity, the smallness of external stimuli can mean the smallness of the shear strain $\gamma(t)$ or the shear rate $\dot{\gamma}(t)$. If the material is purely viscous, the superposition principle holds even if $\gamma(t)$ is very large. If the material is purely elastic, the superposition principle holds even if $\dot{\gamma}(t)$ is large. In both cases, the superposition principle holds as long as the stress created by the deformation is small. Therefore the condition for the superposition principle to hold is that the stress is small.

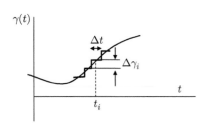

Fig. 9.3 Any time-dependent strain $\gamma(t)$ is expressed by the superposition of many step strains of magnitude $\Delta\gamma_i$ applied at time t_i. The stress response for the strain $\gamma(t)$ can be calculated by superposing the response for such step strains.

$$\sigma(t) = \sum_i G(t - t_i)\Delta\gamma_i = \sum_i G(t - t_i)\dot\gamma(t_i)\Delta t \qquad (9.6)$$

In the limit $\Delta t \to 0$, this gives

$$\sigma(t) = \int_{-\infty}^{t} dt' G(t - t')\dot\gamma(t') \qquad (9.7)$$

Viscoelasticity which satisfies the superposition principle is called linear viscoelasticity. The linear viscoelasticity of incompressible materials is completely characterized by a single function $G(t)$. If $G(t)$ is given, the stress response for arbitrary strain can be calculated.

For example, consider that a shear flow of constant shear rate $\dot\gamma$ is started at $t = 0$. The stress at time t is calculated as

$$\sigma(t) = \int_{0}^{t} dt' G(t - t')\dot\gamma = \dot\gamma \int_{0}^{t} dt' G(t') \qquad (9.8)$$

The shear stress increases monotonically, and approaches a constant value $\sigma(\infty)$. The steady state viscosity η_0 is defined by the ratio $\sigma(\infty)/\dot\gamma$. According to eq. (9.8), the steady state viscosity is given by

$$\eta_0 = \int_{0}^{\infty} dt G(t) \qquad (9.9)$$

9.1.3   Complex modulus

A standard way of characterizing the linear viscoelasticity is to measure the stress response for oscillatory strain

$$\gamma(t) = \gamma_0 \cos\omega t \qquad (9.10)$$

According to eq. (9.7), the shear stress for this strain is calculated as

$$\sigma(t) = \int_{-\infty}^{t} dt' G(t - t')[-\gamma_0\omega \sin\omega t']$$
$$= \int_{0}^{\infty} dt' G(t')[-\gamma_0\omega \sin\omega(t - t')]$$
$$= \gamma_0 [G'(\omega) \cos\omega t - G''(\omega) \sin\omega t] \qquad (9.11)$$

where

$$G'(\omega) = \omega \int_{0}^{\infty} dt \; \sin\omega t G(t) \qquad (9.12)$$

$$G''(\omega) = \omega \int_{0}^{\infty} dt \; \cos\omega t G(t) \qquad (9.13)$$

The quantity

$$G^*(\omega) = G'(\omega) + iG''(\omega) \qquad (9.14)$$

is called the complex modulus.[2] This is because eq. (9.11) is written in the following compact form

$$\sigma(t) = \gamma_0 \Re\big[[G'(\omega) + iG''(\omega)]e^{i\omega t}\big] = \gamma_0 \Re\big[G^*(\omega)e^{i\omega t}\big] \qquad (9.16)$$

[2] $G^*(\omega)$ is related to $G(t)$ by

$$G^*(\omega) = i\omega \int_{0}^{\infty} dt G(t)e^{-i\omega t} \qquad (9.15)$$

$G'(\omega)$ and $G''(\omega)$ are, respectively, called the storage modulus and loss modulus, or simply the real part and the imaginary part of the complex modulus.

$G'(\omega)$ represents the elastic response, while $G''(\omega)$ represents the viscous response. Whether a material is fluid or solid is seen in the behaviour of the complex modulus at small ω. To see this, consider that $G(t)$ is given by

$$G(t) = G_e + Ge^{-t/\tau} \qquad (9.17)$$

where G_e is the equilibrium modulus, which is zero for a viscoelastic fluid but is non-zero for a viscoelastic solid. For this $G(t)$, $G'(\omega)$ and $G''(\omega)$ are calculated as follows[3]

$$G'(\omega) = G_e + G\frac{(\omega\tau)^2}{1 + (\omega\tau)^2}, \qquad G''(\omega) = G\frac{\omega\tau}{1 + (\omega\tau)^2} \qquad (9.18)$$

Therefore the asymptotic behaviour of $G^*(\omega)$ as $\omega \to 0$ is given by

$$G^*(\omega) = \begin{cases} G_e & \text{for viscoelastic solid} \\ i\omega\eta_0 & \text{for viscoelastic fluid} \end{cases} \qquad (9.19)$$

where $\eta_0 = G\tau$ is the steady state viscosity defined in eq. (9.9). Typical behaviour of $G'(\omega)$ and $G''(\omega)$ is shown in Fig. 9.4.

9.1.4   Nonlinear viscosity

According to the superposition principle, when a viscoelastic fluid is sheared with constant shear rate $\dot{\gamma}$, the shear stress σ is proportional to $\dot{\gamma}$, and the relation is written in the form of eq. (9.2). Such a linear relation breaks down if the stress (or the shear rate) becomes large.

In general, the shear stress σ in a steady shear flow is a nonlinear function of the shear rate $\dot{\gamma}$. Typical behaviour of $\sigma(\dot{\gamma})$ is shown in Fig. 9.5(a), and the ratio

$$\eta(\dot{\gamma}) = \frac{\sigma(\dot{\gamma})}{\dot{\gamma}} \qquad (9.20)$$

is called the steady state viscosity. The viscosity appearing in eq. (9.2) or eq. (9.9) is the value of $\eta(\dot{\gamma})$ in the limit $\dot{\gamma} \to 0$.

In soft matter, $\eta(\dot{\gamma})$ usually decreases with the increase of $\dot{\gamma}$ as shown by curve (ii) in Fig. 9.5(a). This phenomenon is called shear thinning. In polymeric fluids, $\eta(\dot{\gamma})$ is often approximated by

$$\eta(\dot{\gamma}) = \frac{\eta_0}{1 + (\dot{\gamma}/\dot{\gamma}_c)^n} \qquad (9.21)$$

where n and $\dot{\gamma}_c$ are material parameters. $\dot{\gamma}_c$ is of the order of $1/\tau$, the inverse of the stress relaxation time, and is very small. Therefore, in the flow of polymer fluids, the nonlinear effect is quite important.

In concentrated colloidal suspensions, the shear thinning is very strong, and the flow curve is often represented by curve (iii) in Fig. 9.5(a).

[3] Here we have used
$$\int_0^\infty dt\, e^{-i\omega t} = \lim_{s \to 0} \int_0^\infty dt\, e^{-i\omega t - st}$$
$$= \lim_{s \to 0} \frac{1}{s + i\omega} = \frac{1}{i\omega}$$

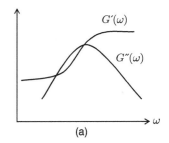

viscoelastic solid

(a)

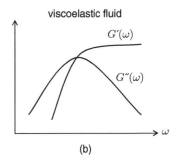

viscoelastic fluid

(b)

Fig. 9.4 Typical behaviour of the storage modulus $G'(\omega)$ and the loss modulus $G''(\omega)$ of viscoelastic materials. (a) Viscoelastic solids; (b) viscoelastic fluids.

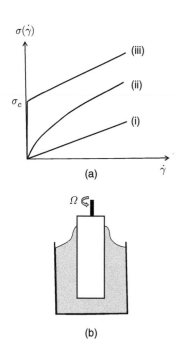

Fig. 9.5 (a) The shear stress σ in steady shear flow is plotted against the shear rate $\dot{\gamma}$ for various materials. (i) Ideal viscous fluid; (ii) shear thinning fluid seen in polymeric liquids; and (iii) a Bingham fluid seen in concentrated colloidal suspensions. (b) Rod climbing effect: when a rod is rotated in a polymeric fluid, the fluid clings to the rod and the surface takes the form shown in the figure.

Such material does not flow if the shear stress σ is less than a critical stress σ_c, and the material starts to flow when σ exceeds σ_c. i.e.,

$$\dot{\gamma} = \begin{cases} 0 & \sigma < \sigma_c \\ \frac{\sigma - \sigma_c}{\eta_B} & \sigma > \sigma_c \end{cases} \tag{9.22}$$

Such a fluid is called a Bingham fluid. (η_B is a material parameter called the Bingham viscosity.)

Nonlinear effects appear in other phenomena. When a polymeric fluid is sheared in a concentric cylinder, the fluid clings to the internal cylinder and often climbs along the cylinder as shown in Fig. 9.5(b). This phenomenon is called the Weissenberg effect, or the rod-climbing effect. Since the climbing takes place independently of the sign of the strain rate $\dot{\gamma}$, the Weissenberg effect is an intrinsically nonlinear effect in a viscoelastic liquid.

9.1.5 Constitutive equations

As demonstrated by the nonlinear effects described above, the relation between stress and deformation of soft matter can be quite complex. As discussed in Section 3.1 and Appendix A, the stress is a tensorial quantity (denoted by the stress tensor $\boldsymbol{\sigma}$), and the deformation is also a tensorial quantity (described by, for example, the deformation gradient tensor $\boldsymbol{E}$). In viscoelastic materials, the stress tensor at time t depends not only on the current value of the deformation gradient $\boldsymbol{E}(t)$, but also on the rate of deformation $\dot{\boldsymbol{E}}(t)$, and, in general, on the previous values of the deformation ($\boldsymbol{E}(t')$ for $t' < t$). Since the relation is generally nonlinear, the relation between the stress and deformation can be quite complex.

Though the relation may be complex, the mapping is unique, i.e., the stress is uniquely determined if the past deformation applied to the material is given. Therefore it is possible to construct a mathematical model for the relation. The equation that determines the stress tensor for a given deformation history is generally called the constitutive equation. If the constitutive equation is given, the flow and deformation of the material can be predicted from solving the equation of motion for the material. This is the basis of continuum mechanics. Further discussions on continuum mechanics and constitutive equations are given in Appendix A. Here we summarize the points which are relevant in the subsequent discussion.

In deriving the constitutive equation, we may assume that the material is deformed uniformly. This is due to the locality principle which states that the stress at any material point depends only on the history of the local deformation that the point has experienced in the past. Therefore we may assume that the material point located at $\boldsymbol{r}_0$ at some reference time t_0 is displaced to the point

$$\boldsymbol{r} = \boldsymbol{E}(t, t_0) \cdot \boldsymbol{r}_0, \qquad \text{or} \qquad r_\alpha = E_{\alpha\beta}(t, t_0)r_{0\beta} \tag{9.23}$$

at time t. The tensor $\boldsymbol{E}(t,t_0)$ represents the deformation gradient tensor introduced in Section 3.1.4. If the deformation is described by eq. (9.23), the velocity of the material point at $\boldsymbol{r} = \boldsymbol{E}(t,t_0)\cdot\boldsymbol{r}_0$ at time t is given by

$$\boldsymbol{v} = \dot{\boldsymbol{E}}(t,t_0)\cdot\boldsymbol{r}_0 \qquad (9.24)$$

where $\dot{\boldsymbol{E}}(t,t_0)$ represents $\partial\boldsymbol{E}(t,t_0)/\partial t$. The velocity gradient tensor $\boldsymbol{\kappa}(t)$ is defined by $\kappa_{\alpha\beta}(t) = \partial v_\alpha/\partial r_\beta$. By eq. (9.24), this can be written as

$$\kappa_{\alpha\beta}(t) = \frac{\partial v_\alpha}{\partial r_\beta} = \frac{\partial v_\alpha}{\partial r_{0\mu}}\frac{\partial r_{0\mu}}{\partial r_\beta} = \dot{E}_{\alpha\mu}(E^{-1})_{\mu\beta} \qquad (9.25)$$

Therefore, the deformation history is specified by the past deformation gradient tensor $\boldsymbol{E}(t',t_0)$, or the past velocity gradient tensor $\boldsymbol{\kappa}(t')$ $(t' < t)$. The constitutive equation determines the stress tensor $\boldsymbol{\sigma}(t)$ for given deformation history.

There are two cases in which the constitutive equation can be uniquely determined experimentally. One is the case of linear viscoelasticity. In this case, the relation between stress and strain is expressed uniquely by the relaxation modulus $G(t)$. Therefore by measuring $G(t)$, or equivalently, by measuring $G'(\omega)$ or $G''(\omega)$, one can obtain the constitutive equation. The other is the case of a purely elastic material for which the stress depends only on the deformation gradient $\boldsymbol{E}$ from the reference state. In this case, the stress is determined by the deformation free energy $f(\boldsymbol{E})$ (see Appendix A.4). The form of $f(\boldsymbol{E})$ is determined by measuring the stress for biaxial stretching, and therefore one can uniquely determine the constitutive equation experimentally.

Apart from such cases, there is no general way of determining the constitutive equation for nonlinear viscoelastic materials. What is normally done is that certain model equations are assumed, and the parameters which appear in the equations are determined experimentally. The model equations are derived either by a phenomenological argument (considering the material symmetry and the compatibility with irreversible thermodynamics) or by molecular models. In what follows, we shall elaborate this latter approach, and show a few examples in which the constitutive equations are derived by molecular models.

9.2 Molecular models

9.2.1 Microscopic expression for the stress tensor

To derive the constitutive equation by molecular models, it is important to know how the stress is expressed microscopically at the molecular level. The macroscopic definition for the stress tensor is given in Fig. A.2. The microscopic expression for the stress tensor is obtained by the argument shown in Fig. 9.6.

To obtain the microscopic expression for the stress tensor at a material point P, we consider a small cubic box centred at P. We divide the box into two parts by a plane normal to the z-axis (see Fig. 9.6(a)), and

Fig. 9.6 (a) Definition of the stress tensor component $\sigma_{\alpha z}$. Consider a plane of area S normal to the z-axis, and let $\boldsymbol{F}$ be the total force that the material above the plane exerts on the material below the plane. Then $\sigma_{\alpha z}$ is defined by $\sigma_{\alpha z} = F_\alpha/S$, i.e., the α-component of the force $\boldsymbol{F}$ per unit area. (b) The molecular origin of the force $\boldsymbol{F}$. If $\boldsymbol{f}_{ij}$ denotes the force that molecule j exerts on molecule i, $\boldsymbol{F}$ is given by $\boldsymbol{F} = \sum_{i \in \text{below}} \sum_{j \in \text{above}} \boldsymbol{f}_{ij}$.

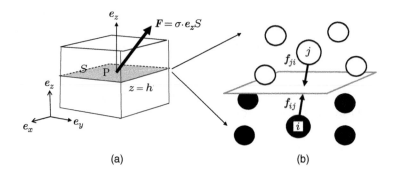

(a)   (b)

[4] Here we are ignoring the contribution to the stress tensor from the momentum transfer caused by the motion of segments across the plane. This contribution is important in gases or liquids of simple molecules, but can be ignored in the slow relaxation of soft matter. In eq. (9.26), $\Theta(x)$ is the step function, i.e., $\Theta(x) = 1$ for $x > 0$ and $\Theta(x) = 0$ for $x < 0$.

define $\boldsymbol{F}$ as the force that the material above the plane in the box exerts on the material below the plane. The stress tensor component $\sigma_{\alpha z}$ is defined by $\sigma_{\alpha z} = F_\alpha/S$, where S is the area of the plane.

Let us assume that the material consists of solute segments (denoted by the spheres in Fig. 9.6(b)) and a solvent of Newtonian fluid of viscosity η_s. Let $\boldsymbol{f}_{ij}$ be the force exerted on segment i by segment j. This force contributes to the stress tensor if segment i is below the plane and segment j is above the plane (see Fig. 9.6(b)). Hence[4]

$$\sigma_{\alpha z}^{(p)} = \frac{1}{S} \sum_{i,j} f_{ij\alpha} \Theta(h - r_{iz}) \Theta(r_{jz} - h) \tag{9.26}$$

where $\boldsymbol{r}_i$ is the position vector of segment i, and the summation is taken over all segments in the box. The superscript (p) is included to denote that it is the stress due to the forces acting between the solute segments.

Now if the system is homogeneous, the stress tensor will not depend on the position h of the plane. Therefore, we may define $\sigma_{\alpha z}^{(p)}$ as the average of (9.26) with respect to h in the region of $0 < h < L$:

$$\sigma_{\alpha z}^{(p)} = \frac{1}{SL} \sum_{i,j} \int_0^L dh\, f_{ij\alpha} \Theta(h - r_{iz}) \Theta(r_{jz} - h)$$

$$= \frac{1}{V} \sum_{i,j} f_{ij\alpha} (r_{jz} - r_{iz}) \Theta(r_{jz} - r_{iz}) \tag{9.27}$$

where $V = SL$ is the volume of the box. Let

$$\boldsymbol{r}_{ij} = \boldsymbol{r}_i - \boldsymbol{r}_j \tag{9.28}$$

be the relative vector between i and j. Equation (9.27) is then written as

$$\sigma_{\alpha z}^{(p)} = \frac{1}{V} \sum_{i,j} f_{ij\alpha} r_{jiz} \Theta(r_{jiz}) = \frac{1}{V} \sum_{i>j} [f_{ij\alpha} r_{jiz} \Theta(r_{jiz}) + f_{jia} r_{ijz} \Theta(r_{ijz})] \tag{9.29}$$

Using the relations $\boldsymbol{f}_{ij} = -\boldsymbol{f}_{ji}$ and $\boldsymbol{r}_{ij} = -\boldsymbol{r}_{ji}$, we can write eq. (9.29) as

$$\sigma_{\alpha z}^{(p)} = \frac{1}{V} \sum_{i>j} [-f_{ij\alpha}r_{ijz}\Theta(r_{jiz}) - f_{ij\alpha}r_{ijz}\Theta(r_{ijz})] = -\frac{1}{V}\sum_{i>j} f_{ij\alpha}r_{ijz} \tag{9.30}$$

Equation (9.30) can be interpreted as follows. The force $\boldsymbol{f}_{ij}$ contributes to the stress if the line segment joining segments i and j intersects the plane S. If the ij pair is distributed uniformly in the box, the probability that such an intersection takes place is Sr_{ijz}/V. Summing up the contribution from all pairs, we have eq. (9.30).

The tensor $\boldsymbol{f}_{ij}\boldsymbol{r}_{ij}$ is called the force dipole. The force dipole is an analogue of the electric dipole in electrostatics. In electrostatics, if two charges q and $-q$ are placed in space at relative vector $\boldsymbol{r}$, the electric dipole moment is given by $q\boldsymbol{r}$. Likewise, if two forces $\boldsymbol{f}$ and $-\boldsymbol{f}$ are acting at points separated by vector $\boldsymbol{r}$, the force dipole moment is given by $\boldsymbol{f}\boldsymbol{r}$. The electric dipole moment is a vector, while the force dipole moment is a tensor. Equation (9.30) indicates that the stress tensor is a force dipole moment per unit volume of the material.

In general, if the system consists of some interacting segments, their contribution to the stress tensor can be written as

$$\sigma_{\alpha\beta}^{(p)} = -\frac{1}{V}\sum_{i>j} \langle f_{ij\alpha}r_{ij\beta}\rangle = -\frac{1}{2V}\sum_{i,j}\langle f_{ij\alpha}r_{ij\beta}\rangle \tag{9.31}$$

where $\langle \cdots \rangle$ represents the average for the distribution of segments.

Equation (9.31) can be written in a slightly different form by use of eq. (9.28):

$$\sigma_{\alpha\beta}^{(p)} = -\frac{1}{2V}\sum_{i,j}\langle f_{ij\alpha}(r_{i\beta} - r_{j\beta})\rangle$$

$$= -\frac{1}{2V}\sum_{i,j}\langle f_{ij\alpha}r_{i\beta}\rangle - \frac{1}{2V}\sum_{i,j}\langle f_{ji\alpha}r_{j\beta}\rangle$$

$$= -\frac{1}{V}\sum_{i}\langle F_{i\alpha}r_{i\beta}\rangle \tag{9.32}$$

where

$$\boldsymbol{F}_i = \sum_j \boldsymbol{f}_{ij} \tag{9.33}$$

represents the total force acting on segment i.[5]

In the present situation, segments are immersed in a Newtonian fluid of viscosity η_s, which also contributes to the stress tensor. Accounting for such a contribution, we finally have the microscopic expression for the stress tensor:

$$\sigma_{\alpha\beta} = -\frac{1}{V}\sum_{i}\langle F_{i\alpha}r_{i\beta}\rangle + \eta_s(\kappa_{\alpha\beta} + \kappa_{\beta\alpha}) - p\delta_{\alpha\beta} \tag{9.35}$$

The last two terms represent the contribution from the solvent.

[5] Equation (9.32) corresponds to the expression for the electric dipole moment $\boldsymbol{P}$ for the collection of point charges q_i located at r_i:

$$\boldsymbol{P} = \sum_i q_i r_i \tag{9.34}$$

Notice that this expression is valid only when $\sum_i q_i = 0$, i.e., the total charge in the system is zero. Likewise, eq. (9.32) is valid only when $\sum_i \boldsymbol{F}_i = 0$. If $\sum_i \boldsymbol{F}_i$ is not equal to zero, diffusion of segments takes place, and the treatment in Chapter 8 is needed.

9.2.2   Stress tensor derived from the Onsager principle

If the time evolution of the system is described by the Onsager principle, the stress tensor can be derived from the Rayleighian by formula (7.42). To see this, consider the geometry shown in Fig. 7.3(c). Here the system is sandwiched between two parallel plates and is sheared. The Rayleighian R of this system involves the position of the top plate $x(t)$ as an external parameter. According to eq. (7.42), the force $F(t)$ needed to displace the top plate at speed $\dot{x}$ is given by

$$F(t) = \frac{\partial R}{\partial \dot{x}} \tag{9.36}$$

Using the shear stress $\sigma_{xy}(t)$, and the shear strain $\gamma(t)$, $F(t)$ and $x(t)$ are written as $F(t) = \sigma_{xy}(t)S$ and $\dot{x}(t) = h\dot{\gamma}$, where S is the area of the top plate, and h is the distance between the plates. Equation (9.36) therefore gives

$$\sigma_{xy}(t) = \frac{1}{Sh} \frac{\partial R}{\partial \dot{\gamma}} = \frac{1}{V} \frac{\partial R}{\partial \dot{\gamma}} \tag{9.37}$$

where $V = Sh$ is the volume of the system.

In general, if the Rayleighian is written as a function of the velocity gradient tensor $\kappa_{\alpha\beta}$, the stress tensor $\sigma_{\alpha\beta}$ is obtained by the formula

$$\sigma_{\alpha\beta}(t) = \frac{1}{V} \frac{\partial R}{\partial \kappa_{\alpha\beta}} \tag{9.38}$$

9.2.3   Viscoelasticity of polymeric fluids

Polymeric fluids (polymer solutions and polymer melts) are generally viscoelastic. The reason for this can be understood by examining Fig. 9.7. Suppose that a step shear is applied at time $t = 0$. When the shear strain is applied, polymer chains are distorted from their equilibrium conformation, which creates a stress (Fig. 9.7(a) to (b)). The stress has the same origin as that in rubbers and gels, i.e., it is due to the elasticity of the polymer chains. Unlike rubbers and gels, the polymer chains in polymeric fluids can recover their equilibrium conformation while the system is macroscopically deformed. As the polymer chains recover their equilibrium conformation, the stress decreases in time, and finally vanishes completely even though the material remains distorted. Therefore, the stress relaxation in polymeric fluids is a direct consequence of the conformational relaxation of polymer chains.

How the relaxation of polymer conformation takes place depends on whether the polymers are entangled or not. If the chains are not entangled, the problem is how a deformed chain placed in a viscous medium recovers the equilibrium conformation. If the chains are entangled, an entirely different thinking is needed. In the following we shall discuss these cases separately.

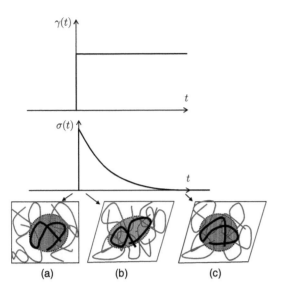

Fig. 9.7 Top: the response of shear stress $\sigma(t)$ of a polymeric fluid when a step shear $\gamma(t)$ is applied. Bottom: molecular conformation at each stage of the stress relaxation process. (a) Before the deformation, polymer molecules are at equilibrium and their average shape is spherical. (b) After the deformation, the polymer is deformed from the spherical shape and creates the restoring stress. (c) As time goes on, the polymer conformation goes back to equilibrium, and the stress relaxes to zero.

9.3 Viscoelasticity of non-entangled polymers

9.3.1 Dumbbell model

We now discuss the viscoelasticity of polymer fluids for which the entanglement effect is not important. This is the case of dilute polymer solutions, or concentrated solutions (or melts) of short polymers. To discuss the conformational dynamics, we consider the simple model shown in Fig. 9.8. Here a polymer molecule is represented by a dumbbell consisting of two segments connected by a spring (Fig. 9.8(b)). The segment represents half of the polymer molecule as in (a). One can consider a more elaborate model which consists of many segments as in (c), but the essential point can be seen in the dumbbell model.

We assume that the medium surrounding the segments is a Newtonian fluid. This assumption is reasonable for dilute polymer solutions, but is not so in polymer melts where the medium is made of other polymers. However, it is known that the assumption works in polymer melts as long as the polymer chains are short.

Let $\boldsymbol{r}_1$ and $\boldsymbol{r}_2$ be the position vectors of the segments. Then the potential energy of the dumbbell is written as

$$U(\boldsymbol{r}_1, \boldsymbol{r}_2) = \frac{k}{2}(\boldsymbol{r}_1 - \boldsymbol{r}_2)^2 \qquad (9.39)$$

where k is the spring constant of the dumbbell. Suppose that the dumbbell is placed in a viscous fluid which is flowing with velocity gradient $\kappa_{\alpha\beta}$. The flow field is given by

$$\boldsymbol{v}(\boldsymbol{r}; t) = \boldsymbol{\kappa}(t) \cdot \boldsymbol{r}, \qquad \text{or} \qquad v_\alpha(\boldsymbol{r}; t) = \kappa_{\alpha\beta}(t) r_\beta \qquad (9.40)$$

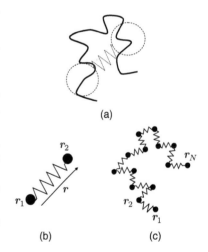

Fig. 9.8 (a) The conformation of a polymer molecule is most crudely represented by the dumbbell shown by the dashed line. (b) The dumbbell model. (c) The Rouse model.

To discuss the conformational change of the dumbbell, we consider the conformational distribution function $\Psi(r_1, r_2, t)$, and derive the time evolution equation for Ψ using the Onsager principle.

When a segment i moves in a flow field, there is an extra contribution to the energy dissipation due to the relative motion between the segment and the surrounding fluid. Most simply, this is given by $\zeta(\dot{r}_i - \kappa \cdot r_i)^2$. Hence the energy dissipation per unit volume is given by

$$\Phi = \frac{n_p}{2} \int dr_1 \int dr_2 \left[\zeta(\dot{r}_1 - \kappa \cdot r_1)^2 + \zeta(\dot{r}_2 - \kappa \cdot r_2)^2 \right] \Psi \qquad (9.41)$$

where n_p is the number of dumbbell molecules in a unit volume.

It is convenient to express r_1 and r_2 in terms of the position of the centre of mass $R = (r_1 + r_2)/2$ and the end-to-end vector $r = r_2 - r_1$ as

$$r_1 = R - \frac{1}{2}r, \qquad r_2 = R + \frac{1}{2}r \qquad (9.42)$$

Equation (9.41) is then written as

$$\Phi = \frac{n_p}{2} \int dR \int dr \left[2\zeta(\dot{R} - \kappa \cdot R)^2 + \frac{1}{2}\zeta(\dot{r} - \kappa \cdot r)^2 \right] \Psi \qquad (9.43)$$

Since the dumbbells are homogeneously distributed in the system, we may write Ψ as $\psi(r, t)$, and ignore the motion of the centre of mass. The energy dissipation function is then written as

$$\Phi = \frac{n_p}{2} \int dr \frac{1}{2}\zeta(\dot{r} - \kappa \cdot r)^2 \psi \qquad (9.44)$$

On the other hand, the free energy is given by the same form as in eq. (7.86)

$$A = n_p \int dr \left[k_B T \psi \ln \psi + U\psi \right] \qquad (9.45)$$

Hence

$$\dot{A} = n_p \int dr [k_B T(\ln \psi + 1) + U]\dot{\psi} \qquad (9.46)$$

Using

$$\dot{\psi} = -\frac{\partial}{\partial r} \cdot (\dot{r}\psi) \qquad (9.47)$$

and integration by parts, eq. (9.46) is written as

$$\dot{A} = n_p \int dr \dot{r}\psi \frac{\partial \tilde{U}}{\partial r} \qquad (9.48)$$

where

$$\tilde{U} = k_B T \ln \psi + U \qquad (9.49)$$

is the effective potential which is the sum of the spring potential and the diffusion potential (see Section 7.3.4). Minimizing $R = \Phi + \dot{A}$ with respect to $\dot{r}$, we have

$$\frac{\zeta}{2}(\dot{r} - \kappa \cdot r) = -\frac{\partial \tilde{U}}{\partial r}, \tag{9.50}$$

or

$$\dot{r} = -\frac{2}{\zeta}\left(k_B T \frac{\partial \ln \psi}{\partial r} + kr\right) + \kappa \cdot r \tag{9.51}$$

Therefore the time evolution equation for $\psi(r,t)$ becomes

$$\frac{\partial \psi}{\partial t} = \frac{\partial}{\partial r} \cdot \left(\frac{2k_B T}{\zeta}\frac{\partial \psi}{\partial r} + \frac{2k}{\zeta}r\psi - \kappa \cdot r\psi\right) \tag{9.52}$$

The stress tensor of the system can be calculated by using eq. (9.38).

$$\sigma_{\alpha\beta}^{(p)} = \frac{\partial R}{\partial \kappa_{\alpha\beta}} = -n_p \int dr\, \frac{\zeta}{2}(\dot{r}_\alpha - \kappa_{\alpha\mu}r_\mu)r_\beta\psi \tag{9.53}$$

By use of eq. (9.51), eq. (9.53) is written as

$$\sigma_{\alpha\beta}^{(p)} = n_p \int dr\, \frac{\partial \tilde{U}}{\partial r_\alpha}r_\beta\psi \tag{9.54}$$

This expression is in agreement with eq. (9.31) since the force acting between the segments in the dumbbell is given by $f = -\partial\tilde{U}/\partial r$. Equation (9.54) is rewritten by use of eq. (9.49) and integration by parts as

$$\sigma_{\alpha\beta}^{(p)} = n_p \int dr\left[k_B T \frac{\partial \psi}{\partial r_\alpha}r_\beta + kr_\alpha r_\beta\psi\right]$$
$$= -n_p k_B T\delta_{\alpha\beta} + n_p k\langle r_\alpha r_\beta\rangle \tag{9.55}$$

Hence the stress tensor is written as

$$\sigma_{\alpha\beta} = n_p k\langle r_\alpha r_\beta\rangle + \eta_s(\kappa_{\alpha\beta} + \kappa_{\beta\alpha}) - p\delta_{\alpha\beta} \tag{9.56}$$

Here we have added the solvent contribution to the stress. The isotropic term $-n_p k_B T\delta_{\alpha\beta}$ has been absorbed in the pressure term $p\delta_{\alpha\beta}$.

Linear viscoelasticity

As a first application of the above equations, let us consider the linear viscoelasticity of the dumbbell system. Consider the shear flow. The velocity field of the shear flow is given by

$$v_x = \dot{\gamma}r_y, \qquad v_y = v_z = 0 \tag{9.57}$$

In this case, eq. (9.52) becomes

$$\frac{\partial \psi}{\partial t} = \frac{\partial}{\partial r} \cdot \left(\frac{2k_B T}{\zeta}\frac{\partial \psi}{\partial r} + \frac{2k}{\zeta}r\psi\right) - \dot{\gamma}\frac{\partial}{\partial r_x}(r_y\psi) \tag{9.58}$$

On the other hand, the shear component of the stress tensor (9.56) is written as

$$\sigma_{xy} = n_p k \langle r_x r_y \rangle + \eta_s \dot{\gamma} \tag{9.59}$$

To calculate the term $\langle r_x r_y \rangle$ in eq. (9.59), we consider the time derivative of $\langle r_x r_y \rangle$. Using eq. (9.58), this is given by

$$
\begin{aligned}
\frac{d}{dt} \langle r_x r_y \rangle &= \int d\boldsymbol{r} \; r_x r_y \frac{\partial \psi}{\partial t} \\
&= \int d\boldsymbol{r} \; r_x r_y \left[\frac{\partial}{\partial \boldsymbol{r}} \cdot \left(\frac{2k_B T}{\zeta} \frac{\partial \psi}{\partial \boldsymbol{r}} + \frac{2k}{\zeta} \boldsymbol{r} \psi \right) - \frac{\partial \dot{\gamma} r_y \psi}{\partial r_x} \right]
\end{aligned}
\tag{9.60}
$$

Using integration by parts for the right-hand side, eq. (9.60) is rewritten as

$$\frac{d}{dt} \langle r_x r_y \rangle = -4 \frac{k}{\zeta} \langle r_x r_y \rangle + \dot{\gamma} \langle r_y^2 \rangle \tag{9.61}$$

In calculating the linear viscoelasticity, we can ignore terms of order $\dot{\gamma}^2$ or higher, and retain terms of first order in $\dot{\gamma}$ only. In this case, the average $\dot{\gamma} \langle r_y^2 \rangle$ may be replaced by $\dot{\gamma} \langle r_y^2 \rangle_0$, where $\langle \cdots \rangle_0$ represents the average for $\dot{\gamma} = 0$, i.e., the average at equilibrium. Since the equilibrium distribution of $\boldsymbol{r}$ is proportional to $\exp(-k\boldsymbol{r}^2/2k_B T)$, we have

$$\langle r_y^2 \rangle_0 = \frac{k_B T}{k} \tag{9.62}$$

Hence eq. (9.61) becomes as follows

$$\frac{d}{dt} \langle r_x r_y \rangle = -\frac{1}{\tau} \langle r_x r_y \rangle + \dot{\gamma} \frac{k_B T}{k} \tag{9.63}$$

where

$$\tau = \frac{\zeta}{4k} \tag{9.64}$$

Solving eq. (9.63), and using eq. (9.56), we have the following expression for the shear stress

$$\sigma_{xy}(t) = n_p k_B T \int_{-\infty}^{t} dt' e^{-(t-t')/\tau} \dot{\gamma}(t') + \eta_s \dot{\gamma}(t) \tag{9.65}$$

Comparing this with eq. (9.7), we have the relaxation modulus

$$G(t) = G_0 e^{-t/\tau} \tag{9.66}$$

where the initial shear modulus G_0 is given by

$$G_0 = n_p k_B T \tag{9.67}$$

This expression for the shear modulus G_0 has a form similar to that in rubber (eq. (3.55)). In the case of rubber, n_p represents the number of subchains per unit volume. The similarity arises from the fact that the stress of both systems has the same origin, i.e., the elasticity of the polymer chains. Since the end-to-end vector of the dumbbell molecule in solutions and the end-to-end vector of the subchain in a rubber behaves in the same way under step deformation, the initial stress becomes the same in both systems.

Constitutive equation

Equations (9.52) and (9.56) can be regarded as the constitutive equation for a solution of dumbbell molecules. For a given velocity gradient $\kappa(t)$, $\psi(r, t)$ is obtained by solving eq. (9.52), and then the stress tensor is calculated by eq. (9.56). Therefore eqs. (9.52) and (9.56) give the stress response of the system for a given velocity gradient. The relation between the stress and the velocity gradient is not apparent in this form. In the case of dumbbell molecules, the relation can be made more explicitly.

To calculate the average $\langle r_\alpha r_\beta \rangle$, we multiply both sides of eq. (9.52) by $r_\alpha r_\beta$ and integrate over r. Using integration by parts, we have

$$\frac{d}{dt}\langle r_\alpha r_\beta \rangle = -\frac{1}{\tau}\langle r_\alpha r_\beta \rangle + \kappa_{\alpha\mu}\langle r_\beta r_\mu \rangle + \kappa_{\beta\mu}\langle r_\alpha r_\mu \rangle + 4\delta_{\alpha\beta}\frac{k_B T}{\zeta} \qquad (9.68)$$

We define the stress due to dumbbell molecules in eq. (9.56) by

$$\sigma^p_{\alpha\beta} = n_p k \langle r_\alpha r_\beta \rangle \qquad (9.69)$$

Equation (9.68) then gives the following equation for $\sigma^p_{\alpha\beta}$

$$\dot{\sigma}^p_{\alpha\beta} = -\frac{1}{\tau}\sigma^p_{\alpha\beta} + \kappa_{\alpha\mu}\sigma^p_{\beta\mu} + \kappa_{\beta\mu}\sigma^p_{\alpha\mu} + \frac{G_0}{\tau}\delta_{\alpha\beta} \qquad (9.70)$$

The constitutive equation (9.70) is known as the Maxwell model in rheology.[6] The Maxwell model is a typical constitutive equation for viscoelastic fluids. It describes the elastic response of viscoelastic fluids and some nonlinear effects (such as the Weissenberg effect). On the other hand, the model cannot describe shear thinning. If we apply eq. (9.70) to steady shear flow, we find that the steady state viscosity is constant.[7] Various modifications have been proposed to remove this deficit (see the Further reading at the end of this chapter).

[6] Equation (9.70) is for homogeneous flow where $\sigma^p_{\alpha\beta}$ does not depend on position r. In the general case where $\sigma^p_{\alpha\beta}$ depends on r, $\dot{\sigma}^p_{\alpha\beta}$ has to be interpreted as the material time derivative, i.e., $\sigma^p_{\alpha\beta}(r, t)\Delta t$ represents the difference of $\sigma^p_{\alpha\beta}$ at time t and $t + \Delta t$ for the same material point

$$\dot{\sigma}^p_{\alpha\beta}(r, t)\Delta t = \sigma^p_{\alpha\beta}(r + v\Delta t, t + \Delta t)$$
$$-\sigma^p_{\alpha\beta}(r, t) \qquad (9.71)$$

i.e.,

$$\dot{\sigma}^p_{\alpha\beta} = \frac{\partial \sigma^p_{\alpha\beta}}{\partial t} + v_\mu \frac{\partial \sigma^p_{\alpha\beta}}{\partial r_\mu} \qquad (9.72)$$

[7] According to eq. (9.70), the steady state viscosity is given by $G_0\tau + \eta_s$, and is independent of the shear rate $\dot{\gamma}$.

9.3.2   Dumbbell model and pseudo-network model

The constitutive equation (9.70) can be written in integral form. It can be shown that the solution of eq. (9.70) is given by

$$\sigma^p_{\alpha\beta} = -\int_{-\infty}^{t} dt' G(t - t')\frac{\partial}{\partial t'}B_{\alpha\beta}(t, t') \qquad (9.73)$$

where $G(t)$ is given by eq. (9.66) and $B_{\alpha\beta}(t, t')$ is defined by

$$B_{\alpha\beta}(t, t') = E_{\alpha\mu}(t, t')E_{\beta\mu}(t, t') \qquad (9.74)$$

Equation (9.73) is a generalization of eq. (9.65).

Suppose that a step deformation $\boldsymbol{E}$ is applied to the Maxwell model at time $t = 0$. According to eq. (9.73), the stress tensor at time t is given by

$$\boldsymbol{\sigma}^p = n_p k_B T \boldsymbol{B} e^{-t/\tau} \tag{9.75}$$

Therefore the initial stress after the deformation is $n_p k_B T \boldsymbol{B}$. Again, this expression is identical to the expression for the stress tensor of Kuhn's model for rubber (see eq. (A.24)). This allows us an interpretation of the constitutive equation of the Maxwell model.

Suppose that the cross-links in rubbers are not permanent, but are constantly generated and destroyed. Let τ be the lifetime of such temporary cross-links. If we apply a step deformation $\boldsymbol{E}$ to the system, we will have an initial stress $n k_B T \boldsymbol{B}$ (n is the number density of subchains in the temporary cross-linked network). As time goes on, the cross-links are destroyed, and the stress decays exponentially. This model gives the stress described by eq. (9.75), and is called the pseudo-network model.

In the dumbbell model, the spring connecting the segments is not destroyed, but it relaxes to the equilibrium state due to the Brownian motion of segments. In both the pseudo-network model and the dumbbell model, the stress is given by the elasticity of the polymer chain and the stress relaxation is a result of the conformational relaxation of the polymer. This concept is also useful for entangled polymers as we shall see later.

9.3.3   Rouse model

In the dumbbell model, the conformation of a polymer chain is represented by two vectors $\boldsymbol{r}_1$ and $\boldsymbol{r}_2$. To describe the polymer conformation in more detail, we can consider the model shown in Fig. 9.8(c), where N segments $\boldsymbol{r}_1, \boldsymbol{r}_2, \ldots, \boldsymbol{r}_N$ are connected by springs. The potential energy of such a model is written as

$$U(\boldsymbol{r}_1, \boldsymbol{r}_2, \ldots) = \frac{k}{2} \sum_{i=2}^{N} (\boldsymbol{r}_i - \boldsymbol{r}_{i-1})^2 \tag{9.76}$$

This model is called the Rouse model. The chain conformation in the Rouse model can be represented by a set of normal coordinates which are linear combinations of $\boldsymbol{r}_i$ and are independent of each other. Accordingly, the Rouse model becomes equivalent to a set of independent dumbbell models. As a result, the relaxation modulus of the Rouse model is given by

$$G(t) = n_p k_B T \sum_{p=1}^{N} \exp(-p^2 t/\tau_R) \tag{9.77}$$

where n_p is the number of polymers in unit volume and τ_R is defined by

$$\tau_R = \frac{\zeta N^2}{2\pi^2 k} \tag{9.78}$$

The relaxation modulus $G(t)$ is equal to $n_p N k_B T$ at $t = 0$, and relaxes in time. For $t \ll \tau_R$, the sum over p in eq. (9.77) can be approximated by an integral over p:

$$G(t) = n_p k_B T \int_0^\infty dp \exp(-p^2 t / \tau_R) = \frac{\sqrt{\pi}}{2} n_p k_B T \left(\frac{\tau_R}{t}\right)^{1/2}, \qquad (t < \tau_R)$$

$$(9.79)$$

Therefore $G(t)$ decays as a power law $t^{-1/2}$ on short time-scales. This characteristic behaviour is also seen in the frequency domain. If we use eq. (9.79) for $G(t)$ in eqs. (9.12) and (9.13), we find that $G'(\omega)$ and $G''(\omega)$ increases in proportion to $\omega^{1/2}$ in the high-frequency region $\omega \tau_R \gg 1$. Such behaviour is indeed observed in polymeric fluids at high frequency.

Using eq. (9.9), the steady state viscosity of the Rouse model is calculated as

$$\eta_0 = \int_0^\infty dt\, G(t) = n_p k_B T \tau_R \frac{\pi^2}{6} = \frac{n_p k_B T \zeta N^2}{12 k} \qquad (9.80)$$

The steady state viscosity η_0 is proportional to $n_p N^2$. If the weight concentration of polymer is constant, $n_p N$ is constant. Therefore, η_0 is proportional to N. This N dependence is indeed observed for polymer melts of small molecular weight where the entanglement effect is not important.

9.4   Viscoelasticity of entangled polymers

9.4.1   Entanglement effect

In polymer solutions, with increasing concentration, the polymers begin to entangle with each other as shown in Fig. 9.9(a). In this situation, the solution becomes quite viscous, and starts to show distinctive viscoelasticity. It has been observed that the viscosity η_0 and the relaxation time τ of polymeric fluids (polymer solutions and polymer melts) increases with the molecular weight M as

$$\eta_0 \propto M^{3.4}, \qquad \tau \propto M^{3.4} \qquad (9.81)$$

This strong molecular weight dependence is caused by the entanglement of polymers.

Polymer molecules cannot pass through each other. This constraint arises from the essential nature of polymers (i.e., as one-dimensional objects embedded in three-dimensional space), and exists for all polymers, independently of polymer species. The effect caused by this constraint is called the entanglement effect.

The entanglement effect is a kinetic effect and does not affect the equilibrium properties.[8] Imagine an ideal situation in which the polymer is represented by a geometrical curve (a one-dimensional object with no thickness). As far as the equilibrium properties are concerned, the entanglement effect is totally negligible: a solution of such objects behaves

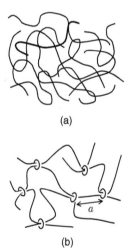

(a)

(b)

Fig. 9.9 (a) Polymer molecules in entangled state. (b) The slip-link model.

[8] In polymers with very special structures such as ring-like polymers or polymers with loops, the entanglement does affect the equilibrium properties.

as an ideal solution even in the entangled state. On the other hand, the molecular motion is drastically slowed down by the entanglements. For example, the relaxation time of a polymer is strongly affected by the entanglements.

Historically, entanglements have been modelled by certain kinds of junctions between polymers. Such junctions are not permanent, and will disappear after some time. This primitive notion is the basis of the pseudo-network model for polymeric fluids.

In the pseudo-network model, polymeric fluids are regarded as a network of polymers connected by entanglement junctions. These entanglement junctions are created randomly, and are destroyed after a certain characteristic lifetime. This model is able to account for the viscoelasticity of polymeric fluids. For example, in the situation shown in Fig. 9.7, when polymeric fluids are deformed, stress appears like in a rubber. However, with the passage of time, the junctions disappear and therefore the stress relaxes.

In the conventional pseudo-network model, the mechanisms for creation and destruction of junctions were not specified. Therefore the model could not explain the relation between the molecular structure and rheological parameters such as eq. (9.81). On the other hand, the creation and destruction dynamics of entanglement junctions can be discussed by the alternative model shown in Fig. 9.9(b). Here, the entanglement junctions are represented by small rings, called slip-links, which confine two polymer chains. In this model, polymer chains are assumed to be able to move through the slip-links freely. When the chain end moves out of a slip-link, the slip-link is destroyed and the entanglement junction disappears. On the other hand, the chain ends will find other partners and create new slip-links. Consequently, the average number of slip-links per chain remains constant. This model is called the slip-link model.

Although the entanglement junctions are represented by small rings in this model, the actual constraint caused by entanglement will not be a truly localized one. Identifying where and how many entanglement points there are in typical situations like that shown in Fig. 9.9(a) requires some kind of ansatz. What is assumed by the slip-link model is that once two polymers start to entangle, their motions are constrained for a certain period of time (from the time of creation of the slip-links to the time of their destruction) and over a certain spatial length (about the mean distance between the slip-links). An important quantity here is the mean distance between neighbouring slip-links. This length a is called the mean distance between entanglements. This is a molecular parameter determined by the rigidity and the local packing of chain segments, and is typically a few nm (i.e., much larger than the atomic length).

The molecular weight of the part of the polymer chain between neighbouring entanglement points is called the entanglement molecular weight and is denoted by M_e. M_e is analogous to M_x, the molecular weight between cross-links in rubber (see eq. (3.56)). When a polymeric fluid is deformed, all entanglement points behave as transient cross-link points in rubber. The elastic modulus of such a transient state is given by

$$G_0 = \frac{\rho R_G T}{M_e} \qquad (9.82)$$

G_0 is called the entanglement shear modulus. Equation (9.82) can be used to determine M_e.

9.4.2   Reptation theory

If a step shear of strain γ is applied to a polymeric fluid, the polymers are deformed, thereby generating the shear stress $G_0\gamma$. This stress decays in time as the polymer conformation relaxes to equilibrium. Let us consider this process in detail.

For simplicity, let us assume that the entanglement points are distributed uniformly along the chain with equal spacing a. If the molecular weight of the polymer is M, the number of entanglement points per chain is given by

$$Z = \frac{M}{M_e} \qquad (9.83)$$

Hence the contour length L of the curve which connects the entanglement points of the chain is given by

$$L = Za = \frac{M}{M_e}a \qquad (9.84)$$

Figure 9.10 shows how the entanglement points of a particular chain are created and destroyed. Suppose that the chain in configuration (a) in Fig. 9.10 moves to the right. Then the slip-link P_0 is evacuated by the polymer and is therefore destroyed. On the other hand, a new slip-link P_{Z+1} is created at the right end of the chain. If the chain moves to the left as in (c), the slip-links P_{Z+1} and P_Z are destroyed, and new slip-links P_0 and P_{-1} are created.

Therefore, the entanglement points are constantly destroyed and created due to the one-dimensional Brownian motion of the chain along the slip-links. Such one-dimensional motion was proposed by de Gennes, and is called reptation.

In the reptation theory, the polymer chain is constrained by the slip-links which are placed along the chain with average separation a. Such constraints by other chains can also be represented by the tube shown in Fig. 9.11. If the tube is assumed to be made of Z segments which have diameter a and length a connected randomly, the tube gives effectively the same constraints as the slip-links. Therefore, the tube model and the slip-link model are equivalent to each other.

To see the main characteristics of the reptation motion, let us assume that the polymer is moving randomly along the tube keeping its length L constant. Let us focus on a certain slip-link P_n in Fig. 9.10(a) which

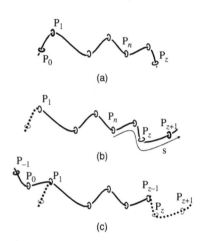

Fig. 9.10 The destruction and creation processes of the entanglement junctions in the slip-link model.

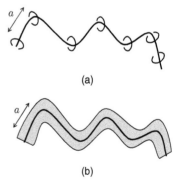

Fig. 9.11 Two equivalent models representing the entanglement effect: (a) slip-link model and (b) tube model.

existed at time $t = 0$, and consider the probability that it has not been
destroyed during the time between 0 and t.

To solve the problem, we take a curvilinear coordinate taken along
the central axis of the tube. The origin of the coordinate is taken at
P_n. Let s be the curvilinear coordinate of the right end of the chain. We
consider the probability $\psi_n(s,t)$ that the entanglement point P_n remains
undestroyed at time t and that the right end of the chain is at s. This
probability satisfies the following diffusion equation

$$\frac{\partial}{\partial t}\psi_n(s,t) = D_c\frac{\partial^2}{\partial s^2}\psi_n(s,t) \tag{9.85}$$

where D_c is the diffusion constant of the polymer along the tube.

The entanglement point P_n is destroyed when it is visited by one of
the two chain ends. Since this happens when s becomes equal to 0 (P_n
is visited by the right end) or L (P_n is visited by the left end), $\psi_n(s,t)$
satisfies the following boundary conditions

$$\psi_n(0,t) = 0, \qquad \psi_n(L,t) = 0 \tag{9.86}$$

Solving the diffusion equation (9.85) under the boundary conditions
(9.86), and the initial condition $\psi_n(s,0) = \delta(s - L + na)$, we have

$$\psi_n(s,t) = \frac{2}{L}\sum_{p=1}^{\infty}(-1)^p\sin\left(\frac{ps\pi}{L}\right)\sin\left(\frac{pna\pi}{L}\right)\exp(-tp^2/\tau_d) \tag{9.87}$$

where

$$\tau_d = \frac{L^2}{\pi^2 D_c} \tag{9.88}$$

is called the reptation time. τ_d represents the lifetime of an entanglement.

According to the Einstein relation (6.35), D_c is related to the friction
constant ζ_c by $D_c = k_BT/\zeta_c$, where ζ_c denotes the force needed to pull
the chain along the tube with unit velocity. This force is proportional
to the molecular weight M of the chain. Therefore D_c depends on M as
$D_c \propto M^{-1}$. On the other hand, L is proportional to M. Therefore, the
reptation time τ_d is proportional to M^3.

The probability that the entanglement point P_n remains at time t is
given by the integral of $\psi_n(s,t)$ for s from 0 to L. If this probability is
averaged for n, we have the probability $\psi(t)$ that an entanglement point
arbitrarily chosen at $t = 0$ still remains undestroyed at time t:

$$\psi(t) = \frac{1}{Z}\sum_{n=1}^{Z}\int_0^L ds\psi_n(s,t) \tag{9.89}$$

Substituting $\psi_n(s,t)$ of eq. (9.87), and replacing the sum over n by an
integral, we finally have the probability $\psi(t)$

$$\psi(t) = \frac{1}{Z}\int_0^Z dn\int_0^L ds\psi_n(s,t) = \frac{8}{\pi^2}\sum_{p=1,3,5,...}\frac{1}{p^2}\exp(-tp^2/\tau_d) \tag{9.90}$$

In the summation on the right-hand side of this equation, the term $p = 1$ has the dominant contribution. Therefore, $\psi(t)$ may be approximated by

$$\psi(t) \simeq \exp(-t/\tau_d) \qquad (9.91)$$

9.4.3   Stress relaxation

Let us now consider the stress relaxation of entangled polymers using the above result. We consider the situation shown in Fig. 9.7. At time $t = 0$, a step shear of strain γ is applied. Immediately after the application of the shear, all entanglement points act as transient cross-links in rubber. Therefore the shear stress $\sigma_{xy}(t)$ at $t = 0$ is given by

$$\sigma_{xy}(0) = G_0\gamma \qquad (9.92)$$

where G_0 is the shear modulus given by eq. (9.82).

As time goes on, these entanglement points gradually disappear due to the Brownian motion of the chain. Since the probability that the entanglement points which existed at $t = 0$ remain intact at a later time t is given by $\psi(t)$, the shear stress at time t is given by

$$\sigma_{xy}(t) = G_0\gamma\psi(t) \qquad (9.93)$$

Therefore, the relaxation modulus is given by

$$G(t) = G_0\psi(t) \qquad (9.94)$$

Notice that $G(t)$ decreases almost as a single exponential relaxation curve with relaxation time τ_d, i.e., the value of $G(t)$ at the longest relaxation time τ_d is about the same as the initial value, i.e., $G(\tau_d) \simeq G(0)$. This behaviour is quite different from that of the Rouse model. In the Rouse model, the value of $G(t)$ at the longest relaxation time τ_R is about $G(0)/N$, which is very small for large N. Indeed the relaxation modulus of entangled polymers (with narrow molecular weight distribution) decays almost with a single relaxation time.

By eqs. (9.9) and (9.94), the viscosity η_0 is given by

$$\eta_0 \simeq G_0\tau_d \qquad (9.95)$$

Since G_0 is independent of the molecular weight M, the viscosity is also proportional to M^3. The exponent 3 is slightly different from what is observed experimentally. The reason for the discrepancy will be discussed later. On the other hand, η_0 is proportional to M in the Rouse model. The difference in the molecular weight dependence between the Rouse model and the reptation model comes from two effects: one is the difference in the relaxation time: τ_R is proportional to M^2, while τ_d is proportional to M^3. The other is the difference in the distribution of the relaxation times. $G(t)$ of the Rouse model has a broad distribution of relaxation times, while $G(t)$ of the reptation model has almost a single relaxation time.

As the magnitude of the shear strain γ increases, there appears a new relaxation mode. When a sufficiently large shear is applied, the mean distance between the entanglement points becomes significantly larger than the equilibrium value a. This is seen as follows. Consider a vector joining neighbouring entanglement points. If this vector is (r_{0x}, r_{0y}, r_{0z}) before the deformation, it will be transformed to

$$r_x = r_{0x} + \gamma r_{0y}, \qquad r_y = r_{0y}, \qquad r_z = r_{0z} \qquad (9.96)$$

after the deformation. Therefore the mean square magnitude of r is calculated as

$$\langle r_x^2 + r_y^2 + r_z^2 \rangle = \langle r_{0x}^2 + r_{0y}^2 + r_{0z}^2 \rangle + 2\gamma \langle r_{0x} r_{0y} \rangle + \gamma^2 \langle r_{0y}^2 \rangle \qquad (9.97)$$

Since $\langle r_{0x}^2 \rangle = \langle r_{0y}^2 \rangle = \langle r_{0z}^2 \rangle = a^2/3$ and $\langle r_{0x} r_{0y} \rangle = 0$, we have

$$\langle r_x^2 + r_y^2 + r_z^2 \rangle = \left(1 + \frac{\gamma^2}{3} \right) a^2 \qquad (9.98)$$

Hence, the distance between the entanglement points increases by the factor[9]

$$\alpha(\gamma) = \left(1 + \frac{\gamma^2}{3} \right)^{1/2} \qquad (9.99)$$

[9] Here $\langle |r| \rangle$ is approximated by $\langle r^2 \rangle^{1/2}$.

Since the polymer is elongated along the tube axis (the path connecting the slip-links), the polymer first shrinks along the tube to recover the equilibrium length L as shown in Fig. 9.12(b) $\to$ (c). Since this relaxation process is not affected by the tube, the relaxation time is given by the Rouse relaxation time τ_R. As the polymer chain shrinks, the length of the polymer along the tube decreases by the factor $1/\alpha(\gamma)$, and the tension acting along the polymer also decreases by the same factor. Therefore the stress decreases by the factor $1/\alpha(\gamma)^2$. The stress at time $t \simeq \tau_R$ becomes

$$\sigma_{xy}(\tau_R) \simeq G_0 \frac{\gamma}{\alpha(\gamma)^2} = G_0 \frac{\gamma}{1 + \gamma^2/3} \qquad (9.100)$$

After the tension along the chain is relaxed, stress relaxation takes place by reptation. Therefore the stress for time $t > \tau_R$ is given by

$$\sigma_{xy}(t) = G_0 \frac{\gamma}{1 + \gamma^2/3} \psi(t) \qquad (9.101)$$

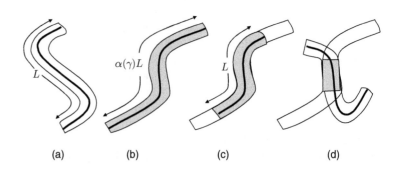

Fig. 9.12 Explanation of stress relaxation by the tube model. (a) Before deformation, the contour length of the tube is L, and the orientation of the tube is isotropic. (b) Immediately after deformation, the contour length of the tube is elongated (becomes $\alpha(\gamma)L$), and the tube is oriented. (c) The contour length recovers the equilibrium value L. (d) The tube gradually becomes unoriented as the tube segments at the ends are replaced by new ones.

(a)   (b)   (c)   (d)

Experimentally, it has been observed that as the shear strain becomes large, the stress first relaxes quickly with relaxation time τ_R and then slowly with relaxation time τ_d.

By extending such considerations to general deformations, a constitutive equation has been obtained for the reptation model (see the reference (2) in Further reading).

9.4.4   Entanglements in real systems

Effect of contour length fluctuation

In the above discussion, the polymer is assumed to have a constant contour length along the tube. In reality, the contour length L is not constant, rather it is fluctuating with time. As shown in Section 3.2, a stretched polymer chain behaves as a linear spring with elastic constant $k = 3k_BT/Nb^2$. Therefore the free energy of a polymer of contour length L is given by

$$U(L) = \frac{3k_BT}{2Nb^2}L^2 - F_{eq}L \qquad (9.102)$$

where F_{eq} is the equilibrium tensile force which arises from the extra entropy of the chain ends: the segments in the middle of the tube are constrained by the tube, while the segments at the ends of the tube are not constrained. Therefore the end segments have larger entropy than the middle segments. This difference in the entropy gives the force F_{eq}. F_{eq} can be expressed in terms of the equilibrium contour length L_{eq} of the tube:

$$F_{eq} = \frac{3k_BT}{Nb^2}L_{eq} \qquad (9.103)$$

Using eq. (9.103), $U(L)$ is written as

$$U(L) = \frac{3k_BT}{2Nb^2}(L - L_{eq})^2 + \text{constant} \qquad (9.104)$$

Therefore, the magnitude of the contour length fluctuation is

$$\langle (L - L_{eq})^2 \rangle = \frac{Nb^2}{3} \qquad (9.105)$$

The tube is made of Z segments of length a which orient randomly in space. Therefore the mean square end-to-end vector of a tube is Za^2. This has to be equal to the mean square end-to-end vector of a polymer, Nb^2. Therefore

$$Za^2 = Nb^2 \qquad (9.106)$$

On the other hand, the equilibrium contour length is given by $L_{eq} = Za$, and the magnitude of the fluctuation is estimated by $\Delta L = \langle (L - L_{eq})^2 \rangle^{1/2}$. Hence

$$\frac{\Delta L}{L_{eq}} \simeq \frac{\sqrt{N}b}{Za} = \frac{1}{\sqrt{Z}} \qquad (9.107)$$

If Z is very large, the fluctuation of the contour length is negligible. In practice, Z is typically between 10 and 100, and the effect of the fluctuations is non-negligible.

As an example of the effect of the contour length fluctuation, consider the reptation time τ_d. In the analysis in Section 9.4.2, it was assumed that for the entanglement point P_1 to be destroyed, the centre of mass of the polymer must move to the right by the distance a. If the contour length is fluctuating, the centre of mass need not move the full distance a: the chain end can slip away from P_1 when the polymer shrinks along the tube by fluctuation. This is true for the other inner entanglement points P_2, P_3, etc. Therefore for all entanglement points which existed at $t = 0$ to disappear, the polymer does not need to travel the full distance L_{eq}: it is enough to diffuse the distance $L_{eq} - \Delta L$. Therefore, the reptation time is better estimated by

$$\tau_d \simeq \frac{(L_{eq} - \Delta L)^2}{D_c} \simeq \tau_d^0 \left(1 - \frac{\Delta L}{L_{eq}}\right)^2 \propto M^3 \left(1 - \sqrt{\frac{M_e}{M}}\right)^2 \quad (9.108)$$

Due to the fluctuation effect, the reptation time becomes less than the ideal time $\tau_d^0 \propto M^3$. The reduction in τ_d is more pronounced as the molecular weight M decreases. Therefore in the plot of $\log \tau_d$ against $\log M$, the slope becomes larger than 3. In fact, the experimentally observed value of the slope is between 3.1 and 3.5.

Effect of chain branching

The reptation motion discussed above does not take place for branched polymers. Consider, for example, the star polymer made of three arms of equal length (see Fig. 9.13). For such a polymer, the branch point of the polymer is essentially fixed at the branch point of the tube: if the polymer branch point moves into one branch of the tube, it will be pulled back to the tube branch point by the tensile forces F_{eq} acting in each arm.

In this situation, the reptation motion is blocked, and the conformational relaxation of the chain can take place by the contraction of the chain along the tube. Consider an arm of the tube which has equilibrium length L_{eq}. If the contour length of the arm shrinks to length a, the whole arm can move out of the old tube and can enter into a new tube.

The contraction of the chain along the tube costs free energy, and therefore the process can be regarded as an activated process. Hence the relaxation time of the branched polymer can be estimated by

$$\tau_{br} = \tau_e \exp(\Delta U/k_B T) \quad (9.109)$$

where $\Delta U = U(a) - U(L_{eq})$ is the difference in the free energy between the contracted state and the equilibrium state. Using eq. (9.104), ΔU is estimated as

$$\Delta U = U(a) - U(L_{eq}) \approx \frac{k_B T}{N b^2} L_{eq}^2 = Z k_B T \quad (9.110)$$

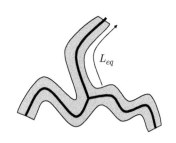

L_{eq}

Fig. 9.13 Tube model for star polymers.

Let M_a be the molecular weight of the arm of the polymer. The relaxation time is then estimated by

$$\tau_{br} = \tau_e \exp(\nu M_a/M_e) \qquad (9.111)$$

where ν is a certain numerical constant. Such an exponential dependence of the relaxation time for star polymers has indeed been observed experimentally.

Many body effect

The theory described above assumes that the slip-links are destroyed only when they are visited by the chain ends. In reality, the slip-links confine two chains, and therefore they are destroyed when one of the chain ends of the two chains visits the slip-link. Therefore, the creation and destruction of slip-links depends on the motion of other chains. Furthermore, the positions of the slip-links are not fixed in space and are fluctuating. Theories have been developed to take into account such effects. By such theories and computer simulations, the dynamics of flexible polymers in the entangled state is now reasonably well understood.

9.5   Rod-like polymers

9.5.1   Solutions of rod-like polymers

In this section we shall consider solutions of rod-like polymers. The viscoelastic properties of a solution of rod-like polymers change dramatically with concentration. Figure 9.14 shows sketches of solutions of rod-like polymers at various concentrations.

In a very dilute solution (a), the polymers can move independently of each other. This solution is essentially a Newtonian fluid, but rod-like polymers give viscoelastic properties to the fluid (as in the case of flexible polymers). This is due to the orientational degrees of freedom of the rod-like polymer.

With the increase of concentration, the solution first changes to the state shown in (b). In this state, the motion of each rod is strongly hindered by other rods, i.e., there is a strong entanglement effect. As a

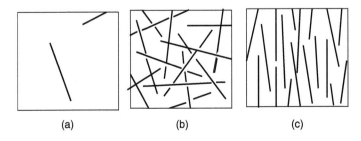

(a)   (b)   (c)

Fig. 9.14 Solutions of rod-like polymers. (a) Dilute solution; (b) concentrated solution in the isotropic state; (c) concentrated solution in the nematic state.

consequence, the solution becomes very viscous and starts to show pronounced viscoelasticity. The concentration c^* at which the state changes from (a) to (b) can be estimated by

$$c^* \frac{M}{N_A} R_g^3 \approx 1 \qquad (9.112)$$

where M is the molecular weight, N_A is the Avogadro number, and R_g is the radius of gyration of the polymer (see problem (8.1)). The radius of gyration of a rod-like polymer is much larger than that of a flexible polymer at the same molecular weight. Therefore, rod-like polymers show the entanglement effect at much lower concentration than flexible polymers.

With further increase of the concentration, the solution of rod-like polymers transforms into a nematic liquid crystal phase (see Chapter 5). Since the liquid crystal is an anisotropic fluid, the rheological behaviour changes drastically at this transition.

In the following, we shall consider rigid rod-like molecules of diameter b and length L, and discuss the rheological properties of the solutions at various concentrations. The concentration is denoted by the number density n_p, or the weight concentration $c = n_p M / N_A$ of the molecules.

9.5.2   Viscoelasticity of dilute solutions

Let us first consider dilute solutions of rod-like polymers (solutions with $n_p L^3 \ll 1$). The viscoelasticity of such a system can be discussed by the same method as in Section 9.3.1. Let $\boldsymbol{u}$ be the unit vector parallel to the rod, and let $\psi(\boldsymbol{u}, t)$ be the distribution function. The conservation equation for ψ is written as (see Section 7.5)

$$\dot{\psi} = -\mathcal{R} \cdot (\boldsymbol{\omega} \psi) \qquad (9.113)$$

where $\boldsymbol{\omega}$ is the angular velocity of the rod, and $\mathcal{R}$ is defined by

$$\mathcal{R} = \boldsymbol{u} \times \frac{\partial}{\partial \boldsymbol{u}} \qquad (9.114)$$

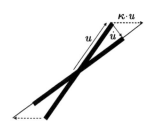

Fig. 9.15 Rotation of a rod-like polymer placed in a velocity field $\boldsymbol{v}(\boldsymbol{r}) = \boldsymbol{\kappa} \cdot \boldsymbol{r}$. In the velocity gradient $\boldsymbol{\kappa}$, the unit vector parallel to the rod changes as $\dot{\boldsymbol{u}} = \boldsymbol{\kappa} \cdot \boldsymbol{u} - (\boldsymbol{u} \cdot \boldsymbol{\kappa} \cdot \boldsymbol{u}) \boldsymbol{u}$. This is written as $\dot{\boldsymbol{u}} = \boldsymbol{\omega}_0 \times \boldsymbol{u}$, where $\boldsymbol{\omega}_0 = \boldsymbol{u} \times (\boldsymbol{\kappa} \cdot \boldsymbol{u})$.

If the rod rotates in a medium at rest, the energy dissipation is given by $(1/2)\zeta_r \boldsymbol{\omega}^2$, where ζ_r is the rotational diffusion constant given by eq. (7.84). If the medium is flowing with velocity gradient $\boldsymbol{\kappa}$, the rod rotates with angular velocity

$$\boldsymbol{\omega}_0 = \boldsymbol{u} \times (\boldsymbol{\kappa} \cdot \boldsymbol{u}) \qquad (9.115)$$

The reason for this is explained in Fig. 9.15. Therefore the energy dissipation function is written as

$$\Phi = \frac{n_p}{2} \int d\boldsymbol{u} \left[\zeta_r (\boldsymbol{\omega} - \boldsymbol{\omega}_0)^2 + w(\boldsymbol{\kappa}, \boldsymbol{u}) \right] \psi \qquad (9.116)$$

where $w(\boldsymbol{\kappa}, \boldsymbol{u})$ represents the energy dissipation in the system when the rod rotates with optimal angular velocity $\boldsymbol{\omega}_0$. Such a term exists since each point on the rod cannot completely follow the medium flow due to the rigid shape constraint of the rod. The actual form of w can be calculated by Stokesian hydrodynamics. The calculation gives the following energy dissipation function

$$\Phi = \frac{n_p}{2} \int d\boldsymbol{u} \left[\zeta_r (\boldsymbol{\omega} - \boldsymbol{\omega}_0)^2 + \frac{\zeta_r}{2} (\boldsymbol{u} \cdot \boldsymbol{\kappa} \cdot \boldsymbol{u})^2 \right] \psi \qquad (9.117)$$

On the other hand, the free energy term $\dot{A}$ is given by the same form as in eq. (7.87). As a result the Rayleighian is given by

$$R = \frac{n_p}{2} \int d\boldsymbol{u}\, \psi \left[\zeta_r (\boldsymbol{\omega} - \boldsymbol{\omega}_0)^2 + \frac{\zeta_r}{2} (\boldsymbol{u} \cdot \boldsymbol{\kappa} \cdot \boldsymbol{u})^2 \right] + n_p k_B T \int d\boldsymbol{u}\, \psi \boldsymbol{\omega} \cdot \mathcal{R}(\ln \psi) \qquad (9.118)$$

The extremum condition $\delta R / \delta \boldsymbol{\omega} = 0$ gives

$$\zeta_r (\boldsymbol{\omega} - \boldsymbol{\omega}_0) = -k_B T \mathcal{R}(\ln \psi) \qquad (9.119)$$

or

$$\boldsymbol{\omega} = -D_r \mathcal{R}(\ln \psi) + \boldsymbol{\omega}_0 \qquad (9.120)$$

where $D_r = k_B T / \zeta_r$. From eqs. (9.113), (9.115), and (9.120), the time evolution equation becomes

$$\frac{\partial \psi}{\partial t} = D_r \mathcal{R}^2 \psi - \mathcal{R} \cdot (\boldsymbol{u} \times \boldsymbol{\kappa} \cdot \boldsymbol{u} \psi) \qquad (9.121)$$

On the other hand, the stress tensor is given by

$$\begin{aligned}
\sigma_{\alpha\beta}^{(p)} = \frac{\partial R}{\partial \kappa_{\alpha\beta}} &= n_p \int d\boldsymbol{u}\, \psi \left[\zeta_r (\boldsymbol{\omega} - \boldsymbol{\omega}_0) \cdot \frac{\partial \boldsymbol{\omega}}{\partial \kappa_{\alpha\beta}} + \frac{1}{2} \zeta_r \boldsymbol{u} \cdot \boldsymbol{\kappa} \cdot \boldsymbol{u} u_\alpha u_\beta \right] \\
&= n_p \int d\boldsymbol{u}\, \psi \left[-\zeta_r [\boldsymbol{u} \times (\boldsymbol{\omega} - \boldsymbol{\omega}_0)]_\alpha u_\beta + \frac{1}{2} \zeta_r \boldsymbol{u} \cdot \boldsymbol{\kappa} \cdot \boldsymbol{u} u_\alpha u_\beta \right]
\end{aligned} \qquad (9.122)$$

The first term in the integral can be rewritten by use of eq. (9.120) as

$$\begin{aligned}
-\int d\boldsymbol{u}\, \psi \zeta_r [\boldsymbol{u} \times (\boldsymbol{\omega} - \boldsymbol{\omega}_0)]_\alpha u_\beta &= k_B T \int d\boldsymbol{u}\, \psi [\boldsymbol{u} \times (\mathcal{R} \ln \psi)]_\alpha u_\beta \\
&= k_B T \int d\boldsymbol{u} [e_{\alpha\mu\nu} u_\mu (\mathcal{R}_\nu \psi) u_\beta] \\
&= -k_B T \int d\boldsymbol{u} [e_{\alpha\mu\nu} \psi \mathcal{R}_\nu (u_\mu u_\beta)] \\
&= k_B T \int d\boldsymbol{u}\, \psi (3 u_\alpha u_\beta - \delta_{\alpha\beta}) \\
&= k_B T \langle 3 u_\alpha u_\beta - \delta_{\alpha\beta} \rangle \qquad (9.123)
\end{aligned}$$

Hence the stress tensor is given by

$$\sigma_{\alpha\beta}^p = n_p k_B T \langle 3 u_\alpha u_\beta - \delta_{\alpha\beta} \rangle + \frac{1}{2} n_p \zeta_r \langle u_\alpha u_\beta u_\mu u_\nu \rangle \kappa_{\mu\nu} \qquad (9.124)$$

The first term on the right-hand side is non-zero if the distribution of $\boldsymbol{u}$ is not isotropic. This term arises from the thermodynamic force which drives the system to the isotropic distribution of the molecules.

As an example, let us consider steady shear flow with shear rate $\dot{\gamma}$. In this case, the shear stress is given by

$$\sigma_{xy} = \sigma_{xy}^p + \eta_s\dot{\gamma} = 3n_p k_B T\langle u_x u_y\rangle + n_p\frac{\zeta_r}{2}\dot{\gamma}\langle u_x^2 u_y^2\rangle + \eta_s\dot{\gamma} \qquad (9.125)$$

For small shear rate, $\langle u_x u_y\rangle$ is estimated as $\dot{\gamma}/D_r \simeq \dot{\gamma}\zeta_r/k_B T$ and $\dot{\gamma}\langle u_x^2 u_y^2\rangle \approx \dot{\gamma}\langle u_x^2 u_y^2\rangle_0 \approx \dot{\gamma}$. Therefore the shear stress due to the polymer is estimated as

$$\sigma_{xy}^p \simeq n_p\dot{\gamma}\zeta_r \simeq n_p L^3\eta_s\dot{\gamma} \qquad (9.126)$$

where we have used eq. (7.84).

Equations (9.125) and (9.126) indicate that the viscosity of the solution of rod-like molecules increases by the factor $n_p L^3$ compared with the viscosity η_s of pure solvent. In a dilute solution (where $n_p L^3 < 1$) this viscosity change is small. On the other hand, in a concentrated solution where $n_p L^3 > 1$, the viscosity change is quite dramatic as we shall show in the next section.

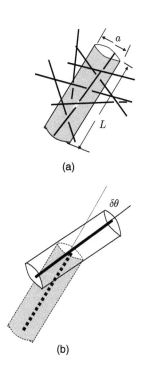

(a)

(b)

Fig. 9.16 Tube model for the state (b) in Fig. 9.14. (a) The tube constraining the motion of a rod-like polymer. (b) The mechanism of rotation of the polymer constrained by the tube.

9.5.3 Viscoelasticity of concentrated solutions in the isotropic phase

In concentrated solutions where $n_p L^3 \gg 1$ is satisfied, the free rotational Brownian motion of rod-like molecules becomes impossible as it is always hindered by other rods (see Fig. 9.14(b)). In this state, the rotational diffusion is strongly slowed down. Let us estimate the effective rotational diffusion constant in the situation shown in Fig. 9.14(b).

Due to the hindrance of surrounding rods, each rod is effectively confined in a tube-like region as shown in Fig. 9.16(a). The diameter a of the tube can be estimated as follows. Consider a cylindrical region of radius r and length L. If the surrounding polymers are placed randomly in space with number density n_p, the average number of polymers which intersect the cylinder is estimated by $n_p L S_c(r)$, where $S_c(r)$ is the surface area of the cylinder ($S_c(r) \simeq 2\pi r L$). When $r \simeq a$, this number should be of the order of 1. Therefore $n_p L S_c(a) \approx 1$, i.e., $n_p a L^2 \approx 1$. This gives

$$a \simeq \frac{1}{n_p L^2} \qquad (9.127)$$

If the polymer stays inside the tube, its direction is essentially confined by the tube. If the polymer moves out of the tube as shown in Fig. 9.16(b), it can change its direction by an angle $\delta\theta \simeq a/L$. The overall rotation of the polymer is attained by the repetition of this process. Therefore the rotational diffusion constant in the entangled state is estimated by

$$D_r \simeq \frac{(\delta\theta)^2}{\tau_d} \simeq \frac{a^2}{L^2\tau_d} \qquad (9.128)$$

Here τ_d represents the time needed for the polymer to move out of the tube. τ_d can be estimated by

$$\tau_d \simeq \frac{L^2}{D_t} \qquad (9.129)$$

where D_t is the translational diffusion constant of the polymer along the tube axis. Since the translational diffusion of the polymer along the tube axis is not hindered, D_t is considered to be equal to the translational diffusion constant in dilute solution. Therefore

$$D_t \simeq \frac{k_BT}{\eta_s L} \qquad (9.130)$$

Therefore τ_d is estimated as

$$\tau_d \simeq \frac{\eta_s L^3}{k_BT} \simeq \frac{1}{D_r^0} \qquad (9.131)$$

where D_r^0 is the rotational diffusion constant in the dilute solution. From eqs. (9.127), (9.128), and (9.131), it follows that

$$D_r \simeq \frac{a^2}{L^2}D_r^0 \simeq \frac{D_r^0}{(n_pL^3)^2} \qquad (9.132)$$

Therefore, in the concentration region where $n_pL^3 > 1$, the rotational diffusion of the polymer is strongly slowed down.

Although the entanglement affects the kinetic coefficients such as D_r strongly, it does not affect the equilibrium properties. Therefore the time evolution of the distribution function $\psi(\boldsymbol{u},t)$ is given by the same diffusion equation (9.121) with the effective diffusion constant D_r. Also the expression for the stress tensor is given by eq. (9.117).

Even though the expression for the stress tensor is not changed, the rheological behaviour is changed dramatically. To see this, let us consider the steady state viscosity. As discussed in Section 9.5.2, each term in eq. (9.125) is estimated as

first term: $\quad n_p k_B T \langle u_x u_y \rangle \simeq n_p k_B T \dfrac{\dot{\gamma}}{D_r} \simeq (n_p L^3)^3 \eta_s \dot{\gamma} \quad$ (9.133)

second term: $\quad n_p \dot{\gamma}\zeta_r^0 \langle u_x^2 u_y^2 \rangle \simeq n_p \dot{\gamma}\zeta_r^0 \simeq n_p L^3 \eta_s \dot{\gamma} \qquad$ (9.134)

In the concentration region of $n_pL^3 > 1$, the first term is much larger than the other terms. If we ignore the other terms, the expression for the stress tensor is given by

$$\sigma_{\alpha\beta} = 3n_p k_B T Q_{\alpha\beta} - p\delta_{\alpha\beta} \qquad (9.135)$$

where $Q_{\alpha\beta}$ is the orientational order parameter tensor introduced in Chapter 5:

$$Q_{\alpha\beta} = \left\langle u_\alpha u_\beta - \frac{1}{3}\delta_{\alpha\beta} \right\rangle \qquad (9.136)$$

Therefore the stress tensor in the concentrated solution is directly related to the orientational order parameter. The order parameter $Q_{\alpha\beta}$ can be calculated by solving the diffusion equation for $\psi(\boldsymbol{u}, t)$. The relaxation time of $Q_{\alpha\beta}$ is of the order of $1/D_r$. Therefore the concentrated solution of rod-like polymer shows pronounced viscoelasticity. The instantaneous elastic modulus G_0, and the relaxation time τ are given by

$$G_0 \simeq n_p k_B T \simeq \frac{cR_G T}{M} \tag{9.137}$$

$$\tau \simeq \frac{1}{D_r} \propto c^2 M^7 \tag{9.138}$$

9.5.4   Viscoelasticity of nematic solutions

If the concentration of the rod-like polymer is increased further, the solution turns into the nematic phase as discussed in Chapter 5. The equilibrium property of the nematic phase can be discussed by the mean field theory described in Section 5.2. The mean field theory gives the following mean field potential

$$U_{mf}(\boldsymbol{u}) = -U u_\alpha u_\beta Q_{\alpha\beta} \tag{9.139}$$

The mean field potential exerts a torque $-\mathcal{R}U_{mf}$ on individual molecules. Therefore the time evolution equation for ψ is now given by

$$\frac{\partial \psi}{\partial t} = D_r \mathcal{R} \cdot [\mathcal{R}\psi + \beta \psi \mathcal{R}U_{mf}(\boldsymbol{u})] - \mathcal{R}(\boldsymbol{u} \times \boldsymbol{\kappa} \cdot \boldsymbol{u}\psi) \tag{9.140}$$

Similarly, the expression for the stress tensor is given by

$$\sigma_{\alpha\beta} = 3n_p k_B T Q_{\alpha\beta} + \frac{1}{2}n_p \langle (\mathcal{R}_\alpha U_{mf})u_\beta \rangle - p\delta_{\alpha\beta} \tag{9.141}$$

Equations (9.140) and (9.141) give the constitutive equation for the nematic phase of rod-like polymers. The rheological properties of the liquid crystalline phase can be discussed using these equations (see reference 2 in Further reading.)

9.6   Summary of this chapter

Soft matter exhibits complex mechanical behaviour called nonlinear viscoelasticity. The mechanical behaviour is characterized by constitutive equations which express the stress for a given history of strain. If the stress level is small, the constitutive equation is completely represented by the linear viscoelastic function, e.g., by the relaxation modulus $G(t)$, or the complex modulus $G^*(\omega)$.

In polymeric fluids (polymer solutions and melts), constitutive equations have been derived by molecular theories. These theories assume that the stress of polymeric materials is directly connected to the deformation of the polymer conformation. When polymeric fluids are deformed, restoring stress is created by the deformed polymers. The stress relaxes as the polymer relaxes to the equilibrium conformation. The relaxation mode depends on whether the polymers are entangled or not. In the non-entangled state, the relaxation is described by the dumbbell model or the Rouse model which assume that the polymer chains are undergoing Brownian motion in a viscous medium. In the entangled state, the relaxation is described by the reptation model in which the polymer chain is undergoing Brownian motion confined by slip-links or tubes.

Solutions of rod-like polymers can be discussed by the same theoretical framework as that for flexible polymers. The stress in this system is directly connected to the orientational order parameter of the polymer, and is therefore obtained by solving the rotational diffusion equation.

Further reading

(1) *Dynamics of Polymeric Liquids*, Volumes 1 and 2, R. Byron Bird, Robert C. Armstrong, Ole Hassager, Wiley (1987).

(2) *The Theory of Polymer Dynamics*, Masao Doi and Sam Edwards, Oxford University Press (1986).

(3) *The Structure and Rheology of Complex Fluids*, Ronald G. Larson, Oxford University Press (1998).

Exercises

(9.1) The relation between the stress $\sigma(t)$ and strain $\gamma(t)$ of a material is often represented by the mechanical models shown in Fig. 9.17. The models consist of elastic elements represented by springs in which $\sigma(t)$ and $\gamma(t)$ are related by $\sigma = G\gamma$ (Fig. 9.17(a)), and viscous elements represented by dashpots in which $\sigma = \eta\dot{\gamma}$ (Fig. 9.17(b)). Answer the following questions.

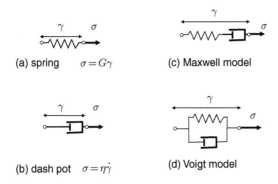

(a) spring $\sigma = G\gamma$

(b) dash pot $\sigma = \eta\dot{\gamma}$

(c) Maxwell model

(d) Voigt model

Fig. 9.17 Mechanical models for viscoelastic materials.

(a) Consider the model shown in Fig. 9.17(c) in which the spring and the dashpot are connected in series. This model is called the Maxwell model. Show that $\sigma(t)$ and $\gamma(t)$ are related by the following differential equation

$$\frac{\dot{\sigma}}{G} + \frac{\sigma}{\eta} = \dot{\gamma} \qquad (9.142)$$

(b) Calculate the relaxation modulus $G(t)$ for the above model.

(c) Calculate $\gamma(t)$ when $\sigma(t)$ is changed as follows

$$\sigma(t) = \begin{cases} 0 & t < 0 \\ \sigma_0 & 0 < t < t_0 \\ 0 & t > t_0 \end{cases} \qquad (9.143)$$

(d) Consider the model shown in Fig. 9.17(d) in which the spring and the dashpot are connected in parallel. This model is called the Voigt model. Calculate the relaxation modulus $G(t)$ of the Voigt model. Also calculate the strain $\gamma(t)$ when the stress is changed as in eq. (9.143).

(e) Construct the mechanical model which gives the relaxation modulus of eq. (9.17).

(9.2) Consider the uniaxial elongation shown in Fig. 3.3. The material point at (x_0, y_0, z_0) is displaced to

$$x = \frac{1}{\sqrt{\lambda}}x_0, \qquad y = \frac{1}{\sqrt{\lambda}}y_0. \qquad z = \lambda z_0 \quad (9.144)$$

The Henkey strain $\epsilon(t)$ is defined by

$$\epsilon(t) = \ln \lambda(t) \qquad (9.145)$$

Answer the following questions.

(a) Show that the velocity at point (x, y, z) is written as

$$v_x = -\frac{1}{2}\dot{\epsilon}x, v_y = -\frac{1}{2}\dot{\epsilon}y, v_z = \dot{\epsilon}z \qquad (9.146)$$

(b) Show that for a Newtonian fluid of viscosity η, the elongational stress $\sigma(t) = \sigma_{zz}(t) - \sigma_{xx}(t)$ is given by

$$\sigma(t) = 3\eta\dot{\epsilon}(t) \qquad (9.147)$$

(c) Suppose that a weight W is hung on the bottom of a cylindrical sample of cross-section A. Show that the elongation of the cylinder is described by

$$\dot{\lambda} = \frac{W}{3A\eta}\lambda^2 \qquad (9.148)$$

Show that λ goes to infinity at a certain time t_∞.

(d) For a viscoelastic fluid, eq. (9.147) is written as

$$\sigma(t) = 3\int_{-\infty}^{t} dt' \, G(t - t')\dot{\epsilon}(t') \qquad (9.149)$$

Obtain the time evolution equation for λ in the same situation discussed above.

(9.3) Consider a non-Newtonian fluid flowing in a channel made of two parallel plates driven by the pressure Δp applied at the end of the channel (see Fig. 9.18). Let $p_x = \Delta p/L$ be the pressure gradient. We take the coordinate as in Fig. 9.18. By symmetry, the velocity of the fluid has an x-component only and is written as $v_x(y)$. Answer the following questions.

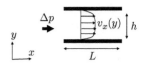

Fig. 9.18

(a) Show that the shear stress $\sigma_{xy}(y)$ at position y is given by

$$\sigma_{xy} = \left(\frac{h}{2} - y\right)p_x \qquad (9.150)$$

(Hint: Use the force balance equation (A.12).)

(b) The shear stress σ_{xy} is related to the velocity gradient $\dot{\gamma} = dv_x/dy$ by $\sigma_{xy} = \dot{\gamma}\eta(\dot{\gamma})$. Suppose that the steady state viscosity of the fluid is given by

$$\eta(\dot{\gamma}) = \eta_0\left(\frac{\dot{\gamma}}{\dot{\gamma}_0}\right)^{-n} \qquad (9.151)$$

where η_0, $\dot{\gamma}_0$, and n are constant. Obtain the velocity profile of the fluid in the channel.

(9.4) Consider the steady shear flow in which the velocity field is given by

$$v_x = \dot{\gamma}y, \qquad v_y = v_z = 0 \qquad (9.152)$$

Solve eq. (9.70), and obtain all components of the stress tensor.

(9.5) Solve the constitutive equation (9.70) when the elongational flow of constant elongational rate $\dot{\epsilon}$ is started at $t = 0$. The velocity field is given by

$$v_x = -\frac{1}{2}\dot{\epsilon}x, \qquad v_y = -\frac{1}{2}\dot{\epsilon}y, \qquad v_z = \dot{\epsilon}z \quad (9.153)$$

(9.6) Calculate the complex modulus $G^*(\omega)$ for the Rouse model, and show that for $\omega\tau_R \gg 1$, $G'(\omega)$ and $G''(\omega)$ become equal to each other and are proportional to $\omega^{1/2}$.

(9.7) Consider a solution of rod-like particles of length $L = 0.5\,\mu m$, diameter $d = 5\,nm$ in water (viscosity $1\,mPa \cdot s$). Estimate the viscosity of the solution for the volume fractions $\phi = 0.01\%, 0.1\%, 1\%$, and 10%.

Ionic soft matter

10

Many soft matter systems have ionic groups which dissociate in aqueous environments. Examples are shown in Fig. 10.1. (a) Colloidal particles in water usually have charges on the surface. (b) Water soluble polymers (especially biopolymers, such as proteins and DNA, etc.) also have ionic groups which dissociate in water and give charges to the polymer. Such a polymer is called polyelectrolyte. (c) Surfactants which have ionic head groups form micelles that have charges on the surface.

As shown in these examples, ionic soft matter consists of macro-electrolytes which produce macro-ions (big ions of the form of particles, polymers, or micelles) which have many ions attached to them, and many small ions which can move around in the solvent. Ions attached to the macro-ions are called fixed ions (meaning ions fixed to backbones), and the other ions which can move around are called mobile ions. In usual electrolytes, all ions are equally mobile, while in macro-electrolytes, a large fraction of ions are bound to the backbone structure and are less mobile. This gives special features to ionic soft matter.

The properties of ionic soft matter depend strongly on the ionic condition in the solvent, i.e., the pH (the concentration of H^+ ions) and salinity (the concentration of salts). For example, colloids of charged particles aggregate when salt is added to the system. Swollen gels made of polyelectrolytes shrink when the pH or ion concentration of the solvent are changed. The ionic state can also be controlled by an electric field. In ionic soft matter, the diffusion and flow of ions and solvents are coupled with electric fields. These properties give us more opportunities in controlling the state of the soft matter and make them more attractive.

In this chapter, we shall discuss these distinctive features of ionic soft matter.

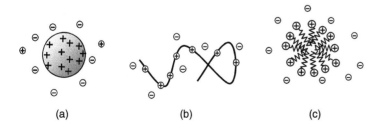

(a)   (b)   (c)

Fig. 10.1 Examples of ionic soft matter. (a) Charged colloid; (b) charged polymer (polyelectrolyte); (c) charged micelle.

10.1   Dissociation equilibrium

10.1.1   Dissociation equilibrium in simple electrolytes

Materials which dissociate into ions when dissolved in solvent are called electrolytes. The solvent is usually water, but other solvents are known. In the following, we shall assume that the solvent is water.

Consider an electrolyte AB which dissociates into A^+ and B^- in water. The dissociation equilibrium is determined by the condition

$$\mu_{AB} = \mu_{A^+} + \mu_{B^-} \tag{10.1}$$

where μ_{AB}, μ_{A^+}, μ_{B^-} are the chemical potentials of AB (undissociated electrolyte), A^+, and B^-, respectively.

We assume that the concentration of electrolyte in the solution is small,[1] and use dilute solution theory for the chemical potentials. In this case, the chemical potential μ_i of solute i can be written as a function of the number density n_i in the following way

$$\mu_i = \mu_i^0(T, P) + k_B T \ln n_i \tag{10.2}$$

where $\mu_i^0(T, P)$ is a constant which does not depend on n_i, but does depend on temperature T and pressure P. Equations (10.1) and (10.2) give

$$\frac{n_{A^+} n_{B^-}}{n_{AB}} = \tilde{K}_{AB} \tag{10.3}$$

where

$$\tilde{K}_{AB} = e^{\beta(\mu_{AB}^0 - \mu_{A^+}^0 - \mu_{B^-}^0)} \tag{10.4}$$

is called the dissociation constant. Notice that $\tilde{K}_{AB}$ has units of number density $[\mathrm{m}^{-3}]$. In the study of electrolytes, it is customary to denote the number density in units of $[\mathrm{mol\ dm}^{-3}]$. We shall denote the number density of species i written in units of $[\mathrm{mol\ dm}^{-3}]$ as $[i]$. Equation (10.3) is therefore written as

$$\frac{[A^+][B^-]}{[AB]} = K_{AB} \quad [\mathrm{mol\ dm}^{-3}] \tag{10.5}$$

The dissociation constant K_{AB} is often expressed in terms of the 'pK' value defined by

$$\mathrm{pK}_{AB} = -\log_{10} K_{AB} \tag{10.6}$$

Among various ions dissolved in water, the ions H^+ and OH^- are special since they are produced by the dissociation of the major component, water. The dissociation equilibrium for water is written as

$$[H^+][OH^-] = K_w \quad [(\mathrm{mol\ dm}^{-3})^2] \tag{10.7}$$

K_w is called the ionic product of water and is equal to $10^{-14}\ [(\mathrm{mol\ dm}^{-3})^2]$.

[1] The number of water molecules in $1\,\mathrm{dm}^3$ of pure water is $10^3/18 = 56\,\mathrm{mol}$. The molar fraction of H^+ ions in a strong acid of pH $= 2$ is $10^{-2}/56 = 0.018\%$. The molar fraction of Na^+ in sea water is ca. 1%.

In pure water, $[H^+]$ is equal to $[OH^-]$, and is therefore equal to 10^{-7} [mol dm^{-3}]. The pH is defined by

$$pH = -\log_{10}[H^+] \tag{10.8}$$

Electrolytes are classified into three types, acid (which produces H^+ by dissociation), base (which produces OH^-), and salt (which produces neither H^+ nor OH^-).

Consider the dissociation of acid HA:

$$HA \leftrightarrow H^+ + A^- \tag{10.9}$$

Let α be the dissociation fraction. Then

$$[A^-] = \alpha[HA]_0; \qquad [HA] = (1 - \alpha)[HA]_0 \tag{10.10}$$

where $[HA]_0$ is the concentration of added acid. Equation.(10.5) is then written as

$$\frac{[H^+][A^-]}{[HA]} = \frac{\alpha}{1-\alpha}[H^+] = K_{HA} \tag{10.11}$$

This gives the relation

$$\log_{10}\frac{1-\alpha}{\alpha} = pK_{HA} - pH \tag{10.12}$$

The dissociation of acid is suppressed as pH decreases, i.e., as the solution becomes more acidic.

10.1.2 Dissociation equilibrium in macro-electrolytes

Variation of net charge of macro-ions

Macro-electrolytes have many dissociative groups. In water, macro-electrolytes produce two kinds of ions. One is the fixed ions which are attached to the backbone structure and cannot move freely. The other is the mobile ions which can move around in the solvent. Mobile ions which have the opposite sign of charge to that of fixed ions are called counter-ions, and those having the same sign of charge are called co-ions.

Quite often, macro-electrolytes have both acidic and basic groups. Upon dissociation, acidic groups leave negative fixed ions and basic groups leave positive fixed ions. Since the basic groups dissociate at low pH, and the acidic groups dissociate at high pH, the net charge of the macro-ion changes from positive to negative as the pH of the solution increases (see Fig. 10.2). The point at which the net charge of the macro-ion becomes zero is called the point of zero charge (pzc). This can be measured by observing the motion of the macro-ions in an electric field. Across the pzc, the direction of the motion of the macro-ion is reversed.

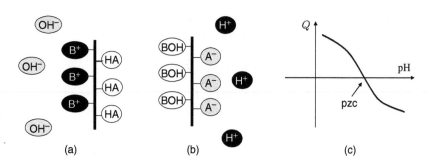

Fig. 10.2 Dissociation of polyelectrolyte which has both acidic and basic groups. (a) At low pH, basic groups dissociate. (b) At high pH, the acidic groups dissociate. (c) The net charge Q of a macro-ion is plotted against pH.

(a)   (b)   (c)

Interaction between the dissociated groups

In solutions of simple electrolytes, the dissociation behaviour can be well predicted by eq. (10.5) which is derived from dilute solution theory. However, in solutions of macro-electrolytes, this theory does not work even in the dilute limit since the local concentration of fixed ions in macro-electrolytes is not low. For example, in polyelectrolytes, the distance between the neighbouring ions is limited by the backbone chains as shown in Fig. 10.3(a). In this situation, the interaction between the fixed ions becomes quite important.

Consider a pair of neighbouring dissociative groups bound to a polymer chain. If both groups dissociate, the fixed ions will have a Coulombic interaction energy $e_0^2/(4\pi\epsilon\ell)$, where e_0 is the elementary charge, ϵ is the dielectric constant of water ($\epsilon = 80\epsilon_0 = 80 \times 8.854 \times 10^{-12}\,\mathrm{F/m}$), and ℓ is the distance between the neighbouring fixed ions. This energy has to be accounted for in the free energy of the dissociated state. If this energy is accounted for, the dissociation constant $\hat{K}_{AB}$ will be modified by a factor

$$e^{-e_0^2/(4\pi\epsilon\ell k_B T)} \tag{10.13}$$

Therefore simultaneous dissociation of closely separated ionic groups is strongly suppressed: if one dissociates, the chance that the other dissociates becomes very small. In other words, the dissociation of ionic groups in macro-ions is strongly correlated. Such a correlation exists for ion pairs separated by less than the length

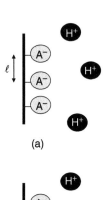

(a)

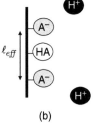

(b)

Fig. 10.3 Correlation effect in the dissociation of neighbouring ionic groups. Dissociation of all three groups at the same time as in (a) rarely happens since such a state has a large Coulomb energy. (b) The average distance ℓ_{eff} between the dissociated groups cannot be less than the Bjerrum length ℓ_B.

$$\ell_B = \frac{e_0^2}{4\pi\epsilon k_B T} \tag{10.14}$$

The length ℓ_B is called the Bjeruum length. In water, the Bjeruum length is 0.74 nm.

Due to this correlation effect, the line charge density of polyelectrolytes (the net charge per length along the polymer backbone) cannot be indefinitely large: the effective line charge density has an upper bound of about e_0/ℓ_B.

Counter-ion condensation

The existence of the upper bound of the line charge density can be explained in another way. If the charge density of the polymer is high, the counter-ions are strongly attracted by the polymer chain and form a sheath of counter-ions around the chain. Accordingly the net charge density cannot exceed the upper limit e_0/ℓ_B. In this explanation, the effect is called the counter-ion condensation.

Both explanations, the dissociation correlation, or the counter-ion condensation, indicate that the degree of dissociation α of polyelectrolyte is much less than that of simple electrolytes.

10.2 Ionic gels

10.2.1 Free ion model for ionic gels

If charged polymers or charged colloids form a network in solvent, they become ionic gels. The simplest model for ionic gels is the free ion model, which is the same as the free electron model for metals. In this model, the fixed ions form a continuous background charge, and the mobile ions can move freely. Let n_i be the number density of mobile ions of species i having charge e_i, and let n_b and e_b be the number density and the charge of the fixed ions.[2] We assume that the concentrations of ions are low, so that their chemical potentials are given by

$$\mu_i = \mu_i^0 + k_B T \ln n_i + e_i \psi \qquad (10.15)$$

where ψ is the electric potential.

[2] The subscript 'b' means 'bound' or 'backbone'.

Notice that the ion density n_i, and the potential ψ in eq. (10.15) are quantities averaged over a sufficiently large volume. Locally, their value fluctuates as discussed in Section 8.1.2, but the fluctuations can be ignored if the average is taken over a length-scale larger than the correlation length. In a later section, we shall show that this correlation length is the Debye length κ^{-1} given by eq. (10.43). Therefore, the following discussion is valid only for macroscopic quantities which are defined for length-scales larger than the Debye length κ^{-1}.

10.2.2 Charge neutrality condition

In electrical conductors which involve mobile charge carriers (mobile ions, or mobile electrons), there should be no net charge in the bulk at equilibrium, i.e., the charge density must be zero everywhere in the bulk:

$$\sum_i e_i n_i + e_b n_b = 0 \qquad (10.16)$$

This condition is called the charge neutrality condition.

The charge neutrality condition holds for the macroscopic charge density, but it does not hold for the local charge density

$$\hat{\rho}_e(\boldsymbol{r}) = \sum_i e_i \hat{n}_i(\boldsymbol{r}) + e_b \hat{n}_b(\boldsymbol{r}) \qquad (10.17)$$

where $\hat{n}_i(\boldsymbol{r})$ and $\hat{n}_b(\boldsymbol{r})$ are defined by eq. (8.3). The local charge density $\hat{\rho}_e$ can be plus or minus, but if $\hat{\rho}_e(\boldsymbol{r})$ is averaged over a length-scale sufficiently larger than the Debye length κ^{-1}, it must become zero.

The charge neutrality condition is important in both equilibrium and non-equilibrium properties. Indeed, the macroscopic electric field ψ is determined by this condition alone. This will be demonstrated in the following.

10.2.3   Donnan equilibrium

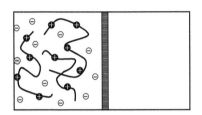

Fig. 10.4 Donnan equilibrium between a polymer solution and pure solvent.

Let us consider the situation shown in Fig. 10.4. A container is divided into two chambers by a semi-permeable membrane which prevents polymer passing through, but allows other components, including solvent and small ions, to pass through. Now suppose that the left chamber is filled with polyelectrolyte solution (with no salt added), and the right chamber is filled with pure water. In this situation, small ions dissociated from the polyelectrolyte in the left chamber do not diffuse into the right chamber even if they can go through the membrane freely. This is due to the charge neutrality condition: if small ions diffuse into the right chamber, the left chamber becomes positively charged (since the polymers have to stay in the left chamber), and the charge neutrality condition is violated. Therefore the small ions are forced to stay in the left chamber.

The imbalance in the solute concentration between the two chambers creates an osmotic pressure. The osmotic pressure acting on the membrane is given by the number density of free ions dissociated from the polyelectrolyte. Since this density is equal to the density of fixed ions, n_b,[3] the osmotic pressure of the polyelectrolyte solution is given by

[3] Here we are assuming that all ions in the left chamber are monovalent, and that the solution is salt free.

$$\Pi = n_b k_B T \qquad (10.18)$$

This is much larger than the osmotic pressure of usual non-electrolytic polymer solutions, where the osmotic pressure is proportional to the number density of polymer chains (n_{chain}). In polyelectrolyte solutions, the osmotic pressure is given by the number density of dissociated free ions n_b, which is much larger than n_{chain} since n_b is of the order of $N n_{chain}$.

The above argument is for salt free solutions. If salt is added to the solution, a portion of the small ions can move to the right chamber, but their concentrations in the two chambers still cannot be equal due to the charge neutrality condition.

The equilibrium distribution of small ions in the two chambers can be determined as follows. Let n_{ip} be the number density of ions of species i in the polyelectrolyte solution (left chamber) and let n_{is} be that in the simple electrolyte solution (right chamber). Let $\Delta\psi$ be the difference of the electric potential in the left chamber relative to that in the right chamber. The equilibrium conditions for the mobile ions in the two chambers are written as

$$\mu_i^0 + k_B T \ln n_{ip} + e_i \Delta\psi = \mu_i^0 + k_B T \ln n_{is} \qquad (10.19)$$

Equation (10.19) gives

$$n_{ip} = n_{is} e^{-\beta e_i \Delta\psi} \qquad (10.20)$$

Equation (10.20) and the charge neutrality condition (10.16) determine n_{ip} and $\Delta\psi$ for given values of n_b and n_{is}.

Consider the situation that all mobile ions are monovalent (having a charge of e_0 or $-e_0$). The charge neutrality condition in the right chamber gives $n_{+s} = n_{-s}$, which will be written as n_s. Equation (10.20) then gives

$$n_{+p} = n_s e^{-\beta e_0 \Delta\psi}, \qquad n_{-p} = n_s e^{\beta e_0 \Delta\psi} \qquad (10.21)$$

If the fixed ions in the polymer are positive, the charge neutrality condition in the left chamber is written as

$$n_b + n_{+p} = n_{-p} \qquad (10.22)$$

The solutions of the above set of equations are

$$n_{+p} = \frac{1}{2}\left(-n_b + \sqrt{n_b^2 + 4n_s^2}\right), \qquad n_{-p} = \frac{1}{2}\left(n_b + \sqrt{n_b^2 + 4n_s^2}\right)$$
$$(10.23)$$

Therefore the electric potential difference $\Delta\psi$ and the osmotic pressure difference $\Delta\Pi$ are given by

$$\begin{aligned}
\Delta\psi &= \frac{k_B T}{e_0} \ln\left(\frac{n_{-p}}{n_s}\right) \\
&= \frac{k_B T}{e_0} \ln\left[\sqrt{1 + \frac{1}{4}\left(\frac{n_b}{n_s}\right)^2} + \frac{n_b}{2n_s}\right] \qquad (10.24)
\end{aligned}$$

$$\Delta\Pi = k_B T[n_{+p} + n_{-p} - 2n_s] = k_B T\left(\sqrt{4n_s^2 + n_b^2} - 2n_s\right) \quad (10.25)$$

This equilibrium solution for the ion distribution is called the Donnan equilibrium, and the potential $\Delta\psi$ is called the Donnan potential.[4]

In the limit of $n_b \gg n_s$, the situation essentially becomes equal to the salt free case. The osmotic pressure is given by eq. (10.18), and increases with the increase of n_b. On the other hand, the electric potential $\Delta\psi$ depends on n_b only weakly

$$\Delta\psi = \frac{k_B T}{e_0} \ln\left(\frac{n_b}{n_s}\right) \qquad (10.26)$$

Since $\ln(n_b/n_s)$ increases little with the increase of n_b, the electric potential inside the polymer solution is about $k_B T/e_0 = 25\,\mathrm{mV}$ even if n_b/n_s

[4] Here we have assumed that μ_i^0 are equal to each other in the two chambers. In reality, μ_i^0 may depend on the polymer concentration since the dielectric constant of the solution changes with polymer concentration. If μ_i^0 are different in the two solutions, there may be non-zero potential difference $\Delta\psi$ even if the polymers have no fixed ions.

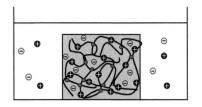

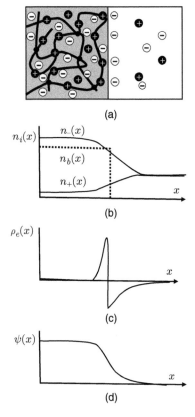

Fig. 10.5 Swelling equilibrium of a polyelectrolyte gel in an electrolyte solution.

Fig. 10.6 (a) Interface between a polyelectrolyte gel and an electrolyte solution. (b) Concentration profiles of fixed ions $n_b(x)$, positive free ions $n_+(x)$, and negative free ions $n_-(x)$. (c) Charge density $\rho_e(x) = e_0(n_b + n_+ - n_-)$. (d) Electric potential $\psi(x)$.

is very large. This is again a manifestation of the correlation effect of the bound charges.

As the salt concentration n_s in the right chamber increases, both the Donnan potential and the osmotic pressure decreases. In the limit of $n_s \gg n_b$, eqs. (10.24) and (10.25) are written as

$$\Delta\psi = \frac{k_B T}{e_0} \frac{n_b}{2n_s} \tag{10.27}$$

$$\Delta\Pi = \frac{n_b^2}{4n_s} k_B T \tag{10.28}$$

10.2.4   Swelling of a polyelectrolyte gel

The situation discussed above is more naturally realized in polyelectrolyte gels. Consider a polyelectrolyte gel immersed in a solution (see Fig. 10.5). In Section 3.4.2, we have shown that the swelling of a gel is determined by the balance of the osmotic pressure and the elastic restoring force of the polymer network (see eq. (3.75)). If the gel is made of polyelectrolytes, the free ions give a large contribution to the osmotic pressure. Therefore, the swelling equilibrium of a polyelectrolyte gel is determined by the following equation

$$\Pi_{sol}(\phi) + \Pi_{ion}(n_b, n_s) = G_0 \left(\frac{\phi}{\phi_0}\right)^{1/3} \tag{10.29}$$

where $\Pi_{ion}(n_b, n_s)$ is the osmotic pressure difference between the inside and outside of the gel due to the mobile ions. (Notice that n_b in eq. (10.29) changes with ϕ as $n_b = n_{b0}\phi/\phi_0$.)

Due to the large osmotic pressure, polyelectrolyte gels swell much more strongly than neutral gels. In fact, water absorbing gels used for diapers are made of polyelectrolyte. The power of water absorbance, however, decreases as the outer solution becomes salty.

10.3   Ion distribution near interfaces

10.3.1   Electric double layer

In the theory of Donnan equilibrium, the Donnan potential $\Delta\psi$ is introduced to make the chemical potential of mobile ions equal in the gel phase and in the outer phase. The origin of the Donnan potential can be seen if we study the ion distribution near the interface separating the two regions. Figure 10.6(a) shows a schematic picture of the interface between the polyelectrolyte gel and the electrolyte solution. Figure 10.6(b) shows the spatial distribution of fixed ions ($n_b(x)$), counter-ions ($n_-(x)$), and co-ions ($n_+(x)$). In the macroscopic theory, it is assumed that $n_+(x)$ and $n_-(x)$ change discontinuously at the interface. In reality, they change continuously as in Fig. 10.6(b).

Figure 10.6(c) shows the spatial distribution of the charge density

$$\rho_e(x) = e_0[n_b(x) + n_+(x) - n_-(x)] \tag{10.30}$$

Far from the interface, $\rho_e(x)$ is zero, but near the interface, $\rho_e(x)$ changes as shown in Fig. 10.6(c). The density of fixed ions $n_b(x)$ changes sharply at the interface, while the densities of mobile ions, $n_+(x)$ and $n_-(x)$, change more gradually. Therefore $\rho_e(x)$ has a positive peak at the gel side, and a negative peak at the solution side. Due to the charge neutrality condition, the total charge in the interfacial area is zero, i.e.,

$$\int_{-\infty}^{\infty} dx \rho_e(x) = 0 \tag{10.31}$$

but the dipole moment

$$P_e = \int_{-\infty}^{\infty} dx x \rho_e(x) \tag{10.32}$$

is non-zero. Such a charge distribution is called an electric double layer. The electric double layer has zero charge, but has non-zero dipole moment.

Across the electric double layer, the electric potential ψ changes sharply. The difference in the electric potential $\Delta\psi = \psi(-\infty) - \psi(\infty)$ is related to the dipole moment P_e by

$$\Delta\psi = -\frac{P_e}{\epsilon} \tag{10.33}$$

This is shown as follows.

In general, the electric potential $\psi(x)$ is related to the charge density $\rho_e(x)$ by the Poisson equation:

$$\epsilon \frac{d^2\psi}{dx^2} = -\rho_e(x) \tag{10.34}$$

The solution of eq. (10.34) is given by[5]

$$\psi(x) = -\frac{1}{2\epsilon} \int_{-\infty}^{\infty} dx' |x - x'| \rho_e(x') \tag{10.35}$$

The Donnan potential $\Delta\psi$ is given by the potential difference between the two points x_1 and x_2 taken far from the interface ($x_1 < 0$, $x_2 > 0$). By eq. (10.35), this is written as

$$\Delta\psi = -\frac{1}{2\epsilon} \int_{-\infty}^{\infty} dx' (x' - x_1)\rho_e(x') + \frac{1}{2\epsilon} \int_{-\infty}^{\infty} dx' (x_2 - x')\rho_e(x') \tag{10.36}$$

By the charge neutrality condition (10.31), eq. (10.36) is written as

$$\Delta\psi = -\frac{1}{\epsilon} \int_{-\infty}^{\infty} dx' x' \rho_e(x') = -\frac{P_e}{\epsilon} \tag{10.37}$$

[5] In one dimension, the principal solution of the Laplace equation $d^2\psi/dx^2 = -\delta(x)$ is $\psi(x) = -|x|/2$ since $d|x|/dx = \Theta(x) - \Theta(-x)$ and $d^2|x|/dx^2 = 2\delta(x)$.

10.3.2 Poisson–Boltzmann equation

The actual form of the ion distribution $n_i(x)$ and the electric field $\psi(x)$ at the interface can be obtained by assuming that eq. (10.19) is valid everywhere, i.e.,

$$\mu_i^0 + k_B T \ln n_i(x) + e_i \psi(x) = \mu_i^0 + k_B T \ln n_{is} \tag{10.38}$$

where n_{is} is the ion density in the solution far from the interface. Equation (10.38) gives

$$n_i(x) = n_{is} e^{-\beta e_i \psi(x)} \tag{10.39}$$

Equations (10.34) and (10.39) then give

$$\epsilon \frac{d^2 \psi}{dx^2} = -e_b n_b(x) - \sum_i e_i n_{is} e^{-\beta e_i \psi} \tag{10.40}$$

This equation, obtained by combining the Boltzmann distribution for ions and the Poisson equation in electrostatics, is called the Poisson–Boltzmann equation. The Poisson–Boltzmann equation determines the potential profile created by fixed ions in a sea of free ions.

10.3.3 Debye length

The Poisson–Boltzmann equation is a nonlinear equation for ψ, and can be solved only for special cases. A commonly made approximation is the linearization approximation (or the Debye approximation) for which the exponential function $e^{-\beta e_i \psi(x)}$ is replaced by a linear function $1 - \beta e_i \psi(x)$. This approximation is valid if $\beta e_i \psi(x)$ is small. In reality, $\beta e_i \psi(x)$ is usually not so small, but the approximation is often used since the result is qualitatively correct in most cases.

If the linearization approximation is used, the charge density of free ions can be written as

$$\rho_e^{free}(x) = \sum_i e_i n_{is} e^{-\beta e_i \psi} = \sum_i e_i n_{is} - \sum_i \beta e_i^2 n_{is} \psi \tag{10.41}$$

The first term on the right-hand side is equal to zero due to the charge neutrality of the bulk solution. Equation (10.40) is then written as

$$\frac{d^2 \psi}{dx^2} - \kappa^2 \psi = -\frac{e_b n_b(x)}{\epsilon} \tag{10.42}$$

where

$$\kappa^2 = \frac{\beta \sum e_i^2 n_{is}}{\epsilon} \tag{10.43}$$

If there is a point charge at the origin, the linearized Poisson–Boltzmann equation becomes

$$\frac{d^2 \psi}{dx^2} - \kappa^2 \psi = -\frac{e_0}{\epsilon} \delta(x) \tag{10.44}$$

The solution of this equation is given by

$$\psi(x) = -\frac{e_0}{2\kappa\epsilon} e^{-\kappa|x|} \tag{10.45}$$

Therefore the effect of the point charge decays exponentially. The length κ^{-1} is called the Debye length. The Debye length represents the distance over which the effect of charge persists; beyond the Debye length, the effect of charge is not observable.

The Debye length is determined by the concentration of free ions. If an electrolyte AB which produces monovalent ions A^+ and B^- is added to water, the Debye length is given by

$$\kappa^{-1} = \sqrt{\frac{\epsilon k_B T}{2ne_0^2}} \tag{10.46}$$

where n represents the number density of A^+ (or B^-) ions in the solution. Equation (10.46) can be written by use of the mean distance $\ell = (2n)^{1/3}$ between the free ions, and the Bjerrum length ℓ_B as

$$\kappa^{-1} = \frac{\ell_B}{\sqrt{4\pi}} \left(\frac{\ell}{\ell_B}\right)^{3/2} \tag{10.47}$$

or by the molar concentration of A^+ ions

$$\kappa^{-1} = \frac{0.3}{\sqrt{[A^+]}} \quad [\text{nm}] \tag{10.48}$$

In pure water $[H^+]$ is 10^{-7} [mol dm^{-3}], and the Debye length is about $1\,\mu m$.

10.3.4 Charge neutrality condition and the Poisson–Boltzmann equation

In the Debye approximation, the electric potential is determined by eq. (10.42). The term $\kappa^2\psi$ represents the charge density created by the redistribution of free ions. Equation (10.42) can be solved by a Fourier transform

$$\psi_k = \int_{-\infty}^{\infty} dx \psi(x) e^{ikx} \tag{10.49}$$

as

$$\psi_k = \frac{\rho_{bk}}{\epsilon(k^2 + \kappa^2)} \tag{10.50}$$

where ρ_{bk} is the Fourier transform of $e_b n_b(x)$, the density of the bound charges. The Fourier transform of the total charge density is given by

$$\rho_{ek} = \rho_{bk} - \epsilon\kappa^2\psi_k = \rho_{bk}\frac{k^2}{k^2 + \kappa^2} \tag{10.51}$$

Hence ρ_{ek} becomes very small if $|k| \ll \kappa$. This indicates that the macroscopic charge density (the charge density averaged over a length larger κ^{-1}) is essentially zero in electrolyte solution. This is the reason for the charge neutrality condition.

If the charge neutrality condition is imposed, the electric potential can be determined without solving the Poisson equation. It can be shown that the solution of the Poisson–Boltzmann equation (10.40) always agrees with the macroscopic analysis which uses the charge neutrality condition (see problem (10.4)).

10.3.5   Ion distribution near charged surfaces

The Poisson–Boltzmann equation is useful in studying the ion distribution near a charged surface. Let us consider the situation shown in Fig. 10.7(a), where a charged wall is placed at $x = 0$ in an electrolyte solution. In this case, $n_b(x)$ in eq. (10.40) or (10.42) is zero for $x > 0$, and $\psi(x)$ satisfies

$$\frac{d^2\psi}{dx^2} + \kappa^2\psi = 0 \qquad (10.52)$$

The effect of the wall charge is given by the boundary condition. Let q be the surface charge density (charge per unit area). Then the boundary condition at $x = 0$ becomes

$$\epsilon\frac{d\psi}{dx} = -q \qquad (10.53)$$

The solution of eq. (10.52) for this boundary condition is

$$\psi(x) = \frac{q}{\epsilon\kappa}e^{-\kappa x} \qquad (10.54)$$

Therefore the surface potential is given by

$$\psi_s = \psi(0) = \frac{q}{\epsilon\kappa} \qquad (10.55)$$

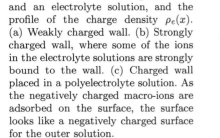

Fig. 10.7 Schematic pictures of the interfaces between a charged surface and an electrolyte solution, and the profile of the charge density $\rho_e(x)$. (a) Weakly charged wall. (b) Strongly charged wall, where some of the ions in the electrolyte solutions are strongly bound to the wall. (c) Charged wall placed in a polyelectrolyte solution. As the negatively charged macro-ions are adsorbed on the surface, the surface looks like a negatively charged surface for the outer solution.

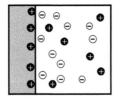
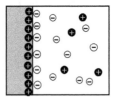
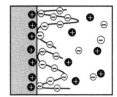
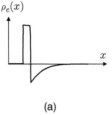
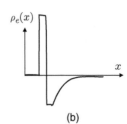
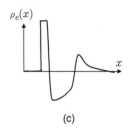

(a)   (b)   (c)

According to eq. (10.55), the electric potential at the surface relative to the bulk is proportional to the charge density q. This indicates that the electric double layer can be regarded as a capacitor with capacitance $\epsilon\kappa$ per unit area.

The electric double layer model gives us a simple picture for the charged surface in electrolyte solution. On the other hand, various subtle problems arise in applying this model to real surfaces. The problems come from the fact that both the surface charge density q and the surface potential ψ_s are not well-defined quantities. For example, to define the surface charge density, one needs to identify which ions belong to the surface and which belong to the solution. Also to define the surface potential, one needs to identify very precisely (i.e., on the atomic scale) where the surface is. To resolve these issues, one has to consider the details of the molecular (or even the electronic) structure at the surface.

Despite such problems, the electric double layer is a useful concept in interpreting the equilibrium and non-equilibrium properties of ionic soft matter as we shall show in the following.

10.3.6   Inter-surface potential between charged surfaces

If two charged surfaces having the same sign of charges are placed parallel to each other at a distance h (see Fig. 10.8), one expects that the surfaces will repel each other due to the Coulombic repulsion. Although this expectation is correct in many cases, it is not always correct. The forces acting between charged surfaces placed in an electrolyte solution is rather subtle. Indeed two surfaces with the same sign of charges can attract each other (see problem (10.5)). This is because there are ions which contribute to the inter-surface force.

To calculate the inter-surface force, let us consider a plane located at x parallel to the surface (see Fig. 10.8). Let $f(x)$ be the force (per unit area) that the material on the left side of the plane exerts on the material on the right side of the plane.

As discussed in Appendix A, the force acting in a continuum generally consists of two parts. One is the surface force which represents the short-range force, and the other is the bulk force which represents the long-range force (like gravity and the Coulombic force). In the present situation, the short-range force is given by the pressure $p(x)$, and the long-range force is given by the Coulombic force.

According to electrostatics, the Coulombic force acting between two charged planes is constant, independent of their separation, and is given by $\rho_1\rho_2/2\epsilon$ per unit area, where ρ_1 and ρ_2 are the charge densities of each plane. The Coulombic force acting between the charges in the left-hand side of the plane and those in the right-hand side of the plane is given by the summation of these forces. Therefore

$$f_{Coulomb}(x) = \frac{1}{2\epsilon}\int_{-\infty}^{x}dx_1\int_{x}^{\infty}dx_2\rho_e(x_1)\rho_e(x_2) \qquad (10.56)$$

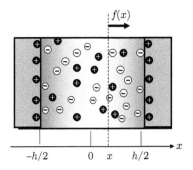

Fig. 10.8 Interaction between charged surfaces separated by distance h.

Using the Poisson equation (10.34), eq. (10.56) is written as

$$f_{Coulomb}(x) = \frac{\epsilon}{2} \int_{-\infty}^{x} dx_1 \int_{x}^{\infty} dx_2 \frac{d^2\psi}{dx_1^2} \frac{d^2\psi}{dx_2^2} = -\frac{\epsilon}{2} \left(\frac{d\psi}{dx}\right)^2 \qquad (10.57)$$

Hence the total force acting across the plane at x is given by

$$f(x) = p(x) - \frac{\epsilon}{2} \left(\frac{\partial\psi}{\partial x}\right)^2 \qquad (10.58)$$

To obtain the pressure, we use the fact that the chemical potential of the solvent is constant everywhere in the system. Now the chemical potential of the solvent is given by (see eqs. (2.27) or (2.81)),

$$\mu_s(x) = \mu_s^0 + v_s[p(x) - \Pi(x)] \qquad (10.59)$$

where v_s is the volume of a solvent molecule, and $\Pi(x)$ is the osmotic pressure at x. $\Pi(x)$ is expressed in terms of the ion concentration as

$$\Pi(x) = k_B T \sum_i n_i(x) \qquad (10.60)$$

The condition that $\mu_s(x)$ is equal to the chemical potential outside of the gap gives the following equation

$$p(x) - \Pi(x) = p_{out} - \Pi_{out} \qquad (10.61)$$

Hence

$$p(x) = p_{out} + \Pi(x) - \Pi_{out} \qquad (10.62)$$

Let n_{is} be the number density of ion i in the outer solution. Then $n_i(x)$ is given by $n_{is} e^{-\beta e_i \psi(x)}$. Hence the pressure is given by eqs. (10.60) and (10.62) as

$$p(x) = p_{out} + k_B T \sum_i n_{is} \left[e^{-\beta e_i \psi(x)} - 1\right] \qquad (10.63)$$

and the total force is

$$f(x) = k_B T \sum_i n_{is} \left[e^{-\beta e_i \psi(x)} - 1\right] - \frac{\epsilon}{2} \left(\frac{\partial\psi}{\partial x}\right)^2 \qquad (10.64)$$

where we have set p_{out} equal to zero. One can confirm that if ψ satisfies the Poisson–Boltzmann equation (10.40), $f(x)$ is independent of x.

Let us calculate the inter-surface force $f(x)$ for the situation shown in Fig. 10.8. We take the origin of the coordinate at the middle of the two charged surfaces. Since the potential is symmetric with respect to the plane at $x = 0$, i.e., $\psi(x) = \psi(-x)$, $d\psi/dx$ is zero at $x = 0$. If we evaluate the force $f(x)$ at $x = 0$, eq. (10.64) gives

$$f_{int} = k_B T \sum_i n_{is} \left[e^{-\beta e_i \psi(0)} - 1\right] \qquad (10.65)$$

where we have written $f(0)$ as f_{int}.

To proceed with the calculation, we use the linearization approximation. The solution of eq. (10.52) under the boundary conditions $d\psi/dx = \pm q/\epsilon$ at $x = \pm h/2$ is given by

$$\psi(x) = \frac{q \cosh \kappa x}{\epsilon \kappa \sinh(\kappa h/2)} \tag{10.66}$$

In the linearization approximation, $\psi(0)$ is small, and eq. (10.65) may be approximated as

$$f_{int} = k_B T \sum_i n_{is} \left[-\beta e_i \psi(0) + \frac{1}{2} [\beta e_i \psi(0)]^2 \right] = \frac{1}{2} \beta \psi(0)^2 \sum_i e_i^2 n_{is} \tag{10.67}$$

Using eqs. (10.43) and (10.67), we have

$$f_{int} = \frac{1}{2} \epsilon \kappa^2 \psi(0)^2 = \frac{q^2}{2\epsilon \sinh^2(\kappa h/2)} \tag{10.68}$$

The force can also be expressed in terms of the surface potential $\psi_s = \psi(h/2)$ as

$$f_{int} = \frac{\epsilon \kappa^2 \psi_s^2}{2 \cosh^2(\kappa h/2)} \tag{10.69}$$

For $\kappa h > 1$, eq. (10.68) and eq. (10.69) are approximated as

$$f_{int} = \frac{1}{2} \frac{q^2}{\epsilon} e^{-\kappa h} = \frac{1}{2} \epsilon \kappa^2 \psi_s^2 e^{-\kappa h} \tag{10.70}$$

Therefore the inter-surface potential decreases exponentially with the increase of the gap distance h.

As discussed above, the inter-surface force f_{int} is the sum of the osmotic pressure (the first term on the right-hand side of eq. (10.64)), and the electrostatic force (the second term in the same equation). At the middle plane where the electric field $E = -d\psi/dx$ is zero, the force f_{int} is entirely determined by the osmotic pressure.

The mechanism that gives a repulsive osmotic pressure between the charged surfaces is the same as that which causes the swelling of a polyelectrolyte gel. To satisfy the charge neutrality condition, the density of counter-ions in the gap region is larger than that in the outer solution. The confinement of the counter ions in the gap gives the repulsive force between the walls.

The fact that the inter-surface force between the charged surfaces is not a simple Coulombic force, and involves the osmotic contribution gives an unexpected result. In the case that the surface potentials ψ_s of the two surfaces are not equal to each other, the inter-surface force f_{int} can be attractive even though the charges of the two surfaces have the same sign (see problem (10.5)).

10.4 Electrokinetic phenomena

10.4.1 Electrokinetic phenomena in gels and colloids

We have discussed the equilibrium state of ionic soft matter. From now on, we shall discuss the non-equilibrium state, i.e., where ions and solvent are flowing under an electric or other fields.

When an electric field is applied to ionic soft matter, the free ions start to move. Their motion generally induces a motion of solvent, as particle motion induces solvent flow (see the discussion in Section 8.2). In the bulk of simple electrolyte solutions, such motion does not take place, since the sum of the electric forces acting on the free ions is zero (due to the charge neutrality condition). On the other hand, in ionic soft matter, the electric field usually induces the flow of solvent, due to the large asymmetry between the macro-ions and counter-ions. For example, in the case of ionic gels, the sum of the forces acting on the free ions is non-zero since there are charges bound to the gel network. The phenomena caused by the coupling between ionic motion and the solvent flow are generally called electrokinetic phenomena.

Several examples of electrokinetic phenomena are shown in Fig. 10.9.

(a) Electro-osmosis (Fig. 10.9(a)): When an electric potential difference $\Delta\psi$ is applied across an ionic gel, solvent starts to move through the gel. This phenomenon, called electro-osmosis, is caused by the free ions which move under the electric field and drag the surrounding solvent.

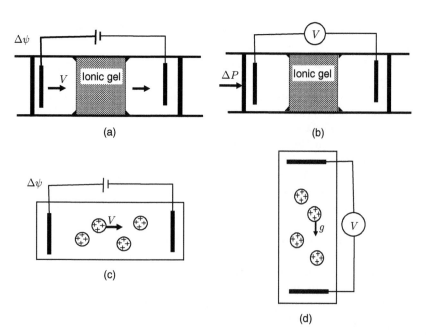

Fig. 10.9 Electrokinetic effect observed in ionic gels ((a) and (b)), and in charged colloidal suspensions ((c) and (d)). (a) Electro-osmosis; (b) streaming potential; (c) electrophoresis; and (d) sedimentation potential. See the text for details.

(b) Streaming potential (Fig. 10.9(b)): When the solvent is forced to move across an ionic gel driven by a pressure difference ΔP, there appears an electric potential difference $\Delta\psi$ between the two chambers. This potential is called the streaming potential. The streaming potential is created to maintain the charge neutrality condition in the two chambers. Without the streaming potential, free ions are convected from one chamber to the other by the solvent flow, thereby violating the charge neutrality condition.

(c) Electrophoresis (Fig. 10.9(c)): When an electric field is applied to a charged particle in an electrolyte solution, the particle starts to move. This phenomenon is called electrophoresis. Though this phenomenon looks simple, it again requires careful analysis since there are free ions which exert forces on the fluid and on the particles.

(d) Sedimentation potential (Fig. 10.9(d)): When colloidal particles sediment under gravity (or centrifugal forces), an electric potential $\Delta\psi$ is created along the direction of gravity. This potential is called the sedimentation potential.

In the following we shall discuss these phenomena using the Onsager principle.

10.4.2   Electromechanical coupling in ionic gels

First we consider phenomena (a) and (b). We consider the situation that an ionic gel (with negative fixed charges) is fixed in space and subject to a pressure gradient $p_x = \partial p/\partial x$ and an electric field $E_x = -\partial\psi/\partial x$. Under these fields, solvent and ions move. To simplify the analysis, we ignore the deformation of the gel network (assuming that the elastic modulus of the gel network is large).

Let $\boldsymbol{v}$ be the velocity of the solvent, and $\boldsymbol{v}_i$ be the velocity of ion i. The energy dissipation function is written as

$$\Phi = \frac{1}{2}\int d\boldsymbol{r}\left[\xi\boldsymbol{v}^2 + \sum_i n_i\zeta_i(\boldsymbol{v}_i - \boldsymbol{v})^2\right] \qquad (10.71)$$

The first term in the integrand represents the energy dissipation due to the motion of solvent relative to the gel network, and the second term is that due to the motion of ions relative to the solvent. The energy dissipation function may include other terms such as $\boldsymbol{v}_i^2$ and $\boldsymbol{v}_i\boldsymbol{v}_j$, but they are ignored here.

The free energy of the system is written as

$$A = \int d\boldsymbol{r}\left[\sum_i k_BTn_i\ln n_i + \sum_i n_ie_i\psi\right] \qquad (10.72)$$

The first term in the integrand is the usual mixing entropy, and the second term represents the potential energy of the ions. A comment is needed about this term. Since n_i satisfies the charge neutrality condition

$$\sum_i n_i e_i + n_b e_b = 0 \tag{10.73}$$

the term is independent of n_i, and one may think that it can be dropped in the free energy expression. However the term is needed in the analysis since the term represents the constraint of the charge neutrality condition. In an electrolyte solution, the electric potential ψ appears as a Lagrangean multiplier to represent the charge neutrality condition (just like the pressure p appears in hydrodynamics to represent the incompressible condition).

The conservation laws for solvent and ions can be written as

$$\nabla \cdot \boldsymbol{v} = 0, \qquad \dot{n}_i = -\nabla \cdot (n_i \boldsymbol{v}_i) \tag{10.74}$$

The Rayleighian is therefore constructed by the standard procedure, and the result is

$$R = \frac{1}{2} \int d\boldsymbol{r} \left[\xi \boldsymbol{v}^2 + \sum_i n_i \zeta_i (\boldsymbol{v}_i - \boldsymbol{v})^2 \right]$$
$$+ \int d\boldsymbol{r} \sum_i n_i \boldsymbol{v}_i \cdot \nabla \left[k_B T \ln n_i + e_i \psi \right] + \int d\boldsymbol{r} \, \boldsymbol{v} \cdot \nabla p \tag{10.75}$$

The last term comes from the incompressible condition $\nabla \cdot \boldsymbol{v} = 0$.

The Onsager principle therefore gives the following equations:

$$\xi \boldsymbol{v} + \sum_i \zeta_i n_i (\boldsymbol{v} - \boldsymbol{v}_i) = -\nabla p \tag{10.76}$$

$$\zeta_i (\boldsymbol{v}_i - \boldsymbol{v}) = -k_B T \nabla \ln n_i - e_i \nabla \psi \tag{10.77}$$

To simplify the analysis further, we consider the situation that the ion concentration is uniform in the system. This is true in the bulk of the ionic gel (but not true near the boundary). If such an assumption is made, eq. (10.77) is replaced by

$$\zeta_i (\boldsymbol{v}_i - \boldsymbol{v}) = -e_i \nabla \psi \tag{10.78}$$

Hence the second term on the left-hand side of eq. (10.76) is written as

$$\sum_i \zeta_i n_i (\boldsymbol{v} - \boldsymbol{v}_i) = \left(\sum_i e_i n_i \right) \nabla \psi = -e_b n_b \nabla \psi \tag{10.79}$$

where the charge neutrality condition (10.73) has been used. Equation (10.76) is then written as

$$\boldsymbol{v} = -\frac{1}{\xi} \nabla p + \frac{n_b e_b}{\xi} \nabla \psi \tag{10.80}$$

The electric current $\boldsymbol{j}_e$ is given by

$$\boldsymbol{j}_e = \sum_i n_i e_i \boldsymbol{v}_i \tag{10.81}$$

which is rewritten using eqs. (10.78) and (10.73) as

$$\boldsymbol{j}_e = \left(\sum_i n_i e_i\right)\boldsymbol{v} - \sum_i \frac{n_i e_i^2}{\zeta_i}\nabla\psi$$

$$= \frac{n_b e_b}{\xi}\nabla p - \left[\frac{(n_b e_b)^2}{\xi} + \sum_i \frac{n_i e_i^2}{\zeta_i}\right]\nabla\psi \qquad (10.82)$$

Equations (10.80) and (10.82) can be written in matrix form

$$\begin{pmatrix} \boldsymbol{v} \\ \boldsymbol{j}_e \end{pmatrix} = \begin{pmatrix} -\kappa & \lambda \\ \lambda & -\sigma_e \end{pmatrix} \begin{pmatrix} \nabla p \\ \nabla\psi \end{pmatrix} \qquad (10.83)$$

where

$$\kappa = \frac{1}{\xi}, \qquad \lambda = \frac{n_b e_b}{\xi}, \qquad \sigma_e = \sum_i \frac{n_i e_i^2}{\zeta_i} + \frac{(n_b e_b)^2}{\xi} \qquad (10.84)$$

Notice that the coefficient matrix in (10.83) is symmetric. This is another example of the Onsager reciprocal relation.

Equation (10.83) describes the electrokinetic phenomena shown in Fig. 10.9(a) and (b). In the situation of Fig. 10.9(a), the pistons in the two chambers are assumed to move freely and therefore ∇p is equal to zero. This implies that the solvent flows with velocity $\lambda\nabla\psi$. In the situation of Fig. 10.9(b), the electric current $\boldsymbol{j}_e$ must be zero,[6] and therefore the streaming potential is given by

$$\nabla\psi = \frac{\lambda}{\sigma_e}\nabla p \qquad (10.85)$$

[6] If there is an electric current flowing across the gel, the charge neutrality condition is violated since the electrodes are connected to a potentiometer.

10.4.3 Kinetic equation for ion distribution in electrolyte solutions

In the situation shown in Fig. 10.9(a) and (b), the solvent is flowing through a gel network almost uniformly, and the energy dissipation is mainly caused by the relative motions between the components. In the situation shown in Fig. 10.9(c) and (d), solvent is flowing around charged particles. In this case, it is necessary to account for the energy dissipation caused by the velocity gradient $\nabla\boldsymbol{v}$. This has already been discussed in Chapter 8; the energy dissipation function is written as

$$\Phi = \frac{1}{2}\int d\boldsymbol{r}\left[\sum_i n_i\zeta_i(\boldsymbol{v}_i - \boldsymbol{v})^2 + \frac{1}{2}\eta\left(\nabla\boldsymbol{v} + (\nabla\boldsymbol{v})^t\right) : \left(\nabla\boldsymbol{v} + (\nabla\boldsymbol{v})^t\right)\right] \qquad (10.86)$$

Repeating the same calculation as in the previous section, we have the following set of equations, which are similar to eqs. (8.54) and (8.55):

$$\zeta_i(\boldsymbol{v}_i - \boldsymbol{v}) = -\nabla(k_B T \ln n_i + e_i\psi) \qquad (10.87)$$

$$\eta\nabla^2\boldsymbol{v} + \sum_i n_i\zeta_i(\boldsymbol{v}_i - \boldsymbol{v}) = \nabla p \qquad (10.88)$$

The electric potential ψ can be determined by the charge neutrality condition or by the Poisson equation

$$\epsilon \nabla^2 \psi = -\sum n_i e_i - \rho_b \qquad (10.89)$$

where ρ_b represents the density of the fixed charges. If eq. (10.89) is used, one can discuss the flow of solvent and ions within the double layer. The set of equations (10.87)–(10.89) represents the dynamic version of the Poisson–Boltzmann equation.

10.4.4   Motion of ions in a narrow channel

As an application of the dynamic Poisson–Boltzmann equation, we consider the problem shown in Fig. 10.10(a). An electrolyte solution is flowing in a capillary driven by an electric field $E = -\partial\psi/\partial x$ and pressure gradient $p_x = \partial p/\partial x$. In this situation, the solvent velocity $\boldsymbol{v}$, and the ion velocity $\boldsymbol{v}_i$ are in the x-direction and only depend on the y-coordinate:

$$\boldsymbol{v} = (v_x(y), 0, 0), \qquad \boldsymbol{v}_i = (v_{ix}(y), 0, 0) \qquad (10.90)$$

The y-component of eq. (10.87) gives

$$\frac{\partial}{\partial y}[k_B T \ln n_i(y) + e_i \psi(x, y)] = 0 \qquad (10.91)$$

Hence

$$n_i(y) = n_i^0 e^{-\beta e_i \psi(x,y)} \qquad (10.92)$$

Therefore the electric potential ψ satisfies the same Poisson–Boltzmann equation at equilibrium. The solution is then written as

$$\psi(x, y) = \psi_{eq}(y) - Ex \qquad (10.93)$$

where $\psi_{eq}(y)$ is the equilibrium solution discussed in Section 10.3.5.

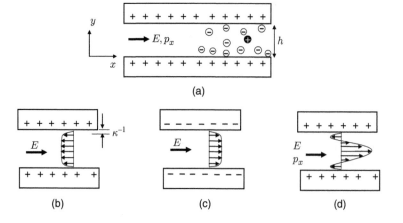

Fig. 10.10 Flow of electrolyte solution in a narrow channel between charged walls. The flow is induced by a pressure gradient p_x and an electric field E.

The x-component of eq. (10.88) becomes

$$\eta \frac{d^2 v_x}{dy^2} + \sum_i n_i(y)\zeta_i[v_{ix}(y) - v_x(y)] = p_x \qquad (10.94)$$

The x-component of eq. (10.87) gives

$$\zeta_i[v_{ix}(y) - v_x(y)] = e_i E \qquad (10.95)$$

Hence eq. (10.94) can be written as

$$\eta \frac{d^2 v_x}{dy^2} + \sum_i n_i(y)e_i E = p_x \qquad (10.96)$$

Using the Poisson equation ($\epsilon d^2\psi/dy^2 = -\sum_i n_i(y)e_i$), eq. (10.96) is written as

$$\eta \frac{d^2 v_x}{dy^2} - \epsilon E \frac{d^2\psi}{dy^2} = p_x \qquad (10.97)$$

The solution of this equation is

$$v_x = -\frac{1}{2\eta}p_x y(h-y) - \frac{\epsilon E}{\eta}(\psi_s - \psi_{eq}(y)) \qquad (10.98)$$

where ψ_s is the electric potential at the surface (i.e., $\psi_s = \psi_{eq}(0)$), and we have used the boundary condition $v_x(0) = v_x(h) = 0$. On the other hand, the current density is given by

$$j_{ex}(y) = \sum_i n_i(y)e_i v_{ix}(y)$$

$$= \left(\sum_i n_i(y)e_i\right)v_x(y) + \sum_i \frac{n_i(y)e_i^2}{\zeta_i}E$$

$$= -\frac{1}{2\eta}\left(\sum_i n_i(y)e_i\right)p_x y(h-y)$$

$$- \frac{\sum_i n_i(y)e_i \epsilon E}{\eta}(\psi_s - \psi_{eq}(y)) + \sum_i \frac{n_i(y)e_i^2}{\zeta_i}E$$

$$(10.99)$$

We now discuss some characteristic aspects of the above solution.

Velocity profile

In the usual capillary, h is much larger than the Debye length. Therefore we first consider the limit of $\kappa h \gg 1$. In this case, $\psi_{eq}(y)$ in eq. (10.98) is almost zero except near the capillary wall. If there is no pressure gradient (i.e., if $p_x = 0$), the velocity is almost constant. Figure 10.10(b) and (c) shows the velocity profile in this situation; the velocity increases sharply with the increase of the distance from the wall, and quickly approaches a constant value

$$v_s = -\frac{\epsilon \psi_s}{\eta} E \qquad (10.100)$$

Notice that this velocity change takes place in a very small length, of the order of the Debye length κ^{-1}. On a macroscopic scale (in which the Debye length κ^{-1} is ignored), the velocity changes discontinuously at the surface: the fluid slips at the surface.

Note that the electric field affects the velocity near the boundary, but does not affect the velocity in the bulk. Again this is a result of the charge neutrality condition. Since the charge density of the bulk fluid is zero, the electric field exerts no forces on the bulk, but does exert forces on the fluid in the electric double layer.

If there is a pressure gradient p_x, the velocity profile has a superimposed parabolic profile as shown in Fig. 10.10(d).

Onsager's reciprocal relation

The total flux of solvent J_s and electric current J_e is given by

$$J_s = \int_0^h dy\ v_x(y), \qquad J_e = \int_0^h dy\ j_{ex}(y) \qquad (10.101)$$

Using eqs. (10.98) and (10.99), we can write the expressions for J_s and J_e in the same form as eq. (10.83):

$$\begin{pmatrix} J_s \\ J_e \end{pmatrix} = - \begin{pmatrix} L_{11} & L_{12} \\ L_{21} & L_{22} \end{pmatrix} \begin{pmatrix} \nabla p \\ \nabla \psi \end{pmatrix} \qquad (10.102)$$

It can be shown that the matrix $\{L\}$ is symmetric ($L_{12} = L_{21}$). Therefore, the phenomena of electro-osmosis and streaming potential are also seen in capillaries.

10.4.5   Electrophoresis of charged particles

We now consider the electrophoresis of colloidal particles. Consider a spherical particle of radius R having surface charge density q moving in an electrolyte solution under an applied electric field E_{ext}.

If the outer solution is a neutral fluid, the particle velocity V can be calculated easily. The force exerted by the electric field on the particle is $4\pi R^2 q E^\infty$. In steady state, this force is balanced with the viscous drag $6\pi\eta RV$. Hence

$$V = \frac{4\pi R^2 q E^\infty}{6\pi \eta R} = \frac{2qR}{3\eta} E_{ext} \qquad (10.103)$$

The charge density q can be represented by the surface potential ψ_s by $\psi_s = qR/\epsilon$. Hence the velocity of the particle is given by

$$V = \frac{2\epsilon\psi_s}{3\eta} E_{ext} \qquad (10.104)$$

Notice that this expression is independent of the size of the particle.

Equation (10.104) is correct if the particle is placed in a non-ionic solvent (or solvent with negligible ion concentration). In reality, the particle is usually surrounded by an electrolyte solution, and therefore the electric force and the hydrodynamic drag acting on the particle will be quite different from this simple argument. To calculate the velocity in such a situation, one has to solve the set of equations described in Section 10.4.3. This problem is a hard mathematical problem, but has been studied by many people, and the results are well documented in the literature (see reference 2 in Further reading). Here we shall limit the analysis to the special (but important) case that the Debye length κ^{-1} is negligibly small compared with the particle size R.

In this situation, we may assume that the bulk of the electrolyte solution is a neutral fluid with viscosity η and conductivity σ_e. The fluid velocity v therefore satisfies the usual Stokes equation

$$\eta \nabla^2 v = \nabla p, \qquad \nabla \cdot v = 0 \tag{10.105}$$

The electric potential ψ also satisfies the Laplace equation

$$\nabla^2 \psi = 0 \tag{10.106}$$

The boundary condition for ψ is

$$\psi(r) = -E^\infty \cdot r, \qquad |r| \to \infty \tag{10.107}$$
$$n \cdot \nabla \psi = 0. \qquad |r| = R \tag{10.108}$$

where n is the unit vector normal to the particle surface. Equation (10.108) represents the condition that the electric current $j_e = -\sigma_e \nabla \psi$ normal to the surface must be zero. The solution of eq. (10.106) under these boundary condition is given by

$$\psi = -E^\infty \cdot r \left[1 - \frac{1}{2} \left(\frac{R}{r} \right)^3 \right] \tag{10.109}$$

The effect of the electric field appears in the boundary condition; due to the boundary condition (10.108), the electric field is parallel to the boundary. Therefore, the problem of the electrophoresis of a sphere becomes locally the same as the problem of electro-osmosis in a channel (see Fig. 10.11). Let V be the velocity of the particle. In usual fluids, the non-slip boundary condition imposes the condition that the fluid velocity at the boundary is equal to V. In the present problem, the fluid velocity is not equal to V due to the large velocity gradient which exists in the electric double layer. Ignoring the thickness of the double layer, the boundary condition at the surface is written as

$$v - V = -\frac{\epsilon \psi_s}{\eta} \nabla \psi \tag{10.110}$$

Equation (10.110) indicates that the tangential component of the fluid velocity is discontinuous (having a slip velocity given by eq. (10.100)), while the normal component of the fluid velocity is continuous.

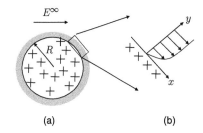

(a)   (b)

Fig. 10.11 Electrophoresis of a charged particle.

The Stokes equation (10.105) can be solved for the boundary condition (10.110), and the particle velocity is determined by the condition that the total force acting on the particle is zero. Such a calculation gives

$$V = \frac{\epsilon \psi_s}{\eta} E^\infty \qquad (10.111)$$

It is interesting that this velocity is not very much different from that of simple estimation. (This is due to the fact that in both limits of $\kappa R \ll 1$ and $\kappa R \gg 1$, the electrophoretic velocity becomes independent of κR.) Equation (10.111) is known to be valid for any shapes of particles as long as the particle surface is homogeneously charged.

The flow velocity is now given by

$$v = \frac{1}{2} \left(\frac{R}{r} \right)^3 \left[\frac{3rr}{r^2} - I \right] V \qquad (10.112)$$

This is quite different from the Stokes solution for the flow field created by the particle motion in usual (non-electrolyte) fluids. In usual fluids, the fluid moves with the particle (the stick boundary condition), and the velocity decays as $v(r) \propto 1/r$ (see Appendix A). On the other hand, in electrophoresis, the fluid slips at the surface of the particle, and the velocity decays much faster as $v(r) \propto 1/r^3$. As a result, the interaction between the particles in electrophoresis is very weak: in the electrophoresis of many particles, particles move almost independently of each other.

10.5   Summary of this chapter

Ionic soft matter (polyelectrolytes, charged colloids, and charged micelles) consists of macro-ions which have many fixed charges and small mobile ions. The dissociation of ionic groups in macro-ions is strongly suppressed due to the Coulombic interaction between the dissociated groups.

The charges of macro-ions are screened by the counter-ions which accumulate around the macro-ions. Due to this effect, the charge density, averaged over a length-scale larger than the Debye length κ^{-1}, is essentially zero. This fact, called the charge neutrality condition, determines the macroscopic quantities such as the average density of small ions in charged gels, and the electric potential difference between the inside and outside of the gel. The distribution of ions on length-scales smaller than the Debye length is determined by the Poission–Boltzmann equation.

The motion of mobile ions in ionic soft matter is generally coupled with the motion of the solvent. This coupling creates various electrokinetic phenomena such as electro-osmosis, electrophoresis, streaming potential, and sedimentation potential. They can be treated by the Onsager principle.

Further reading

(1) *An Introduction to Interfaces and Colloids: The Bridge to Nanoscience*, John C. Berg, World Scientific (2009).

(2) *Electrokinetic and Colloid Transport Phenomena*, Jacob H. Masliyah and Subir Bhattacharjee, Wiley-Interscience (2006).

Exercises

(10.1) Plot the dissociation fraction α of CH_3COOH and NH_4OH as a function of PH. Use pK $= 4.6$ for CH_3COOH and pK $= 4.7$ for NH_4OH.

(10.2) Consider a polymer which contains N_A acidic groups and N_B basic groups, each producing monovalent mobile ions. Let K_A and K_B be their dissociation constants. Calculate the pH at which the total charge of the polymer becomes zero assuming that the dissociation of these groups takes place independently of each other.

(10.3) Consider two compartments of equal volume separated by a semi-permeable membrane which is permeable for anions only. First both compartments are filled with pure water. When sodium chloride is added in one compartment, one comparment becomes acidic while the other becomes basic. Explain the reason for this, and calculate the pH of each compartment for a given number density of added sodium chloride.

(10.4) Solve the linearized Poission–Boltzmann equation (10.42) for a gel interface assuming that the density of the fixed charge is given by

$n_b(x) = n_b\Theta(-x)$. Show that the Donnan potential is given by eq. (10.27).

(10.5) Consider the inter-surface force acting between two surfaces having surface potentials ψ_A and ψ_B. Answer the following questions.

(a) Solve eq. (10.42) under the boundary conditions $\psi(0) = \psi_A$ and $\psi(h) = \psi_B$.

(b) Show that the inter-surface force is given by

$$f_{int} = \frac{\epsilon \kappa^2}{2} \frac{2\psi_A\psi_B \cosh(\kappa h) - \psi_A^2 - \psi_B^2}{\sinh^2(\kappa h)}$$

(10.113)

(c) Show that if $\psi_A \neq \psi_B$, the inter-surface force can be attractive in a certain range of h even if the potentials of the two surfaces have the same sign.

(10.6) Show that $f(x)$ defined by eq. (10.64) is independent of x if $\psi(x)$ satisfies the Poisson–Boltzmann equation (10.40).

(10.7) Calculate the solvent flux J_s and the electric current J_e in eq. (10.101), and obtain the matrix $\{L\}$ in eq. (10.102). Show that it is symmetric.

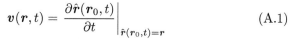

A Continuum mechanics

A.1 Forces acting in a material

Flow and deformation of materials is generally discussed in the framework of continuum mechanics. Continuum mechanics regards a material as an object consisting of points, called material points, and describes the flow and deformation of the material by the motion of these points.

Let $\hat{\boldsymbol{r}}(\boldsymbol{r}_0, t)$ be the position of the material point P, at time t, which was located at position $\boldsymbol{r}_0$ in a certain reference state. The velocity of the point P at time t is given by $\partial \hat{\boldsymbol{r}}(\boldsymbol{r}_0, t)/\partial t$. This can be expressed as a function of the current position $\boldsymbol{r}$

$$\boldsymbol{v}(\boldsymbol{r}, t) = \left. \frac{\partial \hat{\boldsymbol{r}}(\boldsymbol{r}_0, t)}{\partial t} \right|_{\hat{\boldsymbol{r}}(\boldsymbol{r}_0, t) = \boldsymbol{r}} \tag{A.1}$$

The function $\boldsymbol{v}(\boldsymbol{r}, t)$ is called velocity field, and is the central quantity in fluid mechanics. If the velocity field $\boldsymbol{v}(\boldsymbol{r}, t)$ in the past is given, $\hat{\boldsymbol{r}}(\boldsymbol{r}_0, t)$ is calculated by integrating eq. (A.1).

To obtain the equation of motion for the material point P, we consider a small region $\mathcal{V}_P$ enclosing the point P (see Fig. A.1). The mass of this region is given by ρV (ρ and V being the density and the volume of the region), and the equation of motion is written as

$$\rho V \frac{\partial^2 \hat{\boldsymbol{r}}(\boldsymbol{r}_0, t)}{\partial t^2} = \boldsymbol{F}_{tot} \tag{A.2}$$

where $\boldsymbol{F}_{tot}$ is the total force acting on the region $\mathcal{V}_P$.

There are two kinds of forces acting on the region. One is the body force which is proportional to the volume V of the region $\mathcal{V}_P$. Gravity is an example of this kind of force, and can be written as $\rho V \boldsymbol{g}$ ($\boldsymbol{g}$ being the gravitational acceleration vector). The other is the force acting at the outer surface of the region. This force can be expressed by the stress tensor $\boldsymbol{\sigma}$, which is elaborated in the next section.

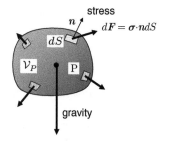

Fig. A.1 Forces acting on a region $\mathcal{V}_P$ surrounding the material point P.

A.2 Stress tensor

The components of the stress tensor are related to the force acting across a plane considered in the material. Consider a plane of area dS normal to the z-axis (see Fig. A.2). The molecules above the plane exert forces on the molecules below the plane. Since the molecular interaction is short

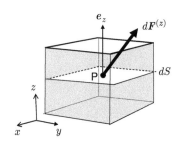

Fig. A.2 Definition of the stress tensor.

ranged, this force $d\boldsymbol{F}^{(z)}$ is proportional to the area dS. The α-component of the force, $dF_\alpha^{(z)}$, per unit area defines the αz-component of the stress tensor, i.e.,

$$dF_\alpha^{(z)} = \sigma_{\alpha z}dS \tag{A.3}$$

In general, $\sigma_{\alpha\beta}$ is defined as the α-component of the force acting across a plane of unit area normal to the β-axis.

From this definition, it can be shown that the force acting across a plane of area dS normal to the unit vector $\boldsymbol{n}$ can be expressed by $\sigma_{\alpha\beta}$ as

$$d\boldsymbol{F} = \boldsymbol{\sigma}\cdot\boldsymbol{n}dS \quad \text{or} \quad dF_\alpha = \sigma_{\alpha\beta}n_\beta dS \tag{A.4}$$

This can be proven as follows. Let $\sigma_{\alpha\beta}$ be the stress tensor at a point P. Let us consider a small tetrahedron ABCD enclosing the point P (see Fig. A.3). The three surfaces ABD, BCD, and CAD are assumed to be normal to the x-, y-, and z-axes, respectively. The forces acting on the surfaces ABD, BCD, and CAD are expressed by $\sigma_{\alpha\beta}$. For example, the force acting on the surface CAD is given by $-\sigma_{\alpha z}dS_z$,[1] where dS_z is the area of the surface CAD. dS_z can be expressed in terms of the area dS of the surface ABC and the unit vector $\boldsymbol{n}$ normal to it as $dS_z = n_z dS$. Similarly, the forces acting on the surfaces ABD and BCD are given by $-\sigma_{\alpha x}n_x dS$ and $-\sigma_{\alpha y}n_y dS$, respectively. If $d\boldsymbol{F}$ denotes the force acting on the surface ABC, the equation of motion for the tetrahedron becomes

$$\rho V \frac{\partial^2 \hat{r}_\alpha}{\partial t^2} = -\sigma_{\alpha x}n_x dS - \sigma_{\alpha y}n_y dS - \sigma_{\alpha z}n_z dS + dF_\alpha + \rho V g_\alpha \tag{A.5}$$

The first four terms on the right-hand side are of the order of the surface area of the tetrahedron, while the left-hand side and the last term on the right-hand side are of the order of the volume of the tetrahedron. In the limit $V \to 0$, the left-hand side becomes negligibly small compared with the right-hand side. Therefore, in the limit $V \to 0$, eq. (A.5) gives

$$-\sigma_{\alpha x}n_x dS - \sigma_{\alpha y}n_y dS - \sigma_{\alpha z}n_z dS + dF_\alpha = 0 \tag{A.6}$$

which is eq. (A.4).

According to eq. (A.4), the total force acting on the surface of the region $\mathcal{V}_P$ in Fig. A.1 is written as

$$F_\alpha = \int dS\, \sigma_{\alpha\beta}n_\beta \tag{A.7}$$

By the divergence theorem,[2] the right-hand side can be written as

$$F_\alpha = \int d\boldsymbol{r}\frac{\partial\sigma_{\alpha\beta}}{\partial r_\beta} \tag{A.9}$$

Therefore the equation of motion is written as

$$\rho V\frac{\partial^2 \hat{r}_\alpha(\boldsymbol{r}_0, t)}{\partial t^2} = \int d\boldsymbol{r}\frac{\partial\sigma_{\alpha\beta}}{\partial r_\beta} + \rho V g_\alpha \tag{A.10}$$

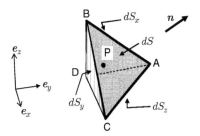

Fig. A.3 Tetrahedron enclosing the material point P.

[1] The stress on the plane CAD is slightly different from the stress at P, but the difference becomes negligible when the size of the tetrahedron goes to zero.

[2] This indicates the following identity for any function $f(\boldsymbol{r})$

$$\int d\boldsymbol{r}\frac{\partial f}{\partial r_\alpha} = \int dS\, f n_\alpha \tag{A.8}$$

In the limit $V \to 0$, this can be written as

$$\rho \frac{\partial^2 \hat{r}_\alpha(\boldsymbol{r}_0, t)}{\partial t^2} = \frac{\partial \sigma_{\alpha\beta}}{\partial r_\beta} + \rho g_\alpha \tag{A.11}$$

In the dynamics of soft matter, the inertia term and the gravitational term are often negligibly small. In this case, eq. (A.11) becomes the following force balance equation

$$\frac{\partial \sigma_{\alpha\beta}}{\partial r_\beta} = 0 \tag{A.12}$$

A.3   Constitutive equations

Equation (A.11) holds for any material (rubber, plastics, polymer solutions, etc). The deformation behaviour of these materials is quite different. The particular properties of the material appear in other equations called constitutive equations. Constitutive equations give the stress tensor for a given deformation of the material.

The stress tensor at a material point P depends on the local history of the deformation that P has experienced. The history of deformation for the point P can be expressed by the deformation gradient tensor defined by

$$E_{\alpha\beta}(t, t_0) = \frac{\partial \hat{r}_\alpha(\boldsymbol{r}_0, t)}{\partial r_{0\beta}} \tag{A.13}$$

This is the same tensor introduced in Section 3.1.4, where t_0 denotes the time for the reference state. Alternatively, the local history of deformation at P can be expressed by the history of the velocity gradient tensor

$$\kappa_{\alpha\beta}(t) = \frac{\partial v_\alpha(\boldsymbol{r}, t)}{\partial r_\beta} \quad \text{at} \quad \boldsymbol{r} = \hat{r}(\boldsymbol{r}_0, t) \tag{A.14}$$

It is easy to show that $E_{\alpha\beta}(t, t_0)$ and $\kappa_{\alpha\beta}(t)$ are related by (see eq. (9.25))

$$\frac{\partial}{\partial t} E_{\alpha\beta}(t, t_0) = \kappa_{\alpha\mu}(t) E_{\mu\beta}(t, t_0) \tag{A.15}$$

Therefore if one of $E_{\alpha\beta}(t, t_0)$ and $\kappa_{\alpha\beta}(t)$ is known, the other can be calculated. The constitutive equations are the equations which determine $\sigma_{\alpha\beta}(t)$ for given values of $E_{\alpha\beta}(t', t_0)$ or $\kappa_{\alpha\beta}(t')$ in the past.

Given the constitutive equations, the equation of motion (A.11) can be solved for all material points, and the flow and deformation of materials is uniquely determined. This is the principle of continuum mechanics.

A.4   Work done to the material

In considering the constitutive equation, it is useful to consider the work needed to cause a deformation in the material.

Consider a material at mechanical equilibrium. Suppose that we deform the material slightly, and displace the point at r to $r+\delta u(r)$. The work we do to a small region V by such a deformation is given by

$$\delta W = \int dS \; \sigma_{\alpha\beta} n_\beta \delta u_\alpha(r) \qquad (A.16)$$

where we have assumed that the body force $\rho V g_\alpha$ is negligible. The right-hand side is rewritten by using the divergence theorem as

$$\delta W = \int dr \; \frac{\partial}{\partial r_\beta}(\sigma_{\alpha\beta}\delta u_\alpha) = \int dr \; \left(\frac{\partial \sigma_{\alpha\beta}}{\partial r_\beta}\delta u_\alpha + \sigma_{\alpha\beta}\frac{\partial \delta u_\alpha}{\partial r_\beta} \right) \qquad (A.17)$$

The first term in the integrand is zero due to the mechanical equilibrium condition (A.12). For a small region, eq. (A.17) is written as

$$\delta W = V \sigma_{\alpha\beta}\delta\epsilon_{\alpha\beta} \qquad (A.18)$$

where

$$\delta\epsilon_{\alpha\beta} = \frac{\partial \delta u_\alpha}{\partial r_\beta} \qquad (A.19)$$

is the strain for the hypothetical deformation.

As an application of eq. (A.18), let us consider a rubbery material for which the free energy is written as a function of E as $f(E)$ per unit volume (see eq. (3.48)). If the system is at equilibrium, the work done to the material by the hypothetical deformation is equal to the change of the free energy. Therefore

$$\delta f(E) = \sigma_{\alpha\beta}\delta\epsilon_{\alpha\beta} \qquad (A.20)$$

Now, if the material is deformed by the strain $\delta\epsilon$ (i.e., if the point at r is displaced to $r + \delta\epsilon \cdot r$), the deformation gradient $E_{\alpha\beta}$ changes as[3]

$$\delta E_{\alpha\beta} = \delta\epsilon_{\alpha\mu}E_{\mu\beta} \qquad (A.21)$$

Hence the free energy $f(E)$ changes as

$$\delta f(E) = \frac{\partial f(E)}{\partial E_{\alpha\beta}}\delta E_{\alpha\beta} = \frac{\partial f(E)}{\partial E_{\alpha\beta}}\delta\epsilon_{\alpha\mu}E_{\mu\beta} \qquad (A.22)$$

Equations (A.20) and (A.22) give

$$\sigma_{\alpha\beta} = E_{\beta\mu}\frac{\partial f(E)}{\partial E_{\alpha\mu}} - p\delta_{\alpha\beta} \qquad (A.23)$$

The last term comes from the fact that since the rubber is incompressible, the hypothetical strain $\delta\epsilon_{\alpha\beta}$ has to satisfy the constraint $\delta\epsilon_{\alpha\alpha} = 0$.[4] If the free energy is given by eq. (3.48), the stress tensor is given by

$$\sigma_{\alpha\beta} = GE_{\alpha\mu}E_{\beta\mu} - p\delta_{\alpha\beta} \qquad (A.24)$$

In the case of shear deformation shown in Fig. 3.3(b), $E_{\alpha\beta}$ is given by eq. (3.51). Hence the shear stress σ_{xy} is calculated as

$$\sigma_{xy} = G\gamma \qquad (A.25)$$

---

[3] This is because

$$\delta E_{\alpha\beta} = \frac{\partial \delta u_\alpha}{\partial r_{0\beta}}$$

$$= \frac{\partial \delta u_\alpha}{\partial r_\mu}\frac{\partial r_\mu}{\partial r_{0\beta}} = \delta\epsilon_{\alpha\mu}E_{\mu\beta}$$

[4] For incompressible materials, the isotropic part of the stress tensor cannot be determined uniquely by constitutive equations. The isotropic stress depends on the external forces applied to the material.

In the case of uniaxial elongation, $E_{\alpha\beta}$ is given by eq. (3.58). Hence eq. (A.24) now gives

$$\sigma_{xx} = \frac{G}{\lambda} - p, \qquad \sigma_{zz} = G\lambda^2 - p \qquad (A.26)$$

Since there is no force acting on the material surface normal to the x-direction, σ_{xx} is equal to zero. Therefore p is equal to G/λ, and eq. (A.26) gives

$$\sigma_{zz} = G\left(\lambda^2 - \frac{1}{\lambda}\right) \qquad (A.27)$$

This is equivalent to eq. (3.61).

A.5   Ideal elastic material

The stress tensor of an isotropic elastic material can be written in a general form if the deformation of the material is small. We assume that the material is isotropic and force free in the reference state and that the displacement vector $\boldsymbol{u} = \boldsymbol{r} - \boldsymbol{r}_0$ is small. We define the tensor $\epsilon_{\alpha\beta}$ by

$$\epsilon_{\alpha\beta} = \frac{\partial u_\alpha}{\partial r_\beta} \qquad (A.28)$$

[5] The stress tensor in the reference state can have an isotropic component $-p_0\delta_{\alpha\beta}$, where p_0 is the pressure at the reference state. Here we are taking p_0 as the reference of the pressure and have ignored the term $-p_0\delta_{\alpha\beta}$.

In the reference state, the stress tensor is zero.[5] If a small strain is applied, a small stress which is linear in the applied strain appears. For an isotropic material, such stress is written as

$$\sigma_{\alpha\beta} = C_1\epsilon_{\gamma\gamma}\delta_{\alpha\beta} + C_2\epsilon_{\alpha\beta} + C_3\epsilon_{\beta\alpha} \qquad (A.29)$$

where C_1, C_2, and C_3 are elastic constants. Now C_2 must be equal to C_3. This is seen as follows. Consider that the material in the reference state is rotated around the unit vector $\boldsymbol{n}$ by an angle $\delta\psi$. The rotation displaces the material point $\boldsymbol{r}_0$ to $\boldsymbol{r} = \boldsymbol{r}_0 + \boldsymbol{n} \times \boldsymbol{r}_0\delta\psi$, and gives

$$\boldsymbol{u} = \boldsymbol{n} \times \boldsymbol{r}_0\delta\psi, \quad \text{and} \quad \epsilon_{\alpha\beta} = e_{\alpha\beta\gamma}n_\gamma\delta\psi \qquad (A.30)$$

where $e_{\alpha\beta\gamma}$ is the Levi-Civita symbol defined in Section 7.5.2. According to eq. (A.29), the stress is given by

$$\sigma_{\alpha\beta} = (C_2 - C_3)e_{\alpha\beta\gamma}n_\gamma\delta\psi \qquad (A.31)$$

On the other hand, the stress should remain unchanged under rotation. Therefore C_2 must be equal to C_3.

It is convenient to express the stress tensor in the following form

$$\sigma_{\alpha\beta} = K\epsilon_{\gamma\gamma}\delta_{\alpha\beta} + G\left(\epsilon_{\alpha\beta} + \epsilon_{\beta\alpha} - \frac{2}{3}\epsilon_{\gamma\gamma}\delta_{\alpha\beta}\right) \qquad (A.32)$$

K and G represent the bulk modulus and the shear modulus defined in Section 3.1.3. Equation (A.32) represents the constitutive equation for the ideal elastic material, or the Hookean solid. The free energy of deformation for such a material is given by

$$f(\epsilon) = \frac{K}{2}\epsilon_{\alpha\alpha}^2 + \frac{G}{4}\left(\epsilon_{\alpha\beta} + \epsilon_{\beta\alpha} - \frac{2}{3}\epsilon_{\gamma\gamma}\delta_{\alpha\beta}\right)^2 \tag{A.33}$$

For the shear deformation shown in Fig. 3.3(b), the displacement vector is given by

$$u_x = \gamma y, \qquad u_y = u_z = 0 \tag{A.34}$$

Equation (A.32) then gives the shear stress

$$\sigma_{xy} = G\gamma \tag{A.35}$$

For uniaxial elongation, the displacement vector is given by

$$u_x = \epsilon_x x, \qquad u_y = \epsilon_x y, \qquad u_z = \epsilon_z z \tag{A.36}$$

Hence eq. (A.32) gives

$$\sigma_{xx} = \sigma_{yy} = K(2\epsilon_x + \epsilon_z) + \frac{2}{3}G(\epsilon_x - \epsilon_z) \tag{A.37}$$

$$\sigma_{zz} = K(2\epsilon_x + \epsilon_z) + \frac{4}{3}G(\epsilon_z - \epsilon_x) \tag{A.38}$$

In the situation shown in Fig. 3.3(c), the stress components σ_{xx} and σ_{yy} are zero, while σ_{zz} is equal to the applied load σ, i.e.,

$$K(2\epsilon_x + \epsilon_z) + \frac{2}{3}G(\epsilon_x - \epsilon_z) = 0 \tag{A.39}$$

$$K(2\epsilon_x + \epsilon_z) + \frac{4}{3}G(\epsilon_z - \epsilon_x) = \sigma \tag{A.40}$$

Solving this equation for ϵ_x and ϵ_z, we have

$$\epsilon_x = -\frac{3K - 2G}{18KG}\sigma, \qquad \epsilon_z = \frac{3K + G}{9KG}\sigma \tag{A.41}$$

Young's modulus E and Poisson's ratio ν of the material are defined by

$$\sigma = E\epsilon_z, \qquad \nu = -\frac{\epsilon_x}{\epsilon_z} \tag{A.42}$$

Equation (A.41) therefore gives

$$E = \frac{9KG}{3K + G} \qquad \nu = \frac{3K - 2G}{2(3K + G)} \tag{A.43}$$

For the constitutive equation (A.32), the force balance equation (A.12) is written as

$$\left(K + \frac{1}{3}G\right)\frac{\partial^2 u_\beta}{\partial r_\alpha \partial r_\beta} + G\frac{\partial^2 u_\alpha}{\partial r_\beta^2} = 0 \tag{A.44}$$

If the material is incompressible, the stress tensor is given by

$$\sigma_{\alpha\beta} = G\left(\epsilon_{\alpha\beta} + \epsilon_{\beta\alpha} - \frac{2}{3}\epsilon_{\gamma\gamma}\delta_{\alpha\beta}\right) - p\delta_{\alpha\beta} \tag{A.45}$$

and the force balance equation (A.12) is written as

$$G\nabla^2 \boldsymbol{u} = \nabla p \tag{A.46}$$

This equation has to be solved under the incompressible condition

$$\nabla \cdot \boldsymbol{u} = 0 \tag{A.47}$$

Equation (A.44) (or eqs. (A.46) and (A.47)) determine the displacement field $\boldsymbol{u}(\boldsymbol{r})$ of isotropic elastic materials at mechanical equilibrium.

A.6   Ideal viscous fluid

In an ideal viscous fluid, the stress $\sigma_{\alpha\beta}$ depends on the current value of the velocity gradient $\kappa_{\alpha\beta}$ only. If the stress is assumed to be a linear function of $\kappa_{\alpha\beta}$, $\sigma_{\alpha\beta}$ is written as

$$\sigma_{\alpha\beta} = D_1\kappa_{\gamma\gamma}\delta_{\alpha\beta} + D_2\kappa_{\alpha\beta} + D_3\kappa_{\beta\alpha} - p\delta_{\alpha\beta} \tag{A.48}$$

where D_1, D_2, and D_3 are material constants. By the same argument as above, D_3 is shown to be equal to D_2. Also, the term D_1 can be omitted since $\kappa_{\gamma\gamma}$ is identically zero for an incompressible fluid. Therefore the constitutive equation for an ideal viscous fluid is written as

$$\sigma_{\alpha\beta} = \eta\left(\kappa_{\alpha\beta} + \kappa_{\beta\alpha}\right) - p\delta_{\alpha\beta} \tag{A.49}$$

where the constant η is the viscosity defined in Section 3.1.2. A fluid described by such a constitutive equation is called a Newtonian fluid.

For a Newtonian fluid, the force balance equation (A.12) becomes

$$\eta\nabla^2 \boldsymbol{v} = \nabla p \tag{A.50}$$

Equation (A.50) and the incompressible condition

$$\nabla \cdot \boldsymbol{v} = 0 \tag{A.51}$$

determine $\boldsymbol{v}$ and p under proper boundary conditions. This equation is called the Stokes equation.

As an application of the Stokes equation, consider a spherical particle of radius a moving in a fluid with velocity $\boldsymbol{U}$. The fluid velocity satisfies the following boundary condition at the surface of the sphere

$$\boldsymbol{v}(\boldsymbol{r}) = \boldsymbol{U} \quad \text{at} \quad |\boldsymbol{r}| = a \tag{A.52}$$

and $\boldsymbol{v}(\boldsymbol{r}) \to 0$ for $|\boldsymbol{r}| \to \infty$. The solution of the above equation is

$$\boldsymbol{v}(\boldsymbol{r}) = \frac{3a}{4r}\left[\left(\boldsymbol{I} + \frac{\boldsymbol{rr}}{r^2}\right) + \frac{a^2}{r^2}\left(\frac{\boldsymbol{I}}{3} - \frac{\boldsymbol{rr}}{r^2}\right)\right] \cdot \boldsymbol{U} \tag{A.53}$$

$$p(\boldsymbol{r}) = \frac{3\eta a}{2r^3}\boldsymbol{r} \cdot \boldsymbol{U} \tag{A.54}$$

The force acting on the particle is thus calculated by

$$\boldsymbol{F}_f = \int dS \boldsymbol{\sigma} \cdot \boldsymbol{n} \tag{A.55}$$

This finally gives the following expression for the hydrodynamic drag for the sphere

$$\boldsymbol{F}_f = -6\pi \eta a \, \boldsymbol{U} \tag{A.56}$$

B Restricted free energy

B.1 Systems under constraint

According to statistical mechanics, the free energy of a system at equilibrium can be calculated if the Hamiltonian of the system is known. The Hamiltonian is a function of the set of generalized coordinates $(q_1, q_2, \ldots, q_f)$ and momenta $(p_1, p_2, \ldots, p_f)$. For simplicity we shall denote the $2f$ set of variables by Γ,

$$\Gamma = (q_1, q_2, \ldots, q_f, p_1, p_2, \ldots, p_f) \tag{B.1}$$

Then the Hamiltonian is written as

$$H(\Gamma) = H(q_1, q_2, \ldots, q_f, p_1, p_2, \ldots, p_f) \tag{B.2}$$

Given $H(\Gamma)$, the free energy of the system at temperature T is calculated by

$$A(T) = -\frac{1}{\beta} \ln \left[\int d\Gamma \; e^{-\beta H(\Gamma)} \right] \tag{B.3}$$

where $\beta = 1/k_B T$.

Equation (B.3) represents the free energy of the system which is at true equilibrium under a given temperature T.

In thermodynamics, we often consider the free energy of a system which is at equilibrium under some hypothetical constraints. Such a free energy is called a restricted free energy. For example, in discussing the phase separation of solutions, we consider the system consisting of two homogeneous solutions (each having concentration ϕ_1, ϕ_2 and volume V_1, V_2), and then minimize the total free energy $F_{tot}(\phi_1, \phi_2, V_1, V_2; T)$ with respect to ϕ_i and V_i to obtain the true equilibrium state. The free energy $F_{tot}(\phi_1, \phi_2, V_1, V_2; T)$ is an example of the restricted free energy.

The restricted free energy appears many times in this book. For example, in Section 3.2.3, we considered the free energy $U(\mathbf{r})$ of a polymer chain whose end-to-end vector is fixed at $\mathbf{r}$. In Section 5.2.3, we considered the free energy $F(S; T)$ of a liquid crystalline system in which the order parameter is fixed at S. In Section 7.2.1, we considered the free energy $A(x)$ for the non-equilibrium state specified by variable x. In this appendix, we discuss the restricted free energy from the general viewpoint of statistical mechanics.

Let us consider a system in which the values of certain physical quantities $\hat{x}_i(\Gamma)$ $(i = 1, 2, \ldots, n)$ are fixed at x_i, i.e.,

$$x_i = \hat{x}_i(\Gamma) \qquad (i = 1, 2, \ldots, n) \tag{B.4}$$

The free energy of such a system is given by

$$A(x, T) = -\frac{1}{\beta} \ln \left[\int d\Gamma \ \Pi_{i=1}^{n} \delta(x_i - \hat{x}_i(\Gamma)) \ e^{-\beta H(\Gamma)} \right] \tag{B.5}$$

where x stands for $(x_1, x_2, \ldots, x_n)$. The constraints can be accounted for by introducing the hypothetical potential

$$U_c(\Gamma, x) = \frac{1}{2} \sum k_i (\hat{x}_i(\Gamma) - x_i)^2 \tag{B.6}$$

where k_i is a large positive constant which gives a penalty for the deviation of the quantity $\hat{x}_i(\Gamma)$ from the given value x_i. Using the formula

$$\delta(x) = \lim_{k \to \infty} \left(\frac{k}{2\pi}\right)^{1/2} e^{-kx^2} \tag{B.7}$$

the free energy of the constrained system can be written as

$$A(x, T) = -\frac{1}{\beta} \ln \left[\int d\Gamma \ e^{-\beta[H(\Gamma) + U_c(\Gamma, x)]} \right] \tag{B.8}$$

where the coefficient $\Pi_i (k_i \beta / 2\pi)^{1/2}$ has been omitted since it only gives terms independent of x_i.

B.2   Properties of the restricted free energy

From eq. (B.5), it is easy to derive the following properties of the restricted free energy.

(1) If the system is at equilibrium without the constraints, the probability that the physical quantity $\hat{x}_i(\Gamma)$ has the value x_i is given by

$$P(x) \propto e^{-\beta A(x, T)} \tag{B.9}$$

(2) The free energy of the system at equilibrium is given by

$$A(T) = -\frac{1}{\beta} \ln \left[\int dx \ e^{-\beta A(x, T)} \right] \tag{B.10}$$

Equations (B.9) and (B.10) indicate that $A(x, T)$ is playing the role of the Hamiltonian $H(\Gamma)$ in the new phase space specified by $x = (x_1, x_2, \ldots, x_n)$.

According to eq. (B.9), the most probable state is the state x which minimizes $A(x, T)$. This corresponds to the thermodynamic principle that the true equilibrium state is the state which minimizes the restricted free energy $A(x, T)$.

To fix the values of the physical quantities at x, external forces need to be applied to the system. For example, to keep the end-to-end vector of a polymer chain at the fixed value r, a force $\hat{f}$ needs to be applied at the chain ends (see the discussion in Section 3.2.3). The force needed to keep $\hat{x}_i(\Gamma)$ at x_i is given by

$$\hat{F}_i(\Gamma, x) = \frac{\partial U_c(\Gamma, x)}{\partial x_i} \tag{B.11}$$

The average of $\hat{F}_i(\Gamma, x)$ is given by

$$F_i(x) = \left\langle \hat{F}_i(\Gamma, x) \right\rangle = \frac{\int d\Gamma \frac{\partial U_c(\Gamma, x)}{\partial x_i} e^{-\beta[H(\Gamma) + U_c(\Gamma, x)]}}{\int d\Gamma e^{-\beta[H(\Gamma) + U_c(\Gamma, x)]}} = \frac{\partial A(x, T)}{\partial x_i} \tag{B.12}$$

This is the generalization of eq. (3.30).

B.3   Method of constraining force

The restricted free energy $A(x, T)$ can be calculated by eqs. (B.5) or (B.8). There is another way of calculating $A(x, T)$. This method has already been used in Section 3.2.3 to calculate the restricted free energy $U(r)$ of a polymer chain whose end-to-end vector is fixed at r. There, instead of fixing the end-to-end vector at r, we assumed that a constant force f is applied at the chain end. If f is suitably chosen, the average end-to-end vector becomes equal to the given value r.[1] If the relation between f and r is known, the restricted free energy $U(r)$ is obtained by eq. (3.30).

The above method can be generalized as follows. We consider the Hamiltonian

$$\tilde{H}(\Gamma, h) = H(\Gamma) - \sum_i h_i \hat{x}_i(\Gamma) \tag{B.13}$$

where h_i represents the hypothetical external force conjugate to $\hat{x}_i(\Gamma)$. We calculate the free energy $\tilde{A}(h_i, T)$ for the Hamiltonian (B.13)

$$\tilde{A}(h, T) = -\frac{1}{\beta} \ln \left[\int d\Gamma \ e^{-\beta \tilde{H}(\Gamma, h)} \right] \tag{B.14}$$

The average of $\hat{x}_i(\Gamma)$ for given forces $(h_1, h_2, \ldots)$ is calculated as

$$\begin{aligned}
x_i &= \frac{\int d\Gamma \ \hat{x}_i(\Gamma) e^{-\beta \tilde{H}(\Gamma, h)}}{\int d\Gamma e^{-\beta \tilde{H}(\Gamma, h)}} = -\frac{1}{\beta} \frac{\int d\Gamma \frac{\partial \tilde{H}(\Gamma, h)}{\partial h_i} e^{-\beta \tilde{H}(\Gamma, h)}}{\int d\Gamma e^{-\beta \tilde{H}(\Gamma, h)}} \\
&= \frac{\partial \tilde{A}(h, T)}{\partial h_i}
\end{aligned} \tag{B.15}$$

[1] The end-to-end vector of a chain with the force applied is not fixed at r; it fluctuates around r. However if the fluctuation is small compared with the mean value (this is indeed the case for a long chain), we may replace the constraint for r with the force f.

On the other hand, the free energy $\tilde{A}(h,T)$ is equal to the minimum of $A(x,T) - \sum_i h_i x_i$ with respect to x_i. Therefore

$$\frac{\partial}{\partial x_i}\left[A(x,T) - \sum_i h_i x_i\right] = 0 \qquad (\text{B.16})$$

Hence

$$h_i = \frac{\partial A(x,T)}{\partial x_i} \qquad (\text{B.17})$$

and $A(x,T)$ is written as

$$A(x,T) = \tilde{A}(h,T) + \sum_i h_i x_i, \qquad (\text{B.18})$$

Equations (B.17) and (B.18) are the Legendre transform. Therefore, the restricted free energy $A(x,T)$ is obtained either by eq. (B.17), or by (B.18). We now consider two examples of the restricted free energy.

B.4 Example 1: Potential of mean force

First we consider a two-component solution. Let $\boldsymbol{r}_{pi}$ $(i = 1, 2, \ldots, N_p)$ and $\boldsymbol{r}_{si}$ $(i = 1, 2, \ldots, N_s)$ be the position vectors of solute molecules and solvent molecules. The potential energy of the system is written as

$$U_{tot}(\boldsymbol{r}_p, \boldsymbol{r}_s) = \sum_{i>j} u_{pp}(\boldsymbol{r}_{pi} - \boldsymbol{r}_{pj}) + \sum_{i,j} u_{ps}(\boldsymbol{r}_{pi} - \boldsymbol{r}_{sj}) + \sum_{i>j} u_{ss}(\boldsymbol{r}_{si} - \boldsymbol{r}_{sj})$$
$$(\text{B.19})$$

where $u_{pp}(\boldsymbol{r})$, $u_{ps}(\boldsymbol{r})$, and $u_{ss}(\boldsymbol{r})$ are the potentials for solute–solute, solute–solvent, and solvent–solvent interactions.

The effective interaction potential between solute molecules is given by the restricted free energy:

$$A_{pp}(\boldsymbol{r}_p) = -\frac{1}{\beta} \ln\left[\int \Pi_i d\boldsymbol{r}_{si}\, e^{-\beta U_{tot}(\boldsymbol{r}_p, \boldsymbol{r}_s)}\right] \qquad (\text{B.20})$$

This represents the effective interaction between the solute molecules with the effect of the solvent included. The effective interaction $\Delta\epsilon$ introduced in eq. (2.49) is an example of such a potential for the lattice model. The interaction potential $U(h)$ between the colloidal particles introduced in Section 2.5.1 is another example of such a potential.

In the theory of solutions, $A_{pp}(\boldsymbol{r}_p)$ is called the potential of the mean force since the average of the force acting on solute i is given by the partial derivative of $A_{pp}(\boldsymbol{r}_p)$:

$$\boldsymbol{f}_i(\boldsymbol{r}_p) = -\frac{\partial A_{pp}(\boldsymbol{r}_p)}{\partial \boldsymbol{r}_{pi}} \qquad (\text{B.21})$$

This is a special case of eq. (B.12). (The minus sign arises from the difference in the definition of the force.) Notice that even though the original potential U_{tot} is written as a sum of two-body potentials such as $u_{pp}(\boldsymbol{r})$, $u_{ps}(\boldsymbol{r})$, and $u_{ss}(\boldsymbol{r})$, the potential A_{pp} is generally not written as a sum of the two-body potentials.

B.5   Example 2: Landau–de Gennes free energy of liquid crystals

As the next example, let us consider the free energy introduced in Section 5.2.3.

We consider N nematic forming molecules, each pointing along the unit vector $\boldsymbol{u}_i$ $(i = 1, 2, \ldots, N)$. We assume that the scalar order parameter of the system is S, i.e., $\boldsymbol{u}_i$ satisfy the constraint of eq. (5.28). To take into account this constraint, we assume that the Hamiltonian of the system is written as

$$\tilde{H}(\boldsymbol{u}_i, h) = -\frac{h}{N} \sum_i \frac{3}{2}\left(u_{iz}^2 - \frac{1}{3}\right) \tag{B.22}$$

where we have assumed (for simplicity) that the molecules are not interacting with each other (i.e., $U = 0$).

The free energy $\tilde{A}(h)$ can be written as

$$\tilde{A}(h) = -\frac{1}{\beta} \ln\left[\int d\boldsymbol{u}\, e^{-\beta h(3/2N)(u_z^2 - 1/3)}\right]^N$$

$$= \tilde{A}(0) - \frac{N}{\beta} \ln\left\langle e^{-\beta h(3/2N)(u_z^2 - 1/3)}\right\rangle_0 \tag{B.23}$$

where $\tilde{A}(0)$ is the free energy of the system with no constraint, and $\langle \cdots \rangle_0$ is the average for $\boldsymbol{u}$ for an isotropic distribution of molecules. In the following, we set $\tilde{A}(0)$ equal to zero.

The right-hand side of eq. (B.23) can be calculated as a power series with respect to h:

$$\left\langle e^{-\beta h(3/2N)(u_z^2 - 1/3)}\right\rangle_0 = \left\langle\left[1 + \frac{3\beta h}{2N}(u_z^2 - 1/3) + \frac{9\beta^2 h^2}{8N^2}(u_z^2 - 1/3)^2\right]\right\rangle_0$$

$$= 1 + \frac{(\beta h)^2}{10N^2} + \cdots \tag{B.24}$$

Hence

$$\tilde{A}(h) = -\frac{\beta}{10N}h^2 \tag{B.25}$$

Therefore the order parameter S of the system is given by

$$S = -\frac{\partial \tilde{A}(h)}{\partial h} = \frac{\beta}{5N}h \tag{B.26}$$

Hence

$$A(S) = \tilde{A}(h) + hS = \frac{\beta}{10N}h^2 = \frac{5Nk_BT}{2}S^2 \tag{B.27}$$

If there is an interaction between the particles, the interaction energy is written as (see eqs. (5.11) and (5.23))

$$E(S) = -\frac{NU}{2}\langle(\boldsymbol{u} \cdot \boldsymbol{u}')^2\rangle = -\frac{NU}{3}S^2 \tag{B.28}$$

Hence the free energy is given by

$$A(S) = \frac{5Nk_BT}{2}S^2 - \frac{NU}{3}S^2 = \frac{5Nk_B}{2}(T - T_c)S^2 \tag{B.29}$$

where $T_c = 2U/15k_B$. Equation (B.29) is the first term in the expansion of eq. (5.35).

C | Variational calculus

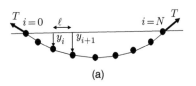

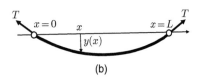

Fig. C.1 (a) A string connecting $N+1$ mass points is hung with a tensile force T. (b) A string with mass density ρ is hung with a tensile force T.

[1] Here δy_0 and δy_N are identically equal to zero since the end particles 0 and N are fixed.

C.1 Partial derivatives of functions

The variational derivative of a functional is an extension of the partial derivative of a function of many variables. We explain this by taking a special problem as an example.

Consider a system consisting of $(N+1)$ mass points (of mass m) each connected by a string of length ℓ. The string is under tension T, and the two end points are fixed at the same height (see Fig. C.1(a)). We consider the mechanical equilibrium of the system under gravity g.

The mechanical equilibrium is the state at which the potential energy of the system is a minimum. Let y_i $(i = 0, 1, \ldots, N)$ be the vertical displacement of the particle i from the line connecting the two end points. The potential energy is written as

$$f(y_0, y_1, \ldots, y_N) = T \sum_{i=0}^{N-1} \left(\sqrt{(y_{i+1} - y_i)^2 + \ell^2} - \ell \right) - \sum_{i=1}^{N-1} mgy_i \quad \text{(C.1)}$$

For simplicity, we consider the case that the displacement $|y_{i+1} - y_i|$ is much less than ℓ. Equation (C.1) is then written as

$$f = \frac{T}{2\ell} \sum_{i=0}^{N-1} (y_{i+1} - y_i)^2 - \sum_{i=1}^{N-1} mgy_i \quad \text{(C.2)}$$

Let $(y_{0,0}, \ldots, y_{N,0})$ be the state of minimum energy. To find the state, we consider the variation of f when y_i is changed slightly from $y_{i,0}$ to $y_{i,0} + \delta y_i$. Since f becomes a minimum at the state $(y_{0,0}, \ldots, y_{N,0})$, the following condition must be satisfied for any δy_i[1]

$$\delta f = f(y_{0,0} + \delta y_0, \ldots, y_{N,0} + \delta y_N) - f(y_{0,0}, y_{10}, \ldots, y_{N,0}) \geq 0 \quad \text{(C.3)}$$

The right-hand side can be expanded with respect to δy_i as

$$\delta f = \sum_{i=0}^{N} \frac{\partial f}{\partial y_i} \delta y_i \quad \text{(C.4)}$$

For this to be positive for any values of $(\delta y_1, \ldots, \delta y_{N-1})$ the following conditions must hold:

$$\frac{\partial f}{\partial y_i} = 0 \qquad i = 1, \ldots, N-1 \quad \text{(C.5)}$$

In the case that f is given by eq. (C.2), eq. (C.5) gives

$$\frac{T}{2\ell}(y_{i+1} - 2y_i + y_{i-1}) - mg = 0, \qquad i = 1, \ldots, N-1 \qquad (C.6)$$

This determines the equilibrium state.

C.2 Functional derivatives of functionals

Let us now consider that the string has a uniform mass density ρ. The state of the string is represented by a function $y(x)$ as in Fig. C.1(b). The potential energy of the system is now written as

$$f[y(x)] = T \int_0^L dx \left[\sqrt{1 + \left(\frac{dy}{dx}\right)^2} - 1 \right] - \int_0^L dx \rho gy(x) \qquad (C.7)$$

For the case $(dy/dx)^2 \ll 1$, eq. (C.7) is written as

$$f[y(x)] = \frac{T}{2} \int_0^L dx \left(\frac{dy}{dx}\right)^2 - \int_0^L dx \rho gy(x) \qquad (C.8)$$

In eqs. (C.7) and (C.8), $f[y(x)]$ represents a scalar, the value of which is determined by the function $y(x)$. Therefore $f[y(x)]$ is regarded as a function of the function $y(x)$, and is called a functional.

The variational calculus is a method of finding the minimum (or maximum) of functionals. The calculation is essentially the same as that for usual functions. The only difference is that the set of variables y_i $(i = 0, 1, \ldots, N)$ is replaced by the function $y(x)$, i.e., the discrete index i is replaced by a continuous variable x.

Let $y_0(x)$ be the state which gives the minimum of $f[y(x)]$. Consider the change of f when $y(x)$ is slightly changed from $y_0(x)$ to $y_0(x)+\delta y(x)$. The change in f is written as

$$\delta f = f[y_0(x) + \delta y(x)] - f[y_0(x)] \qquad (C.9)$$

For small $\delta y(x)$, the right-hand side can be expanded with respect to $\delta y(x)$, and can be written as

$$\delta f = \int_0^L dx \frac{\delta f}{\delta y(x)} \delta y(x) \qquad (C.10)$$

This corresponds to eq. (C.4), where the summation over i is replaced by the integral over x, and the partial derivative $\partial f/\partial y_i$ is replaced by $\delta f/\delta y(x)$. The quantity $\delta f/\delta y(x)$ is called the functional derivative. By the same argument as above, the condition that the state $y_0(x)$ is the state of minimum energy is written as

$$\frac{\delta f}{\delta y(x)} = 0, \quad \text{for} \quad y(x) = y_0(x) \qquad (C.11)$$

Calculation of the functional derivative $\delta f/\delta y(x)$ is straightforward. Let us consider the functional (C.8). For notational simplicity we write $y_0(x)$ as $y(x)$. The variation of $f[y(x)]$ is written as

$$\delta f[y(x)] = \frac{T}{2} \int_0^L dx \left[\left(\frac{dy}{dx} + \frac{d\delta y}{dx} \right)^2 - \left(\frac{dy}{dx} \right)^2 \right] - \int_0^L dx \rho g \delta y \quad \text{(C.12)}$$

Ignoring terms of second order in δy, we have

$$\delta f[y(x)] = T \int_0^L dx \frac{dy}{dx} \frac{d\delta y}{dx} - \int_0^L dx \rho g \delta y \quad \text{(C.13)}$$

The first term on the right-hand side can be written by using integration by parts, as

$$\delta f[y(x)] = T \left[\frac{dy}{dx} \delta y(x) \right]_0^L - T \int_0^L dx \frac{d^2 y}{dx^2} \delta y - \int_0^L dx \rho g \delta y \quad \text{(C.14)}$$

Since the position of the string is fixed at $x = 0$ and $x = L$, $\delta y(x)$ becomes zero at $x = 0$ and at $x = L$. Therefore the first term on the right-hand side of eq. (C.14) becomes zero. Hence

$$\delta f[y(x)] = \int_0^L dx \left[-T \frac{d^2 y}{dx^2} - \rho g \right] \delta y(x) \quad \text{(C.15)}$$

Therefore the functional derivative is given by

$$\frac{\delta f}{\delta y(x)} = -T \frac{d^2 y}{dx^2} - \rho g \quad \text{(C.16)}$$

Let us consider the case that the functional $f[y(x)]$ is written in the following form

$$f[y(x)] = \int_a^b dx F(y(x), y'(x), x) \quad \text{(C.17)}$$

where $y'(x) = dy/dx$, and $F(y, y', x)$ is a certain function which has three arguments, y, y', and x. The functional derivative of such a functional is calculated as

$$\frac{\delta f}{\delta y(x)} = \frac{\partial F}{\partial y} - \frac{d}{dx} \left(\frac{\partial F}{\partial y'} \right) \quad \text{(C.18)}$$

Reciprocal relation

D

In this appendix, we shall present two proofs of the reciprocal relation $\zeta_{ij} = \zeta_{ji}$ for the friction matrix of a particle; one is based on hydrodynamics, and the other is based on Brownian motion theory.

D.1 Hydrodynamic definition of the generalized frictional force

To prove the reciprocal relation by hydrodynamics, we start with the definition of the generalized potential force F_{pi} and the generalized frictional force F_{fi} for particles moving in a viscous fluid subject to a potential $U(x)$.

Suppose that we change the configuration of the particles with velocity $\dot{x}_i$. To cause such a change, we have to do work on the particles. Part of the work is used to change the potential energy $U(x)$. This work is written as

$$W_{rev} = \sum_i \frac{\partial U}{\partial x_i} \dot{x}_i \Delta t = -\sum_i F_{pi} \dot{x}_i \Delta t \qquad (D.1)$$

This defines the potential force conjugate to x_i

$$F_{pi} = -\frac{\partial U}{\partial x_i} \qquad (D.2)$$

The remaining part of the work is dissipated as heat in the viscous fluid. This part is written as

$$W_{irr} = -\sum_i F_{fi} \dot{x}_i \Delta t \qquad (D.3)$$

This defines the frictional force F_{fi} conjugate to x_i

Now when the particle configuration is changed with velocity $\dot{x}_i$, the point located at $\boldsymbol{r}$ on the surface of the particles moves with a certain velocity $\boldsymbol{v}_S(\boldsymbol{r}; x, \dot{x})$. The velocity $\boldsymbol{v}_S(\boldsymbol{r}; x, \dot{x})$ is written as a linear combination of $\dot{x}_i$:

$$\boldsymbol{v}_S(\boldsymbol{r}; x, \dot{x}) = \sum_i \boldsymbol{G}_i(\boldsymbol{r}; x) \dot{x}_i \qquad (D.4)$$

where the function $\boldsymbol{G}_i(\boldsymbol{r}; x)$ is determined by the geometry of the particle.

Using eq. (D.4) as the boundary condition, we can solve the Stokes equation in hydrodynamics for the fluid velocity

$$\eta \frac{\partial^2 v_\alpha}{\partial r_\beta^2} = \frac{\partial p}{\partial r_\alpha}, \qquad \frac{\partial v_\alpha}{\partial r_\alpha} = 0 \qquad (D.5)$$

(see Appendix A), and calculate the hydrodynamic force $\boldsymbol{f}_H(\boldsymbol{r}; x, \dot{x})$ acting on unit area of the particle surface at $\boldsymbol{r}$:

$$f_{H\alpha}(\boldsymbol{r}; x, \dot{x}) = \sigma_{\alpha\beta} n_\beta \qquad (D.6)$$

where $\sigma_{\alpha\beta}$ is the stress tensor

$$\sigma_{\alpha\beta} = \eta \left(\frac{\partial v_\alpha}{\partial r_\beta} + \frac{\partial v_\beta}{\partial r_\alpha} \right) - p\delta_{\alpha\beta} \qquad (D.7)$$

and $\boldsymbol{n}$ is the unit vector normal to the surface.

The work done on the fluid is then given by

$$W_{irr} = -\int dS \boldsymbol{f}_H(\boldsymbol{r}; x, \dot{x}) \cdot \boldsymbol{v}_S(\boldsymbol{r}; x, \dot{x})\Delta t$$

$$= -\int dS \boldsymbol{f}_H(\boldsymbol{r}; x, \dot{x}) \cdot \sum_i \boldsymbol{G}_i(\boldsymbol{r}; x)\dot{x}_i \Delta t \qquad (D.8)$$

Comparing eq. (D.8) with eq. (6.56), we have

$$F_{fi} = \int dS \boldsymbol{f}_H(\boldsymbol{r}; x, \dot{x}) \cdot \boldsymbol{G}_i(\boldsymbol{r}; x) \qquad (D.9)$$

Since the Stokes equation is a linear equation for the fluid velocity $\boldsymbol{v}$, the solution $\boldsymbol{v}$ of the above boundary value problem can be expressed as a linear combination of $\dot{x}_i$. Therefore the frictional forces $\boldsymbol{f}_H$ and F_{fi} are expressed as a linear combination of $\dot{x}_i$. Hence

$$F_{fi} = -\sum_j \zeta_{ij}\dot{x}_j \qquad (D.10)$$

This defines the friction coefficient ζ_{ij}.

D.2   Hydrodynamic proof of the reciprocal relation

Now to prove the reciprocal relation $\zeta_{ij} = \zeta_{ji}$, we consider two situations; one is that the particle configuration is changed with velocity $\dot{x}^{(1)}$, and the other changed with velocity $\dot{x}^{(2)}$. In both situations, the particles are in configuration x. Let $F_{fi}^{(1)}$ and $F_{fi}^{(2)}$ be the frictional forces for each situation. We shall prove the identity

$$\sum_i F_{fi}^{(1)} \dot{x}_i^{(2)} = \sum_i F_{fi}^{(2)} \dot{x}_i^{(1)} \qquad (D.11)$$

The reciprocal relation $\zeta_{ij} = \zeta_{ji}$ follows from this identity.

Let $v^{(a)}(r)$ ($a = 1, 2$) be the velocity field for the two situations. By eqs. (D.10), the left-hand side of eq. (D.11) is written as

$$I \equiv -\sum_i F_{fi}^{(1)} \dot{x}_i^{(2)} = -\int dS \, G_i(r; x) \cdot f_H^{(1)}(r; x, \dot{x}) \dot{x}_i^{(2)}$$

$$= -\sum_i \int dS G_{i\alpha}(r; x) \sigma_{\alpha\beta}^{(1)} n_\beta \dot{x}_i^{(2)} \tag{D.12}$$

Since the fluid velocity satisfies the boundary condition (D.4), $\sum_i G_{i\alpha}(r; x) \dot{x}_i^{(2)}$ is equal to $v_\alpha^{(2)}(r)$ at the particle surface. Hence I is written as[1]

$$I = -\int dS v_\alpha^{(2)} \sigma_{\alpha\beta}^{(1)} n_\beta \tag{D.13}$$

Using the Gauss theorem, the surface integral on the right-hand side of eq. (D.13) can be written as

$$I = \int dr \frac{\partial}{\partial r_\beta} \left(v_\alpha^{(2)} \sigma_{\alpha\beta}^{(1)} \right) = \int dr \left(\frac{\partial v_\alpha^{(2)}}{\partial r_\beta} \sigma_{\alpha\beta}^{(1)} + v_\alpha^{(2)} \frac{\partial \sigma_{\alpha\beta}^{(1)}}{\partial r_\beta} \right) \tag{D.14}$$

The second term in the integrand on the right-hand side of eq. (D.14) vanishes by the Stokes equation (D.5). Since $\sigma_{\alpha\beta}^{(1)} = \sigma_{\beta\alpha}^{(1)}$, the first term on the right-hand side is written as

$$I = \frac{1}{2} \int dr \left(\frac{\partial v_\alpha^{(2)}}{\partial r_\beta} + \frac{\partial v_\beta^{(2)}}{\partial r_\alpha} \right) \sigma_{\alpha\beta}^{(1)} \tag{D.15}$$

Substituting eq. (D.7) and using the incompressible condition $\nabla \cdot v = 0$, we have

$$I = \frac{\eta}{2} \int dr \left(\frac{\partial v_\alpha^{(2)}}{\partial r_\beta} + \frac{\partial v_\beta^{(2)}}{\partial r_\alpha} \right) \left(\frac{\partial v_\alpha^{(1)}}{\partial r_\beta} + \frac{\partial v_\beta^{(1)}}{\partial r_\alpha} \right) \tag{D.16}$$

The expression on the right-hand side is invariant under the exchange of superscripts (1) and (2). Therefore the identity (D.11) has been proven. If we put $\dot{x}_i^{(1)} = \dot{x}_i^{(2)} = \dot{x}_i$ in eq. (D.16), we have

$$-\sum_i F_{fi} \dot{x}_i = \frac{\eta}{2} \int dr \left(\frac{\partial v_\alpha}{\partial r_\beta} + \frac{\partial v_\beta}{\partial r_\alpha} \right)^2 \tag{D.17}$$

The right-hand side of eq. (D.17) is the work done on the fluid. Therefore

$$\frac{\eta}{2} \left(\frac{\partial v_\alpha}{\partial r_\beta} + \frac{\partial v_\beta}{\partial r_\alpha} \right)^2 \tag{D.18}$$

[1] Note that n is a vector pointing from the particle surface to the fluid.

is the energy dissipation rate per unit volume in the fluid. If eq. (6.57) is used for F_{Hi} in eq. (D.17)

$$\sum_{i,j} \zeta_{ij} \dot{x}_i \dot{x}_j = \frac{\eta}{2} \int d\mathbf{r} \left(\frac{\partial v_\alpha}{\partial r_\beta} + \frac{\partial v_\beta}{\partial r_\alpha} \right)^2 \tag{D.19}$$

The right-hand side is non-negative. Therefore eq. (6.61) has been proven.

D.3   Onsager's proof of the reciprocal relation

Next we give a proof based on Brownian motion theory. The proof is based on the ansatz that the Langevin equation (6.58) is valid for an arbitrary potential $U(x)$.

We consider the special case that the particles are subject to a harmonic potential

$$U(x) = \frac{1}{2} \sum_i k_i (x_i - x_i^0)^2 \tag{D.20}$$

and that the particle configuration cannot deviate significantly from the equilibrium configuration $x^0 = (x_1^0, x_2^0, \ldots)$. In this case, eq. (6.58) gives the following linear Langevin equation for the deviation $\xi_i(t) = x_i(t) - x_i^0$:

$$-\sum_j \zeta_{ij}^0 \dot{\xi}_j - k_i \xi_i + F_{ri}(t) = 0 \tag{D.21}$$

where ζ_{ij}^0 represents $\zeta_{ij}(x^0)$, and is a constant independent of $\xi_i(t)$.

For the purpose of proving the relations (6.59) and (6.60), it is sufficient to analyse the short-time behaviour of the time correlation function $\langle \xi_i(t) \xi_j(0) \rangle$. For simplicity of notation, we shall write ζ_{ij}^0 as ζ_{ij} in this appendix. Equation (D.21) is then written as

$$\dot{\xi}_i = -\sum_k (\zeta^{-1})_{ik} [k_k \xi_k - F_{rk}(t)] \tag{D.22}$$

We assume that ξ_i is equal to ξ_{i0} at time $t = 0$. Equation (D.22) is then solved for small Δt as

$$\xi_i(\Delta t) = \xi_{i0} - \Delta t \sum_k (\zeta^{-1})_{ik} k_k \xi_{k0} + \sum_k \int_0^{\Delta t} dt_1 \, (\zeta^{-1})_{ik} F_{rk}(t_1) \tag{D.23}$$

Since the average of $F_{rk}(t)$ is zero, we have

$$\langle \xi_i(\Delta t) \rangle_{\xi_0} = \xi_{i0} - \Delta t \sum_k (\zeta^{-1})_{ik} k_k \xi_{k0}, \quad \text{for} \quad \Delta t > 0 \tag{D.24}$$

Therefore the time correlation becomes

$$\langle \xi_i(\Delta t)\xi_j(0)\rangle = \langle \xi_{i0}\xi_{j0}\rangle - \Delta t \sum_k (\zeta^{-1})_{ik}k_k\langle \xi_{k0}\xi_{j0}\rangle, \quad \text{for} \quad \Delta t > 0$$

$$(\text{D}.25)$$

Since the distribution of ξ_i at equilibrium is proportional to $\exp(-\sum_i k_i\xi_i^2/2k_BT)$, $\langle \xi_{i0}\xi_{k0}\rangle$ is given by

$$\langle \xi_{i0}\xi_{k0}\rangle = \frac{\delta_{ik}k_BT}{k_i} \qquad (\text{D}.26)$$

Equation (D.25) then gives,

$$\langle \xi_i(\Delta t)\xi_j(0)\rangle = k_BT\left[\delta_{ij}k_j^{-1} - \Delta t(\zeta^{-1})_{ij}\right], \quad \text{for} \quad \Delta t > 0 \qquad (\text{D}.27)$$

Therefore the symmetry of the time correlation function $\langle \xi_i(\Delta t)\xi_j(0)\rangle = \langle \xi_j(\Delta t)\xi_i(0)\rangle$ requires the relation $(\zeta^{-1})_{ij} = (\zeta^{-1})_{ji}$, or $\zeta_{ij} = \zeta_{ji}$. This proves the reciprocal relation (6.60).

To prove eq. (6.59), we calculate $\langle \xi_i(\Delta t)\xi_j(\Delta t)\rangle$ using eq. (D.23):

$$\langle \xi_i(\Delta t)\xi_j(\Delta t)\rangle = \langle \xi_{i0}\xi_{j0}\rangle - \Delta t \sum_k (\zeta^{-1})_{ik}k_k\langle \xi_{k0}\xi_{j0}\rangle$$

$$- \Delta t \sum_{k'} (\zeta^{-1})_{jk'}k_{k'}\langle \xi_{i0}\xi_{k'0}\rangle$$

$$+ \sum_{k,k'} \int_0^{\Delta t} dt_1 \int_0^{\Delta t} dt_2 (\zeta^{-1})_{ik}(\zeta^{-1})_{jk'}\langle F_{rk}(t_1)F_{rk'}(t_2)\rangle$$

$$(\text{D}.28)$$

As in the discussion in Section 6.2.2, we assume

$$\langle F_{ri}(t)F_{rj}(t')\rangle = A_{ij}\delta(t-t') \qquad (\text{D}.29)$$

By use of eq. (D.26), eq. (D.28) is written as

$$\frac{\delta_{ij}k_BT}{k_i} = \frac{\delta_{ij}k_BT}{k_i} - \Delta t(\zeta^{-1})_{ij}k_BT\Delta t - \Delta t(\zeta^{-1})_{ji}k_BT\Delta t$$

$$+ \Delta t \sum_{k,k'}(\zeta^{-1})_{ik}(\zeta^{-1})_{jk'}A_{k,k'}\Delta t \qquad (\text{D}.30)$$

Using the reciprocal relation, we have

$$-2(\zeta^{-1})_{ij}k_BT + \sum_{k,k'}(\zeta^{-1})_{ik}(\zeta^{-1})_{jk'}A_{k,k'} = 0 \qquad (\text{D}.31)$$

Solving this for $A_{kk'}$, we have

$$A_{ij} = 2\zeta_{ij}k_BT \qquad (\text{D}.32)$$

This gives eq. (6.59).

Statistical mechanics for material response and fluctuations

E.1   Liouville equation

Statistical mechanics regards any macroscopic system as a dynamical system consisting of molecules which obey Hamilton's equations of motion. As in Appendix B, we denote the set of generalized coordinates $(q_1, q_2, \ldots, q_f)$ and the generalized momenta $(p_1, p_2, \ldots, p_f)$ by Γ:

$$\Gamma = (q_1, q_2, \ldots, q_f, p_1, p_2, \ldots, p_f) \tag{E.1}$$

The microscopic state of the system is completely specified by Γ.

The Hamiltonian of the system is generally written as $H(\Gamma; x)$. Here $x = (x_1, x_2, \ldots)$ denotes the set of external parameters which affect the motion of the molecules, and can be controlled externally. Examples of the external parameters are (i) the position of the piston confining a gas, (ii) the position of a large particle placed in a fluid (see Fig. 6.6), and (iii) the strength of the magnetic field applied to the system (see Fig. 6.9). x can also represent the parameter appearing in the hypothetical potential (B.6).

If the Hamiltonian $H(\Gamma; x)$ is known, the time evolution of the microscopic state can be calculated by solving Hamilton's equations of motion

$$\frac{dq_a}{dt} = \frac{\partial H}{\partial p_a}, \qquad \frac{dp_a}{dt} = -\frac{\partial H}{\partial q_a}, \qquad (a = 1, 2, \ldots, f) \tag{E.2}$$

Let $\psi(\Gamma, t)$ be the probability of finding the system in the state Γ. The time evolution of $\psi(\Gamma, t)$ is given by

$$\begin{aligned}
\frac{\partial \psi}{\partial t} &= -\sum_a \left[\frac{\partial}{\partial p_a}(\dot{p}_a \psi) + \frac{\partial}{\partial q_a}(\dot{q}_a \psi) \right] \\
&= -\sum_a \left[\frac{\partial}{\partial p_a}\left(-\frac{\partial H}{\partial q_a}\psi \right) + \frac{\partial}{\partial q_a}\left(\frac{\partial H}{\partial p_a}\psi \right) \right] \\
&= -L\psi
\end{aligned} \tag{E.3}$$

where L is defined by

$$L = \sum_a \left(\frac{\partial H}{\partial p_a} \frac{\partial}{\partial q_a} - \frac{\partial H}{\partial q_a} \frac{\partial}{\partial p_a} \right) \qquad (E.4)$$

which is called the Liouville operator.

For the time being, we assume that the external parameter x is kept constant. Then eq. (E.3) is formally solved as

$$\psi(\Gamma, t; x) = e^{-tL} \psi_0(\Gamma; x) \qquad (E.5)$$

where $\psi_0(\Gamma; x)$ is the distribution function at $t = 0$.

If the system is at equilibrium at temperature T, the probability that the system is in the state Γ is given by the canonical distribution:

$$\psi_{eq}(\Gamma; x) = \frac{e^{-\beta H(\Gamma; x)}}{\int d\Gamma \; e^{-\beta H(\Gamma; x)}} \qquad (E.6)$$

where

$$\beta = \frac{1}{k_B T} \qquad (E.7)$$

The canonical distribution is a steady state solution for the Liouville equation:

$$e^{-tL} \psi_{eq}(\Gamma; x) = \psi_{eq}(\Gamma; x) \qquad (E.8)$$

E.2 Time correlation functions

Time correlation functions of physical quantities can be expressed by the Liouville operator. Consider the time correlation function of two physical quantities X and Y. Let $\hat{X}(\Gamma; x)$ and $\hat{Y}(\Gamma; x)$ be the values of X and Y in the microscopic state Γ. Then the values of X and Y at time t are written as

$$X(t) = \hat{X}(\Gamma_t; x), \qquad Y(t) = \hat{Y}(\Gamma_t; x) \qquad (E.9)$$

where Γ_t denotes the microscopic state at time t.

We shall consider the time correlation function $\langle X(t)Y(0) \rangle_{eq,x}$ for the system which is at equilibrium under the given external parameter x. If the system is in the microscopic state Γ_0 at time $t = 0$, and if it is in the state Γ at a later time t, the value of $X(t)Y(0)$ is $\hat{X}(\Gamma)\hat{Y}(\Gamma_0)$. (Here the x dependence is dropped for simplicity.) The probability that the system is in the state Γ_0 is $\psi_{eq}(\Gamma_0)$. Hence $\langle X(t)Y(0) \rangle_{eq}$ is written as

$$\langle X(t)Y(0) \rangle_{eq} = \int d\Gamma \int d\Gamma_0 \hat{X}(\Gamma)\hat{Y}(\Gamma_0) G(\Gamma, t|\Gamma_0, 0)\psi_{eq}(\Gamma_0) \qquad (E.10)$$

where $G(\Gamma, t|\Gamma_0, 0)$ represents the probability that the system which was in the state Γ_0 at time $t = 0$ is in the state Γ at time t. This is the solution of the Liouville equation (E.3) under the initial condition

$$G(\Gamma, t|\Gamma_0, 0)|_{t=0} = \delta(\Gamma - \Gamma_0) \qquad (E.11)$$

Therefore, according to eq. (E.5), $G(\Gamma, t|\Gamma_0, 0)$ is given by

$$G(\Gamma, t|\Gamma_0, 0) = e^{-tL}\delta(\Gamma - \Gamma_0) \tag{E.12}$$

Hence

$$\langle X(t)Y(0)\rangle_{eq} = \int d\Gamma \int d\Gamma_0 \hat{X}(\Gamma)\hat{Y}(\Gamma_0)[e^{-tL}\delta(\Gamma - \Gamma_0)]\psi_{eq}(\Gamma_0) \tag{E.13}$$

The integral with respect to Γ_0 gives

$$\langle X(t)Y(0)\rangle_{eq} = \int d\Gamma \hat{X}(\Gamma)e^{-tL}\left[\hat{Y}(\Gamma)\psi_{eq}(\Gamma)\right] \tag{E.14}$$

This is the formal expression for the time correlation function in the equilibrium state.

The symmetric relation (6.7) for the time correlation function can be proven by using eq. (E.14). Let us consider the following variable transformation in the integral of eq. (E.14)

$$q_a \rightarrow q_a, \qquad p_a \rightarrow -p_a \tag{E.15}$$

Such a transformation will be denoted as $\Gamma \rightarrow -\Gamma$. Equation (E.14) is then written as

$$\langle X(t)Y(0)\rangle_{eq} = \int d\Gamma \hat{X}(-\Gamma)e^{-t\hat{L}'}\left[\hat{Y}(-\Gamma)\psi_{eq}(-\Gamma)\right] \tag{E.16}$$

Here L' is given by

$$L' = \sum_a \left(-\frac{\partial H(-\Gamma)}{\partial p_a}\frac{\partial}{\partial q_a} + \frac{\partial H(-\Gamma)}{\partial q_a}\frac{\partial}{\partial p_a}\right) \tag{E.17}$$

Since the Hamiltonian has the property $H(-\Gamma) = H(\Gamma)$, we have

$$L' = -L, \qquad \psi_{eq}(-\Gamma) = \psi_{eq}(\Gamma) \tag{E.18}$$

Using $\hat{X}(-\Gamma) = \epsilon_X \hat{X}(\Gamma)$, $\hat{Y}(-\Gamma) = \epsilon_Y \hat{Y}(\Gamma)$, we finally have

$$\langle X(t)Y(0)\rangle_{eq} = \epsilon_X \epsilon_Y \int d\Gamma \hat{X}(\Gamma)e^{tL}[\hat{Y}(\Gamma)\psi_{eq}(\Gamma)] = \epsilon_X \epsilon_Y \langle X(-t)Y(0)\rangle_{eq} \tag{E.19}$$

which is eq. (6.7)

E.3   Equilibrium responses

Suppose that the system is at equilibrium for a given external parameter x, and that the external parameters are changed slightly at time $t = 0$ as (see Fig. E.1)

$$x \rightarrow x + \delta x \qquad \text{for} \qquad t > 0 \tag{E.20}$$

Then the Hamiltonian changes as

$$H(\Gamma; x) \rightarrow H(\Gamma; x + \delta x) = H(\Gamma; x) + \delta H(\Gamma; x, \delta x) \tag{E.21}$$

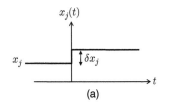

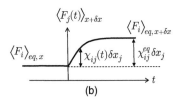

Fig. E.1 (a) Change of external parameter δx_j. (b) Time variation of the average of $\hat{F}_i$ induced by the change of the external parameter.

where

$$\delta H(\Gamma; x, \delta x) = \sum_i \frac{\partial H(\Gamma; x)}{\partial x_i} \delta x_i = -\sum_i \hat{F}_i(\Gamma; x) \delta x_i \qquad (E.22)$$

where $\hat{F}_i(\Gamma; x)$ is defined by

$$\hat{F}_i(\Gamma; x) = -\frac{\partial H(\Gamma; x)}{\partial x_i} \qquad (E.23)$$

$\hat{F}_i(\Gamma; x)$ is called the quantity conjugate to the external parameter x_i. In the following, we shall consider how the average of $\hat{F}_i$ varies in time.

When the system achieves equilibrium under the new parameter $x + \delta x$, the average of $\hat{F}_i$ is easily calculated. In general, the average of $\hat{F}_i(\Gamma; x)$ at equilibrium for a given external parameter x is calculated by

$$
\begin{aligned}
\langle F_i \rangle_{eq,x} &= \int d\Gamma \ \hat{F}_i \psi_{eq}(\Gamma; x) \\
&= -\frac{\int d\Gamma \ \frac{\partial H}{\partial x_i} e^{-\beta H(\Gamma; x)}}{\int d\Gamma \ e^{-\beta H(\Gamma; x)}} \\
&= -\frac{\partial A(x)}{\partial x_i} \qquad (E.24)
\end{aligned}
$$

where $A(x)$ is the free energy of the system

$$A(x) = -\frac{1}{\beta} \ln \left[\int d\Gamma \ e^{-\beta H(\Gamma; x)} \right] \qquad (E.25)$$

Now when the external parameters are changed from x to $x + \delta x$, the equilibrium average of $\hat{F}_i$ is also changed. For small δx, the change can be written as

$$\langle F_i \rangle_{eq,x+\delta x} = \langle F_i \rangle_{eq,x} + \sum_j \chi_{ij}^{eq} \delta x_j \qquad (E.26)$$

The coefficient χ_{ij}^{eq} is given by

$$\chi_{ij}^{eq} = \frac{\partial \langle F_i \rangle_{eq,x}}{\partial x_j} \qquad (E.27)$$

From eqs. (E.24) and (E.27), χ_{ij}^{eq} is calculated as

$$
\begin{aligned}
\chi_{ij}^{eq} &= -\frac{\int d\Gamma \ \frac{\partial^2 H}{\partial x_i \partial x_j} e^{-\beta H}}{\int d\Gamma \ e^{-\beta H}} + \beta \frac{\int d\Gamma \ \frac{\partial H}{\partial x_i} \frac{\partial H}{\partial x_j} e^{-\beta H}}{\int d\Gamma \ e^{-\beta H}} \\
&\quad - \beta \frac{\int d\Gamma \ \frac{\partial H}{\partial x_i} e^{-\beta H} \int d\Gamma \ \frac{\partial H}{\partial x_j} e^{-\beta H}}{\left(\int d\Gamma \ e^{-\beta H} \right)^2} \\
&= \left\langle \frac{\partial F_i}{\partial x_j} \right\rangle_{eq,x} + \beta \left[\langle F_i F_j \rangle_{eq,x} - \langle F_i \rangle_{eq,x} \langle F_j \rangle_{eq,x} \right] \qquad (E.28)
\end{aligned}
$$

Let F_{ri} be the deviation of F_i from the equilibrium value:

$$F_{ri} = \hat{F}_i(\Gamma; x) - \langle F_i \rangle_{eq,x} \qquad \text{(E.29)}$$

Equation (E.28) is then written as

$$\chi_{ij}^{eq} = \left\langle \frac{\partial F_i}{\partial x_j} \right\rangle_{eq,x} + \beta \langle F_{ri} F_{rj} \rangle_{eq,x} \qquad \text{(E.30)}$$

This equation indicates that the response to external perturbations is related to the fluctuations taking place at equilibrium.

E.4 Non-equilibrium responses

We now consider the non-equilibrium response. Before the system achieves equilibrium for the new parameter $x + \delta x$, the average of $\hat{F}_i$ at time t is written as (see Fig. E.1)

$$\langle F_i(t) \rangle_{x+\delta x} = \langle F_i \rangle_{eq,x} + \sum_j \chi_{ij}(t) \delta x_j \qquad \text{(E.31)}$$

The function $\chi_{ij}(t)$ approaches χ_{ij}^{eq} as $t \to \infty$.

To calculate $\chi_{ij}(t)$, we consider the Liouville equation for the Hamiltonian (E.21):

$$\frac{\partial \psi}{\partial t} = -(L + \delta L)\psi \qquad \text{(E.32)}$$

where δL is given by

$$\begin{aligned}
\delta L &= \sum_a \left(\frac{\partial \delta H}{\partial p_a} \frac{\partial}{\partial q_a} - \frac{\partial \delta H}{\partial q_a} \frac{\partial}{\partial p_a} \right) \\
&= -\sum_i \sum_a \delta x_i \left(\frac{\partial \hat{F}_i}{\partial p_a} \frac{\partial}{\partial q_a} - \frac{\partial \hat{F}_i}{\partial q_a} \frac{\partial}{\partial p_a} \right)
\end{aligned} \qquad \text{(E.33)}$$

To solve eq. (E.32), we write the solution

$$\psi(\Gamma, t; x, \delta x) = \psi_{eq}(\Gamma; x) + \delta\psi(\Gamma, t; x, \delta x) \qquad \text{(E.34)}$$

Substituting eq. (E.34) into eq. (E.32) and retaining the linear terms of order δx_i, we have

$$\frac{\partial \delta\psi}{\partial t} = -L\delta\psi - \delta L\psi_{eq} \qquad \text{(E.35)}$$

The solution of this equation can be written as

$$\delta\psi = -\int_0^t dt' e^{-(t-t')L} \delta L \psi_{eq} \qquad \text{(E.36)}$$

By use of eq. (E.6), $\delta L\psi_{eq}$ can be written as

$$\delta L\psi_{eq} = -\beta(\delta LH)\psi_{eq}$$

$$= -\beta\psi_{eq}\sum_{i,a}\delta x_i\left(-\frac{\partial\hat{F}_i}{\partial p_a}\frac{\partial H}{\partial q_a} + \frac{\partial\hat{F}_i}{\partial q_a}\frac{\partial H}{\partial p_a}\right)$$

$$= -\beta\psi_{eq}\sum_i\delta x_i(L\hat{F}_i)$$

$$= -\beta\psi_{eq}\sum_i\delta x_i\hat{\dot{F}}_i \tag{E.37}$$

where $\hat{\dot{F}}_i = L\hat{F}_i$. Hence eq. (E.36) is written as

$$\delta\psi = \beta\sum_i\delta x_i\int_0^t dt'e^{-(t-t')L}\left(\hat{\dot{F}}_i\psi_{eq}\right) \tag{E.38}$$

The average of F_i at time t is calculated as

$$\langle F_i(t)\rangle_{x+\delta x}$$

$$= \int d\Gamma\ \hat{F}_i(\Gamma; x+\delta x)[\psi_{eq}(\Gamma; x) + \delta\psi(\Gamma, t; x, \delta x)]$$

$$= \int d\Gamma\left[\hat{F}_i + \sum_j\frac{\partial\hat{F}_i}{\partial x_j}\delta x_j\right]\left[\psi_{eq} + \beta\sum_j\delta x_j\int_0^t dt'e^{-(t-t')L}\left(\hat{\dot{F}}_j\psi_{eq}\right)\right]$$

$$= \int d\Gamma\left[\hat{F}_i\psi_{eq} + \sum_j\delta x_j\frac{\partial\hat{F}_i}{\partial x_j}\psi_{eq} + \beta\hat{F}_i\sum_j\delta x_j\int_0^t dt'e^{-(t-t')L}\left(\hat{\dot{F}}_j\psi_{eq}\right)\right]$$

$$\tag{E.39}$$

The last term is expressed in terms of the time correlation function using eq. (E.14). Equation (E.39) is then written as

$$\langle F_i(t)\rangle_{x+\delta x} = \langle F_i\rangle_{eq,x} + \sum_j\delta x_j\left\langle\frac{\partial F_i}{\partial x_j}\right\rangle_{eq,x}$$

$$+ \beta\sum_j\delta x_j\int_0^t dt'\langle F_i(t-t')\dot{F}_j(0)\rangle_{eq,x} \tag{E.40}$$

Comparing this with eq. (E.31), $\chi_{ij}(t)$ is obtained as

$$\chi_{ij}(t) = \left\langle\frac{\partial F_i}{\partial x_j}\right\rangle_{eq,x} + \beta\int_0^t dt'\langle F_i(t-t')\dot{F}_j(0)\rangle_{eq,x} \tag{E.41}$$

By using the identity $\langle F_i(t-t')\dot{F}_j(0)\rangle_{eq,x} = \langle F_i(t)\dot{F}_j(t')\rangle_{eq,x}$, the last term is written as

$$\int_0^t dt'\langle F_i(t-t')\dot{F}_j(0)\rangle_{eq,x} = \int_0^t dt'\langle F_i(t)\dot{F}_j(t')\rangle_{eq,x}$$

$$= \langle F_i(t)F_j(t)\rangle_{eq,x} - \langle F_i(t)F_j(0)\rangle_{eq,x}$$

$$\tag{E.42}$$

Hence $\chi_{ij}(t)$ is given by

$$\chi_{ij}(t) = \left\langle \frac{\partial F_i}{\partial x_j} \right\rangle_{eq,x} + \beta \langle F_i(t)F_j(t)\rangle_{eq,x} - \beta \langle F_i(t)F_j(0)\rangle_{eq,x} \quad \text{(E.43)}$$

Using the fluctuating force $F_{ri} = \hat{F}_i(\Gamma, x) - \langle F_i\rangle_{eq,x}$, $\chi_{ij}(t)$ is written as

$$\chi_{ij}(t) = \left\langle \frac{\partial F_i}{\partial x_j} \right\rangle_{eq,x} + \beta \langle F_{ri}(0)_{rj}(0)\rangle_{eq,x} - \beta \langle F_{ri}(t)F_{rj}(0)\rangle_{eq,x} \quad \text{(E.44)}$$

or by eq. (E.30)

$$\chi_{ij}(t) = \chi_{ij}^{eq} - \alpha_{ij}(t) \quad \text{(E.45)}$$

where

$$\alpha_{ij}(t) = \beta \langle F_{ri}(t)F_{rj}(0)\rangle_{eq,x} \quad \text{(E.46)}$$

Therefore eq. (E.31) is written as

$$\langle F_i(t)\rangle_{x+\delta x} = \langle F_i\rangle_{eq,x} + \sum_j \chi_{ij}^{eq}\delta x_j - \sum_j \alpha_{ij}(t)\delta x_j$$

$$= \langle F_i\rangle_{eq,x+\delta x} - \sum_j \alpha_{ij}(t)\delta x_j \quad \text{(E.47)}$$

Equation (E.47) represents the response to a step change of the external parameters. The response for general time variation of the external parameters is calculated by superposing these responses:

$$\langle F_i(t)\rangle_{x+\delta x} = \langle F_i\rangle_{eq,x} + \int_0^t dt' \sum_j \chi_{ij}(t-t')\dot{x}_j(t')$$

$$= \langle F_i\rangle_{eq,x} + \sum_j \int_0^t dt' \left[\chi_{ij}^{eq} - \alpha_{ij}(t-t')\right]\dot{x}_j(t')$$

$$= \langle F_i\rangle_{eq,x} + \sum_j \chi_{ij}^{eq}\delta x_j - \sum_j \int_0^t dt'\alpha_{ij}(t-t')\dot{x}_j(t')$$

$$= \langle F_i\rangle_{eq,x+\delta x} - \sum_j \int_0^t dt'\alpha_{ij}(t-t')\dot{x}_j(t') \quad \text{(E.48)}$$

This gives eqs. (6.69) and (6.70).

E.5   Generalized Einstein relation

As a special case, we consider the situation that x_i stands for the external fields applied to the system. To clarify the situation, we use different symbols h_i to denote the external parameters that represent the strength of the applied field. We assume that the Hamiltonian of the system is written as

$$H(\Gamma; h) = H_0(\Gamma) - \sum_i h_i(t)\hat{M}_i(\Gamma) \qquad (E.49)$$

where $\hat{M}_i$ represents the quantity conjugate to h_i.

We consider the situation that the system was at equilibrium at $t = 0$ with zero external fields, and that step fields h_i are applied for $t \geq \epsilon$ (ϵ being a small positive time), i.e.,

$$h_i(t) = h_i\Theta(t - \epsilon), \qquad \text{or} \qquad \dot{h}_i(t) = h_i\delta(t - \epsilon) \qquad (E.50)$$

We define the response function $\chi_{ij}(t)$ for this situation

$$\langle M_i(t)\rangle_h - \langle M_i(0)\rangle_{eq,0} = \sum_j \chi_{ij}(t)h_j \qquad (E.51)$$

We assume that that the equilibrium average of $\hat{M}_i$ in the absence of the field is zero: $\langle M_i(0)\rangle_{eq,0} = 0$.[1]

For such a situation, eq. (E.48) is written as

$$\langle M_i(t)\rangle_h = \langle M_i(t)\rangle_{eq,h} - \sum_j \int_0^t dt' \, \alpha_{ij}(t - t')h_j\delta(t' - \epsilon)$$

$$= \sum_j \chi_{ij}^{eq}h_j - \sum_j \alpha_{ij}(t)h_j \qquad (E.52)$$

where

$$\alpha_{ij}(t) = \beta\langle M_{ri}(t)M_{rj}(0)\rangle_{eq,0} \qquad (E.53)$$

Since $\langle M_i(0)\rangle_{eq,0}$ is equal to zero, $M_{ri}(t)$ is identical to $M_i(t)$. Therefore $\chi_{ij}(t)$ is written as

$$\chi_{ij}(t) = \beta\langle M_iM_j\rangle_{eq,0} - \beta\langle M_i(t)M_j(0)\rangle_{eq,0}$$

$$= \frac{1}{2k_BT}\langle [M_i(t) - M_i(0)][M_j(t) - M_j(0)]\rangle_{eq,0} \qquad (E.54)$$

This is the generalized Einstein relation which is discussed in Section 6.5.3.

---

[1] This assumption can always be made. If $\langle M_i\rangle_{eq,0}$ is not equal to zero, we can replace $\hat{M}_i$ with $\hat{M}_i' = \hat{M}_i - \langle M_i\rangle_{eq,0}$. Then $\langle \hat{M}_i'\rangle_{eq,0}$ is equal to zero.

F Derivation of the Smoluchowskii equation from the Langevin equation

Here we shall derive the Smoluchowskii equation (7.44) from the Langevin equation (6.39).

Let $P(x, \xi, \Delta t)$ be the probability that the particle located at x at time t moves to the position $x + \xi$ in time Δt. Given $P(x, \xi, \Delta t)$, the distribution of the particle at time $t + \Delta t$ is expressed as

$$\psi(x, t + \Delta t) = \int d\xi \, P(x - \xi, \xi, \Delta t)\psi(x - \xi, t) \tag{F.1}$$

Now according to the Langevin equation (6.39), the displacement ξ for small Δt is given by

$$\xi = V(x)\Delta t + \Delta x_r \tag{F.2}$$

where

$$V(x) = -\frac{1}{\zeta}\frac{\partial U}{\partial x} \tag{F.3}$$

and

$$\Delta x_r = \int_t^{t + \Delta t} dt' v_r(t') \tag{F.4}$$

The mean and the variance of the random variable Δx_r are given by

$$\langle \Delta x_r \rangle = 0, \qquad \langle \Delta x_r^2 \rangle = 2D\Delta t \tag{F.5}$$

Hence

$$\langle \xi \rangle = V\Delta t, \qquad \langle \xi^2 \rangle = 2D\Delta t \tag{F.6}$$

where terms of order Δt^2 or higher are ignored.

To calculate the integral in eq. (F.1), we write

$$\psi(x - \xi, t) = \left[1 - \xi\frac{\partial}{\partial x} + \frac{\xi^2}{2}\frac{\partial^2}{\partial x^2} \right]\psi(x, t) \tag{F.7}$$

$$P(x - \xi, \xi, \Delta t) = \left[1 - \xi \frac{\partial}{\partial x} + \frac{\xi^2}{2} \frac{\partial^2}{\partial x^2} \right] P(x, \xi, t) \qquad \text{(F.8)}$$

The right-hand side of eq. (F.1) can be calculated as an expansion with respect to ξ. For example

$$\int d\xi \; \xi \frac{\partial P(x, \xi, \Delta t)}{\partial x} \psi(x, t) = \psi \frac{\partial}{\partial x} \int d\xi \; \xi P(x, \xi, \Delta t) = \psi \frac{\partial V}{\partial x} \Delta t \quad \text{(F.9)}$$

Summing up such terms, we have

$$\int d\xi \; P(x - \xi, \xi, \Delta t) \psi(x - \xi, t)$$

$$= \psi + \left[-\psi \frac{\partial V}{\partial x} - V \frac{\partial \psi}{\partial x} + D \frac{\partial^2 \psi}{\partial x^2} + 2 \frac{\partial D}{\partial x} \frac{\partial \psi}{\partial x} + \psi \frac{\partial^2 D}{\partial x^2} \right] \Delta t$$

$$\text{(F.10)}$$

This gives the following equation for ψ

$$\frac{\partial \psi}{\partial t} = -\frac{\partial}{\partial x} (V \psi) + \frac{\partial^2}{\partial x^2} (D \psi) \qquad \text{(F.11)}$$

For constant D, eq. (F.11) gives the Smoluchowskii equation (7.44).

Index